Kitchen Design, Installation and Remodeling

Other Residential and Light Construction Books from McGraw-Hill

Bianchina: *Forms & Documents for the Builder*

Bolt: *Roofing the Right Way, 3/e*

Bynum, Woodward, Rubino: *Handbook of Alternative Materials in Residential Construction*

Domel: *Basic Engineering Calculations for Contractors*

Feldman, Feldman: *Construction & Computers*

Frechette: *Accessible Housing*

Gerhart: *Everyday Math for the Building Trades*

Harris: *Noise Control Manual for Residential Buildings*

Hacker: *Residential Steel Design and Construction*

Hutchings: *National Building Codes Handbook*

Jahn, Dettenmaier: *Offsite Construction*

Korejwo: *Bathroom Design, Installation and Remodeling*

Miller, Baker: *Carpentry and Construction, 2/e*

Philbin: *The Illustrated Dictionary of Building Terms*

Philbin: *Painting, Staining and Refinishing*

Powers: *Kitchens: Professional's Illustrated Design & Remodeling Guide*

Powers: *Bathrooms: Professional's Illustrated Design & Remodeling Guide*

Scharff and the Staff of *Roofer* Magazine: *Roofing Handbook*

Scharff and the Staff of *Walls & Ceilings* Magazine: *Drywall Construction Handbook*

Shuster: *Structural Steel Fabrication Practices*

Skimin: *The Technician's Guide to HVAC Systems*

Trellis: *Documents, Contracts and Worksheets for Home Builders*

Vernon: *Professional Surveyor's Manual*

Woodson: *Be a Successful Building Contractor, 2/e*

Dodge Cost Books from McGraw-Hill

Marshall & Swift: *Dodge Unit Cost Book*

Marshall & Swift: *Dodge Repair & Remodel Cost Book*

Marshall & Swift: *Dodge Heavy Construction Unit Cost Book*

Marshall & Swift: *Dodge Electrical Cost Book*

Marshall & Swift: *Home Repair and Remodel Cost Guide*

In order to receive additional information on these or any other McGraw-Hill titles, in the United States please call 1-800-722-4766. Or visit us at www.books.mc-graw-hill.com In other countries, contact your local McGraw-Hill representative.

Kitchen Design, Installation and Remodeling

Pamela L. Korejwo

Leon E. Korejwo (*Illustrator*)

McGraw-Hill
New York San Francisco Washington, D.C. Auckland Bogotá
Caracas Lisbon London Madrid Mexico City Milan
Montreal New Delhi San Juan Singapore
Sydney Tokyo Toronto

McGraw-Hill

A Division of The McGraw·Hill Companies

Copyright © 1998 by The McGraw-Hill Companies, Inc. Printed in the United
States of America. Except as permitted under the United States Copyright Act of 1976,
no part of this publication may be reproduced or distributed in any form or by any
means, or stored in a data base or retrieval system, without the prior written
permission of the publisher.

1 2 3 4 5 6 7 8 9   FGR/FGR   9 0 2 1 0 9 8 7

ISBN 0-07-058070-7 (PBK)
      0-07-058071-5 (HC)

*The sponsoring editor of this book was Zoe G. Foundotos. The editing supervisor was
Sally Glover, and the production supervisor was Sherri Souffrance. It was set in New
Century Schoolbook per the MHT Design by McGraw-Hill's Professional Book Group
composition unit, Hightstown, New Jersey.*

 This book is printed on recycled, acid-free paper containing a
minimum of 50% recycled, de-inked fiber.

Printed and bound by Quebecor / Fairfield.

McGraw-Hill books are available at special quantity discounts to use as premiums
and sales promotions, or for use in corporate training programs. For more information,
please write to the Director of Special Sales, McGraw-Hill, 11 West 19th Street,
New York, NY 10011. Or contact your local bookstore.

Product or brand names used in this book may be trade names or trademarks. Where
we believe that there may be proprietary claims to such trade names or trademarks,
the name has been used with an initial capital or it has been capitalized in the style
used by the name claimant. Regardless of the capitalization used, all such names have
been used in an editorial manner without any intent to convey endorsement of or other
affiliation with the name claimant. Neither the author nor the publisher intends to
express any judgment as to the validity or legal status of any such proprietary claims.

This book is dedicated to the memory of Robert Scharff, who instilled within us an unending fascination and an abiding love for this work. We were given the chance of a lifetime and are truly honored to have had the opportunity to work by his side to discover the path to writing. We thank Bob for the years of knowledge, guidance, faith, and wisdom, which he passed on to us to make this book possible, as well as every book hereafter, to continue his dreams.

This book is also dedicated in memory of one of our best friends, Elmer Nittinger, who dedicated his career to cabinetmaking and kitchen installations. He was truly our first inspiration to the kitchen industry, many years ago.

They are both sadly missed, and although we have been left with the absence of these two great men from our lives, it strengthens us to know that they have been with us every step of the way. They are in our memories daily, and will continue with us on every journey through life.

All of our love and gratitude

p.l.k. & l.e.k.

Contents

Acknowledgments

Organizing a book of this size requires the assistance of many people, corporations, and associations. I would like to thank the following for their help in providing technical data, photographs, forms, and various other materials used to complete this book:

American Home Lighting Institute

American Lighting Association

American Olean Tile Company, Lisa Harlin

Anderson Windows, Inc., Cameron Snyder

Aristokraft, Inc., Melissa M. Crosley

Armstrong World Industries, Jody Harnish

Blanco America

Brandywine Tile & Flooring, Jim McCormick

Broan Mfg. Co., Inc., Keith Lester

Bruce Hardwood Floors, Stephanie Holowak and Maureen King

Elkay Manufacturing Company, Allen Danenberg

Five Star, Lisa Achilles and James Corken

Florida Tile Industries, Inc., Shannon Mitchell

Formica Corporation, Julie Maslov

Four Seasons Sunrooms, Tony Russo

General Electric Co., Dana Fischetti

Halo Lighting-Cooper Industries, Karin Martin

Harris-Tarkett, Inc., Hardwood Flooring

Home Tech Information Systems, Inc., Walter W. Stoeppelworth

Juno Lighting, Inc.

Kitchen & Bath Design News

Kitchen Cabinet Manufacturers Association (KCMA), Janet Titus

Kohler Company, Larry Booth

Leon E. Korejwo, Illustrator

KraftMaid Cabinetry, Inc., Ilana Isakov

David G. Leitheiser, master electrician (owner of the stone)

Mannington Resilient Floors, Joe Fisher

Merillat Industries, Inc., Wendy Knox

National Kitchen & Bath Association, Nick Geragi

National Kitchen Cabinet Association

Perstorp Flooring, Inc., Dave Kroencke

Planit 96

Progress Electric Supply Co.

Sub-Zero Freezer Company, Inc.

The West Big Beaver, Inc.

Thomson Consumer Electrics, Frank McCann

United States Gypsum Company

Wellborn Cabinet, Inc., Kristen Hammond

In addition, thanks to:

- Bill Yatron of Adelphi Kitchens, Robesonia, Pennsylvania, for continuous assistance and advice.
- Jeff Fichthorn and Allen V. Hanson of QuakerMaid Kitchens of Reading, Inc., for always being available to answer questions and provide assistance.
- The following for their help in editing and coordinating the text and art for this book: Carole A. Korejwo, Christopher L. Korejwo, Leon E. Korejwo, and Lauri A. Kohler.
- A special thank you to my illustrator and my father, Leon E. Korejwo, for all of the support, guidance, and knowledge you provided me throughout the completion of this book. Without you, this book would have never been possible.
- To my family, who is always there when it matters, for the big things and the little. Every hour, every moment of every day, I will always love you.
- A special thank you to my grandmother, Dorothy T. Eltz, whose strength and independence has led me to follow in her path. You are truly a hard act to follow. As always, "Everything I know, I learned from my grandmother..."
- One final thank you to April Nolan and Zoe Foundotos, acquisitions editors, as well as the entire editorial staff of McGraw-Hill for their continuous encouragement and support.

Introduction

There has been significant growth in the residential kitchen industry, in both remodeling and new construction. This growth has created the need for knowledgeable and skilled kitchen installers. The complexity of kitchen remodeling projects has increased along with the need for professionalism in installation. Few consumers have the time to be do-it-yourselfers; instead they look for professionals who can handle the entire project for them. Today's customers want accountability, speed, choice, and convenience, as well as quality.

This book was written for the professional kitchen installer/remodeler with knowledge of basic construction skills. It concentrates on the skills and techniques (directing attention to major tasks, rather than an extremely detailed step-by-step process), as well as the basic business concepts, needed for the successful installation of kitchens. The demand for service and quality is more important than ever, and, therefore, the need is greater than ever for installers who are personable, well-trained, technically adept, and professional in every way. A professional installer must have the skills and knowledge to meet the high standards that today's consumers demand.

Installers must be able to competently manage a project from start to finish. They must not only have specific and detailed knowledge of cabinet and countertop installation and carpentry work, but they need to have a thorough understanding of all appliances and equipment items installed in kitchens and a base knowledge of the mechanical systems (electrical, plumbing, HVAC, and cooking equipment ventilation), drywall, painting, wallpapering work, and other associated items, as permitted by local laws and codes. A successful professional installer has mastered the principles of design to make decisions in the field and communicate with the designer, has a clear understanding of safety rules and procedures, and knows how to deal effectively with clients.

The *Kitchen Design, Installation and Remodeling* shows anyone from a nonprofessional craftsman with basic construction knowledge, to a cabinetmaker, to a home construction contractor, the different skills needed to expand into a full business home construction contractor. In addition to the self-employed installer (who needs to know every aspect of the installation business), this book can help someone who has strengths in one area (such as cabinet installation) to recognize where his or her weaknesses may be so that he or she can learn about those areas or successfully subcontract to others.

For already established kitchen installers, the *Kitchen Design, Installation and Remodeling* gives a good overview of all aspects of the business, possibly from a different point of view, so that they may improve their knowledge and skills, as well as address issues of which they may not have been aware.

Written in a clear, easy-to-understand format, this book describes an installation project as it physically progresses and unfolds in a logical process, from initial customer contact, to concept development fabrication, all the way through the installation to the job completion. This approach allows the reader to decide what aspects require increased effort, subcontracting, hiring, or training, to arrive at the level today's customer demands.

This book includes more than 300 fully detailed illustrations to help you visualize the functions necessary for a successful kitchen installation. The photographs and illustrations enhance the text and make it easy for the reader to grasp the principles and materials presented. Checklists, charts, and tables offer quick, instantly usable data.

Chapter 1, "The Kitchen Installation Business," covers the basics of what the kitchen installation business is all about. It includes dealing with the client, customer communication, contracts, proposals, codes and permits, employees, code of ethics, working relations, job safety, and more.

Chapter 2, "Sizing Up the Job and the Customer," discusses the foundation of your business: meeting with the customer to size up the job, scheduling the job, and obtaining the equipment necessary to prepare for the installation project.

Chapter 3, "Basic Principles of Kitchen Design," gives an overview of the basics for kitchen design. For the kitchen to be functional, the designer must take into consideration its location within the home, its size and shape, and the arrangement of its equipment and work centers. Whether or not the kitchen installer is involved with the design process, it is important to master the principles of design so the installer can make decisions in the field and communicate effectively with the designer.

Chapter 4, "Kitchen Extras," is for the installer who may need to assist a client in choosing kitchen extras. This chapter provides many unique ideas.

Chapter 5, "Kitchen Cabinets, Storage, and Countertops," provides the knowledge necessary for assisting the client when choosing cabinets and countertops. Cabinets are a permanent part of any kitchen and therefore should be chosen with care for their quality and lasting value.

Chapter 6, "The Work Center Appliances," provides a thorough understanding of all of the appliances and equipment typically installed in kitchens. The installer should be knowledgeable in appliance sales, customer education, and maintenance of appliances.

Chapter 7, "Developing a Design from Concept to Plans," gives detailed steps to measuring the kitchen and familiarizes the installer with all of the different types of construction drawing sets. Not all kitchen installers will be responsible for preparing kitchen plans; however, all need to be able to read and understand them.

Chapter 8, "Getting the Kitchen Ready," covers the steps necessary to prepare the kitchen for the final installation. The kitchen remodeler must be on the lookout for design and structural traps. For example, work put in by amateurs is often overengineered (and hard to remove) and often ignores codes (which makes it harder to get the job accepted) and standard practices (so studs, pipes, and wires may not be where one would expect them). These steps include the proper removal of appliances, old cabinets, countertops, and flooring. The structural system is explained, including the removal of existing walls or partitions, doors, or windows, as well as preparing for new walls, doors, windows, and flooring.

In many situations, the installer will hire licensed subcontractors for mechanical systems, but the installer needs to have a clear understanding of electrical, plumbing, HVAC, and cooking ventilation to accurately plan, coordinate, and install a kitchen. Chapter 9, "Mechanical Systems," provides that knowledge.

Chapter 10, "Installation of Cabinets, Countertops, and Appliances," provides specific and detailed knowledge of cabinet and countertop construction and installation. The installer needs to be an expert at all of the related skills that pertain to the installation trade.

Chapter 11, "Finishing the Project," covers the basics on finishing products for floors, walls, ceilings, and lighting. If the installer subcontracts the finishing work or has excluded it from the contract altogether, the ultimate responsibility for the client's satisfaction lies with the installer. Being knowledgeable in these areas is important to an installer. In addition, this chapter covers the final cleanup, as well as a post-completion meeting to review the completed installation. The installer needs to supply and coordinate the installation of a wide variety of floor, wall, and ceiling finishes and understand the installation techniques for each of these materials.

1

The kitchen installation business

The kitchen may be the greatest potential attraction to a customer in the decision to purchase a new home or remodel an existing one. What can be accomplished with a kitchen is not something that homebuyers or home-owners may easily visualize, and, therefore, the possibilities may be over-looked. The role of a professional kitchen installer may be the factor that turns a questionable or out-of-date kitchen into a homeowner's dream. What a creative and experienced kitchen professional quickly envisions can turn the potential client into a satisfied buyer. This chapter discusses the place of the kitchen builder in a rapidly expanding business.

What is the kitchen installation business all about?

According to the National Kitchen & Bath Association (NKBA), a leader in set-ting standards in the kitchen and bath industry, consumers spend $27 billion on kitchen remodeling in a good year. Add a million-plus kitchens in new-home con-struction, and you get an idea of the size of this industry. The kitchen indus-try can be large and complicated, with an ever-widening range of products to furnish a room that serves multiple purposes. The functions of other rooms in a house are defined by the room's furnishings, and those furnish-ings can be easily changed or moved around (i.e., a bedroom can be changed to a home office at any time simply by moving the furnishings). The kitchen, on the other hand, must be designed and installed with permanent wiring and plumbing.

Kitchen installations require much more than just the ability to physically install cabinets, fixtures, and appliances. The kitchen builder must combine many skills and trades to merge the needs and budget of their clients with the products available and the physical limitations of the space available (Fig. 1-1). Professional installers are, in many cases, running a business. They must also be able to competently manage a project from start to finish. To ensure that installers can run an efficient and profitable business, they must be equipped with the skills necessary to realistically estimate and price projects, produce professional proposals and documents, and keep accurate records. They must also have the skills to plan and closely supervise an installation schedule.

Today's consumers are looking for the most efficient way to handle their kitchen projects. Few consumers have the time to be do-it-yourselfers; they look for professionals who can handle the entire project for them. They want speed, choice, convenience, high-quality materials, and excellent service and accountability at all stages of the project. A professional installer must have the skills and knowledge to meet the high standards that consumers demand.

Installers do not only work with homeowners. In addition, they provide installation services to kitchen retailers, dealers, designers, contractors, home centers, and lumberyards.

The kitchen marketplace

According to the NKBA, kitchens are sold in three major marketplaces: replacement, remodeling, and new home construction.

Figure 1.1 A kitchen installer must envision all phases of the project, from structure through staining to job completion. (*Leon E. Korejwo*)

Replacement

In the replacement market, homeowners or builders buy cabinets or appliances to replace what has grown old or worn out. Price is usually the main consideration. While design is seldom involved, good salesmanship is important to the kitchen builder who is involved in the replacement business.

The replacement sector includes cabinet resurfacing, which is often offered as a service by a kitchen-specialist firm. Resurfacing, which sometimes includes purchasing new doors and drawer fronts, is a good way to get a new look in a kitchen, although it might preserve an inadequate design.

Remodeling

By far, the largest market is kitchen remodeling. Over the years, the complexity of kitchen remodeling projects has increased and, therefore, has created the need for competent professional installers. Traditional kitchen remodeling, typically cabinet, appliance, and fixture replacement, along with other minor cosmetic changes, has given way to more sophisticated and extensive remodeling projects. In the past, most kitchen remodeling projects were done within the existing kitchen. Now many major projects involve dramatic expansions beyond the existing space (Fig. 1-2). A large amount of remodeling is performed by do-it-yourselfers who actually do it themselves or buy the materials and then hire the work out. Retail home centers are large sources of supplies for do-it-yourself kitchen remodeling. These customers buy cabinets at home centers, building supply stores, or similar outlets that offer stock cabinets, countertop blanks, and a cutting station to cut mitered corners, make sink cutouts, and otherwise work the tops. These sales are over the counter, with little planning or design involved.

In many cases, however, these outlets set up separate, identifiable kitchen departments with separate trained personnel. In such cases, they may offer stock and custom cabinets, as well as professional design service. Many home centers now have computer programs to aid the do-it-yourselfer.

Homeowners, interior designers, and contractors who want professional design and planning for a new kitchen can go to the showroom of a kitchen specialist, preferably one who is a member of the National Kitchen & Bath Association and with certified kitchen designers (CKDs) on staff (Fig. 1-3). High-end remodeling normally is done by professionals at a kitchen specialist's dealership, which is where most kitchen remodeling jobs are sold.

Professional design and planning services also are available from independent designers and architects, if they qualify as kitchen specialists. The independent kitchen designer is a growing factor in design and product specification. Such designers do not have showrooms, but they might take clients around to visit showrooms of other specialists. Professionalism in this specialized field requires detailed knowledge of kitchen products, materials, and techniques. Some architectural or design firms set up separate kitchen departments in which a level of advanced training is maintained.

Figure 1.2 Expanding a kitchen may include an outside addition. (*Leon E. Korejwo*)

Figure 1.3 Kitchen showrooms can help clients plan and visualize their new kitchen. (*Leon E. Korejwo*)

Residential remodeling is now being recognized as a major part of the construction industry. Of all residential remodeling projects, 80 percent were professionally installed. Residential remodeling is expected to climb at a rate of approximately 5-10 percent per year, according to Home Tech Information Systems.

Clients look to the company from which they are buying the product or service to be accountable and to service what they sell. They have a right to expect the most for their investment. As a result, kitchen specialists find it necessary to handle the entire job from inception to installation—that means the total remodeling job if the kitchen is part of a larger project. This way, all aspects of the project can be controlled for quality.

As reported by Walter Stoeppelwerth of Home Tech Information Systems, the building industries are going through total quality management and changing the way they do business. One of the most significant changes is from a top-down to a bottom-up approach to management. The old way was to have each worker handle one small part of the process, with supervisors for every 10 or 15 people to make sure quality was maintained. The new trend is to build quality teams of workers who are responsible for many different jobs. They also have the responsibility for building quality the first time.

A current trend in the remodeling kitchen business is the introduction of lead carpenters or lead installers. This lead worker is assigned the responsibilities for the job from start to finish and is given help on an as-needed basis. Basically, kitchen projects require one or, at most, two people to work effectively.

If the industry is to gain maximum efficiency from installation personnel, these workers must not only be able to do all of the cabinet and countertop installation and carpentry work, but also the electrical, plumbing, mechanical, drywall, painting, wallpapering work, and other associated items, as permitted by local laws and codes. Codes in some areas may limit the areas of work that an installer may perform, and the building codes always prevail. In these instances, subcontractors must be hired.

New home construction

In most cases, new home construction is served by kitchen distributors that keep several lines of stock and semicustom cabinets in their warehouses for builders. Kitchen distributors often serve as independent factory representatives for custom-cabinet producers and act as liaison between a custom manufacturer and both home builders and kitchen specialists in the area. Kitchen dealers also serve specialists in the construction industry who focus on featuring custom cabinetry in their upscale homes. Both kitchen dealers and distributors set up showrooms and offer a professional design service to which the builder may bring a client for visual support in the early planning stages.

Changes in new home construction have also contributed to the growth of the kitchen industry. In the past 10 to 15 years, new home builders have begun to realize that kitchens (as well as bathrooms) are no longer basic utilitarian spaces; they have become the prime selling features of homes. The more than one million new homes being constructed each year represents a substantial portion of the kitchen market. Those involved in new home construction have started to pay more attention not only to the design of kitchens, but also to the quality of the products that are installed, thereby making these new homes more attractive to potential buyers. With the tremendous increase in the variety and styles of cabinets, countertops, and appliances, there is an increasing need for sophistication and technical expertise in kitchen installations.

The NKBA has stated that a significant trend gaining momentum in the new home segment of the industry is that builders, especially tract home builders and developers, are requiring that suppliers of building materials such as roofing, siding, doors, windows, fireplaces, and closet accessories provide the materials on an installed basis. This fact is especially true for kitchen cabinets. This trend will lead directly to an increasing demand for trained kitchen installers in the new home market over the next 10 years.

Builders who use their carpenters to install specialty products (such as kitchen cabinets) have found that, while they may save a small amount initially on labor costs, the additional costs of service or adjustments due to the lack of knowledge of certain features of these products makes the use of a specialized kitchen installer more cost-effective. As discussed later in this chapter, the warranty offered by the installer is also an attractive feature for residential contractors.

Today, the buyer of a kitchen can also be a middleman—a developer or home-builder—rather than the homeowner, with the ultimate purchaser, or end user, being the homeowner. Therefore, the dynamics of dealing with the client are somewhat different.

A number of key competencies have been identified by the NKBA as important to kitchen installation. Depending on your experience or the number of installation projects you have completed, your familiarity with these key competencies will vary. It is incumbent to your professional success to master and use these key competencies.

1. *A basic knowledge of kitchen design.* While you may not be required to do the complete design, you need to master the principles of design so that you can make decisions in the field and communicate effectively with the designer.

2. *A clear understanding of installation and construction terms used in the industry.* The complex nature of this profession requires you to be familiar with a wide array of terms used by installers and the other related trades who work with you.

3. *A thorough understanding of carpentry skills.* Your base of knowledge and experience in carpentry needs to include everything related to the construction of a shell, including the construction of a stud wall and other framing; door and window installation and the construction of openings for these doors and windows; blocking required for cabinet and fixture installation; and general finish trim requirements.

4. *Specific and detailed knowledge of cabinet and countertop construction and installation.* Obviously, this is the basic service that you are providing, and

you need to be an expert at all of the related skills that pertain to the installation trade, including everything from field measurement and survey work to actual layout and installation of cabinetry and equipment.

5. *Familiarity with finish materials.* You are required to supply and coordinate the installation of a wide variety of floor, wall, and ceiling finishes, including vinyl tile and sheet goods, wood, ceramic tile, paint, and wallpaper. You need to understand the installation techniques for each of these materials and their impact on the cabinet and fixture installation.

6. *A thorough understanding of all of the appliances and equipment items that are typically installed in kitchens.* You need to understand standard installation and mounting techniques for these items, as well as their electrical, plumbing, and ventilation requirements.

7. *A basic knowledge of mechanical systems.* Familiarity with all of the various mechanical systems and their impact on your project is valuable. These systems include electrical, plumbing, gas, heating, ventilation, air conditioning, and cooking-equipment exhaust ventilation. The goal here is not to eliminate the need for electricians, plumbers, and other tradespeople, nor to circumvent code requirements, but to have the ability to perform some basic mechanical tasks, including the removal of electrical and plumbing fixtures to permit tear-out of existing kitchens and the installation of fixtures and equipment such as lighting fixtures, range hoods, sinks, or dishwashers where the electrical or plumbing rough-ins have already been completed. A good, basic knowledge of all of these mechanical systems equips you to plan and sequence efficiently the work of the various mechanical trades with your installation work.

8. *Techniques for ceiling construction.* You should know how to install ceiling joists and frame soffits in preparation for drywall, plaster, or other finishes as necessary. When needed, you should also be able to install a suspended ceiling system with acoustical ceiling tiles. Your ability to do this work well significantly enhances your cabinet installation work.

9. *A good general mechanical ability and a clear understanding of how to maintain tools and equipment.* While you may be a cabinet installer first, you need to possess a good general mechanical ability that allows you to deal with all of the side issues that arise in conjunction with cabinet installations. This mechanical ability also includes understanding and using all of the state-of-the-art tools and equipment available to the profession and maintaining this equipment in good working condition.

10. *An understanding of the importance of preinstallation conferences.* The consistent scheduling of preinstallation conferences is important to your success as an installer. You must understand what items require discussion at these meetings and who the key players are.

11. *Strongly developed people skills.* To deal effectively with clients and successfully sell your services, it is vital to be able to get along with people,

including establishing comfortable relationships with clients and working well with others involved in your installation projects.

12. *The practice of good job site management.* An understanding and adherence to effective staging and sequencing of an installation job is extremely important to ensure that the job is completed in a timely and efficient manner. A thorough knowledge of staging and sequencing enables you to provide the client with a realistic schedule for your work.

13. *Adherence to job site and safety recommendations.* To protect yourself, your client, and the client's home and possessions, you must have a clear understanding of established safety rules and procedures and the most effective way to carry out these recommendations.

14. *Familiarity with the legal issues facing the profession.* You are running a business and therefore are subject to laws that regulate and affect your work. In the eyes of the law, ignorance is no excuse. You must be aware of your liability and understand the best ways to protect yourself from litigation.

Ethics

Professional installers must maintain the highest standards of honesty, integrity, and responsibility while conducting business within the profession. You should always strive to uphold these ethical standards in your work, with your fellow professional colleagues, and in your dealings with the public. As a result, you quickly gain a reputation for honesty and integrity. This code of ethics earns high regards from your clients as well as continued work from dealers or designers. A satisfied client is truly the best form of advertising.

The code of ethics, quoted here from the NKBA, applies to all of your professional activities, whenever or wherever they occur. They do not supersede or alter in any way the local or federal regulations that control this profession or the general provision of services to the consumer. Your ethical obligations include the following:

1. Uphold all laws and regulations while conducting your professional activities, and report to the proper authorities those who are knowingly violating established laws.

2. In all of your dealings with the public, represent the profession in a positive light and continually attempt to advance the stature of the profession in the public eye.

3. Always strive to improve your knowledge and skills in the profession and offer your clients the best possible service available.

4. Serve your clients in a timely and competent manner. Promptly acknowledge and respond to all customer complaints and seek mutually agreeable solutions.

5. Uphold the human rights of all of the individuals with whom you deal without regard for race, religion, gender, national origin, age, disability, or sexual orientation.

6. Offer only those warranties or guarantees for products or services that you can actually fulfill.

7. Always provide contracted products and services to the best of your ability, and honor all contractual obligations (unless they are altered or dissolved by the mutual consent of all parties).

8. Always respect your clients' privacy, and keep confidential any sensitive information concerning your dealings with clients and their homes.

9. Only use the terms *insured*, *licensed*, or *bonded* in your advertising if you can provide proof of these to the client.

10. Always be clear in your description of the products and services you are including, as well as those that are specifically excluded or are available at additional charge.

11. Always honestly represent your experience in the profession and the product or service that you are offering to your client.

12. Never attempt to reuse a design prepared for one client for another client.

13. Compensation paid to you by the client for services performed by subcontractors under your supervision should be paid to the subcontractors without delay.

14. Accept additional or side work from a client for whom you are working under contract from a dealer/designer only with permission of that dealer/designer.

15. Free competition among members of the profession is encouraged, but honesty and fairness in the pricing of products and services are your professional responsibility.

Client relationships

In this industry, there are many different types of relationships between the kitchen installer and the customer. These include the turnkey installer working for a kitchen specialist, the installer working as subcontractor to a contractor, installer working for the buy-it-yourself (BIY) homeowner, and the installer working for a home center or lumberyard.

The turnkey installer

The cabinetmaker subcontracted by a kitchen designer who is also acting as general contractor is one type of relationship. Many kitchen specialists are not in the contracting business, but they provide construction management services for the homeowner when requested. In this role, the dealer/designer assists the homeowner in finding a "turnkey" installer who can handle the ordering of all materials and payment of the subcontractors, coordinate the subcontractors, and basically run the job.

In this case, the installer really has two masters. While you are working directly for the clients, and are likely to be paid by them directly, you may also be working under the supervision of the dealer/designer.

Prior to the start of the job, be sure to thoroughly review the plans and investigate the job. If you discover any problems, notify the dealer/designer directly.

It is important to work together to come up with a solution, as well as to convey to all involved parties any revisions.

The installer as subcontractor

The installer as subcontractor to a remodeling contractor or to a contractor who is building a new home is another type of relationship between installers and clients. You may be working as an independent contractor or on an hourly basis. In either case, the general contractor takes responsibility for the total job. However, this does not mean that you are not working for the client. Most importantly, while you are being paid by the contractor, you are working for the client.

Using the bottom-up management approach as recommended by Home Tech Information Systems, you take over the job at the preconstruction conference. The chain of command is then passed from salesperson to installer, and from then on you are responsible for dealing with the homeowner or client, meeting the conditions of the contract, writing up change orders, completing the job to the client's satisfaction, and collecting payments.

The buy-it-yourself homeowner

Many people purchase their own fixtures, cabinets, and appliances, and then hire an installer to handle the installation. These homeowners are basically do-it-yourselfers who do not want to actually undertake the kitchen installation project. However, they are willing to purchase the fixtures, cabinets, and appliances themselves, contract directly with the plumber, electrician, flooring contractor, and heating and air-conditioning mechanic, then hire an installer or remodeling contractor to do the installation. The installer, in this case, is working directly for the homeowner, probably on an hourly or a fixed-price basis. The homeowner may pay the installer a small percentage above actual labor costs to coordinate the efforts of the other subcontractors.

Today, this relationship between the client and installer is popular. Many home centers offer kitchen design services and then sell the cabinets to their customers at competitive prices, roughly equal to what a contractor might pay.

Home center or lumberyard installations

Finally, there is the installer who works for a home center or lumberyard. In this instance, the home center or lumberyard sells the materials to the homeowner and may offer installation service only or all of the subcontractor trades. Although the installer is paid by the home center or lumberyard, the responsibility to satisfy the end user or client still exists.

As described, there are many possible relationships in the kitchen business. Ultimately, however, you are working for the end users, and you must always strive to meet the highest standards of quality to successfully complete the job and therefore receive payment. While the chain of command may vary, the steps to complete a quality installation differ very little.

Dealing with the client

There is much more to this business than just providing a quality installation. You must possess the necessary people skills to deal with clients, whether they are developers, builders, or homeowners. If you are dealing with homeowners, the installation of a kitchen may be the largest construction project they have ever undertaken. As a result, they may have many questions and fears concerning the installation work. The more you can do to develop solutions to these problems and fears, the better off you are.

Maintaining a professional image with clients is of utmost importance at the beginning of an installation. The clients are counting on you to build a kitchen or transform an existing kitchen into the room of their dreams.

Soon after a job starts, your positive image with the client may begin to drop. Remodeling a kitchen is tough for everyone involved. Keep in mind that in many cases you are putting the client's kitchen completely out of commission for quite some time. Obviously, this situation can be very inconvenient. It becomes even worse if the project runs longer than originally anticipated. Every time the job is delayed a day or two waiting for materials or for the plumbing or electrical inspector, the client's anger increases. Obviously, these instances detract from your image.

You must be aware of when you need to meet with the client to help alleviate any problems. You may need to clear the air, explain any problems, and give them a revised schedule for completion. If you are organized and efficient, by the time the installation is complete, your relationship with the client might have risen back to its original level.

It is important to establish a realistic assessment of the time required to complete the installation project. A preconstruction conference is usually required, involving the client, designer, and installer. At this meeting, the client is advised as to what to expect during the installation, and questions concerning the installation are answered. If delays do occur that affect the project schedule, inform the client as soon as possible. Clearly explain the situation and the anticipated length of the delay. The most common complaint by clients is unexplained delays. Also at this meeting, it is important to make the clients feel they are involved with and are an important part of the whole installation process. When questions arise during the installation, you should respond to them immediately and courteously.

Client communication

Establishing good communication

The number one reason relationships between installers and their clients fail is lack of communication between parties. It is imperative for an open line of communication to be established with the client at the preconstruction conference, and it is just as important to continue it throughout the project. Keep the client informed at all times.

Your client has, up to this point, most likely been dealing with a retailer, dealer, or designer who has spent much time listening to their needs and desires for a new kitchen. Now the responsibility for the client and maintaining the established line of communication is shifted to the installer. Provide the same level of open communication as in the design and sales stages of their kitchen project. It is important that the retailer, dealer, or designer clarify to the customers that responsibility is being passed to the installer. It is also important for the initial contact person to express confidence in the installer's abilities. If this transition does not happen, the customers continue to call the dealer or designer with questions and concerns. This situation is neither efficient nor practical since the dealer/designer does not have firsthand experience of what is occurring during the actual installation process. If two or more customers are involved, installers must first determine which one is the designated customer to communicate with.

Because installers are not typically involved in the design and sales of kitchen projects and are not introduced to customers until later, they must make a concerted effort to gain their trust so that the line of communication can be established. The preconstruction conference is the time to do this (see Chap. 2). Encourage the customer to view the relationship with the installer as a partnership. The more customers are made to feel a part of the process, the more comfortable they are in handling project communication—both positive and negative.

Many clients may not be at home during the day when work is being completed, so phone communication has become the standard for answering questions. The telephone works fine for passing small bits of information (i.e., changes in the job schedule); however, explaining large problems and their potential solutions by phone can be difficult. It is much better to meet with the client in person to answer questions or solve problems. Set up an appointment as soon as possible to review or repair any problems. This way, there is much less room for misunderstanding.

An excellent way to communicate between the installer and client is with a memo or on-site communication form (Fig. 1-4a). Let the client know that this form can be used for transmitting questions and concerns. Establish a good place to keep the forms during the installation (Fig. 1-4b). When the customers have a question or concern, they can simply write it down and leave it for the installer to reply. This procedure also provides written verification of unplanned changes or uncalculated cost overruns. Keep in mind that installers must be sensitive to the wide range of languages spoken, and, if necessary, provide installers who can communicate with clients of a different language. It is advisable to size up the client. They may become overly inquisitive or indecisive, causing expensive delays or changes. The person in the field then must be diplomatic while proceeding with the work.

If no one will be at the project site for a day or two, or the project is delayed, make sure the client knows, and clearly explain why. Otherwise, they may assume that their project is being pushed aside in favor of someone else's project. Be honest with them. Getting caught deceiving the client results in loss of trust and creates a negative reputation.

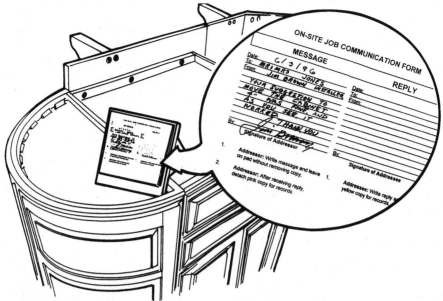

Figure 1.4a Always keep an on-site job communication form in a central location to keep good communication with the client. (*Leon E. Korejwo*)

ON-SITE JOB COMMUNICATION FORM

MESSAGE	REPLY
Date:_____ Message No._____ To:_____ From:_____ _____ _____ _____ _____ _____ _____ _____ By:_____ **Signature or Initials of Addressor**	Date:_____ Message No._____ To:_____ From:_____ _____ _____ _____ _____ _____ _____ _____ By:_____ **Signature or Initials of Addressee**

1. **Addressor:** Write message and leave on pad without removing copy.

1. **Addressee:** Write reply and detach yellow copy for records.

Figure 1.4b An on-site job communication form. (*Home Tech Information Systems, Inc.*)

Patiently and thoroughly answer every question that a client asks, despite how ridiculous it may seem. Remember, installers have been through possibly hundreds of kitchen installations. But for the clients, this one may well be their first. They may have many questions because everything is new to them.

Communication with clients is not necessarily limited to the homeowner. On larger projects, installers may also need to communicate with an architect or designer who may be supervising the overall construction of a project. Contract installers also must deal with retailers, dealers, and designers as clients. The same basic rules apply to all project communications. Be honest, answer their questions fully and patiently, and keep them informed.

It is important to realize that there is a real cost involved in taking the time to effectively communicate with clients. This time must be planned for as a part of the whole installation project.

Dealing with other professionals

Besides dealing with the homeowner, your primary client, you will also be dealing with tradespeople who will contribute to the success of the installation project.

Subcontractors

Because almost all kitchen projects include plumbing and electrical work, and many also require ventilation and HVAC work, subcontractors who specialize in these trades are crucial to the installation. Even if installers could do some of this work themselves, many building codes require licensed professionals to handle these specialty tasks. Installers must build a team of subcontractors who can work together to deliver a project in a quality and timely fashion.

The installer is looking for professional subcontractors who work to meet the schedule, keep the job site clean and organized, handle problems, provide prompt warranty service, and do a top-quality job. Installers want their subcontractors to provide the same level of quality that they provide. The clients do not distinguish between subcontractors and the installer's company. In their eyes, everyone is working for the same company.

It is valuable to spend time recruiting and managing subcontractors. The subcontractor's overall appearance and skill in communication should be evaluated. Installers should also ask for and check out references with other installers and retailers, dealers, and designers on similar projects. This step is especially important when determining plumbers' and electricians' level of experience with similar projects.

Installers should set up monthly meetings with their subcontractors to review the job progress, as well as to go over the subcontractor's performance on these jobs. Their comments and suggestions may lead to more efficient ways of integrating subcontractors into the installer's project. These meetings help keep communication open as well as build team spirit between installers and subcontractors.

Kitchen dealers and designers

Your first contact is with the dealer and designer (who may or may not be the same entity). Your relationship may vary somewhat with the dealer and designer depending on whether you are an installer working as an employee of a kitchen retailer or have been contracted by a dealer or designer to do the installation. Field problems and design interpretations are referred to the dealer or designer for resolution. The clients are most likely to contact the dealer or designer if they have a problem during the installation or are dissatisfied, so many of these issues are relayed to you through the dealer or designer.

Kitchen designers are specialists in kitchens. These individuals are often well-informed about the latest trends in furnishings and appliances, but they may have neither the structural knowledge of the architect nor the aesthetic skill of a good interior designer. Their experience and expertise may be required in special design. They prepare sketches—via a computer simulation or the traditional way—as to what things will look like, draw plans for the contractor, and even supervise the work. If the plan has any structural changes, an architect's services are a good idea and may even be required by code.

If you are working with a kitchen designer, look for a member of the National Kitchen & Bath Association (NKBA) or a certified kitchen designer (CKD). Each association has a code of ethics and a continuing program to inform members about the latest building materials and techniques.

Architect

In some cases, you need to coordinate with an architect. Architects are state-licensed professionals with degrees in architecture. They are trained to create designs that are structurally sound, functional, and aesthetically pleasing. They may be required by code in some areas. Architects also know construction materials, can negotiate bids from contractors, and can supervise the actual work. If stress calculations must be made, architects can make them; other professionals need state-licensed engineers to design the structure and sign the working drawings. If the project involves major structural changes, an architect should be consulted. The architect should visit the project regularly, and may, at times, want to adjust or make changes to the design in the field.

If you are an employee of or under contract with a dealer or designer, these requests for changes should be referred to the dealer or designer with whom the client has a contract. The architect may ask whether a change being considered is feasible or what problems might be encountered if a change were made, and you should provide this information so that an informed decision can be made.

Interior designers

Even if you are working with an architect or kitchen designer, you may wish to call on the services of an interior designer for finishing touches. These experts specialize in the decorating and furnishing of rooms and can offer fresh, innovative ideas and advice. Through their contacts, a client has access to materials and products not available at the retail level.

Contractor or lead carpenter

If the scope of work for the kitchen installation project includes structural modifications or expansion of existing space, a contractor or lead carpenter is a part of the project team. The role of the contractor would be to coordinate the efforts of the subcontractors, and, in some cases, to handle some of the construction work. Often, the contractor plays the role of lead carpenter and actually does most of the construction work.

Plumbers, electricians, and HVAC technicians

Plumbing, electrical, heating, ventilation, and air-conditioning work generally is done by licensed subcontractors (Fig. 1-5). Because these people provide the services that make a kitchen function in which your installation must occur, you need to coordinate closely with them to ensure that the total project runs smoothly and efficiently. Establish as early as possible the place in the schedule for your part of the project relative to these other trades. You may be unable to perform your job before they have done theirs and, as a result, you may be the one who appears to be behind schedule.

Figure 1.5 A subcontractor who specializes in plumbing is crucial to the whole kitchen installation. (*Leon E. Korejwo*)

Code enforcement officials

Finally, you are required to have portions of your work inspected by local code enforcement officials. Often, the contractor (or lead carpenter) arranges for these inspections, so you need to coordinate scheduling and completion of inspections to ensure that your work is not delayed. Code enforcement officials have the authority to delay an unsatisfactory or noncomplying job. If a permit is required and has not been applied for, the official may stop the job or subject a fine. Establish who is responsible for the permit application—the homeowner, contractor, or you. Electrical, plumbing, and disposal may require permits. If enlarging a home is in the plan, local codes may require a zoning change. The municipal zoning officer or building inspector should be consulted if a question arises.

Working relations

Professional appearance

Always keep in mind that personal appearance can say a lot about the way you do your work, as well as the quality of the finished project. A neat and professional appearance suggests to the client that your work will be done neatly and professionally. Torn or soiled clothing and a sloppy appearance may suggest a poor finished product. While this may not actually be the case, a client's first impression is hard to change. Remember, you and the members of your staff are walking advertisements for your company.

Today, in most cases, it is not unusual for both adults in a household to be working outside of the home. Therefore, in many cases, you may be left alone to work in their home. To the client, you are a stranger. As irrelevant as it may seem, a good personal appearance goes a long way to ease their fears about leaving their home. It is advisable to have work shirts, or uniforms, with the name of your company. This presents a neat, organized team look and easily identifies you and your staff. Although it may seem like common sense, you should be clean and well-groomed at all times. However, clients tend to be a bit more forgiving by the end of a long day of work.

In addition, the service vehicle brought to the job site should be clean and in good condition and, if possible, should bear the name of your company. Also, equipment and tools brought into the home should be organized and well maintained, not only for the efficiency of your work, but also to contribute to the client's first impression of you. Another important item to keep in mind is that smoking inside the home greatly increases your liability for the installation project. Do not do it under any circumstances.

Designer awareness

Today's clients can go to any bookstore or newsstand to find magazines full of design ideas and, as a result, are more sophisticated and better informed than ever before. This increased awareness of design options has everyone interested in good design, regardless of the project budget. Interest in personalized design has led to a trend toward highly individualized kitchens. Today's designers are being challenged to create much more than just a functional environment.

For you as an installer, this trend means that there is an increasing level of complexity to the installation of these nontraditional designs. While the designer has the responsibility to follow accepted design principles and good common sense, you will often be asked to create new installation techniques to respond to the designer's intent.

Each installation must be approached as a team effort between the installer, designer, manufacturer, and other tradespeople. You need to be considerate of all parties involved. Be certain to maintain an open line of communication between these parties to ensure the success of the project. It is your skill that makes the design a reality. No matter how carefully a project site has been verified and surveyed, there will, in most cases, be minor adjustments required to transform the design into the finished product. You must assess what changes are needed from the original design intent and work with the designer to develop the correct measures.

When problems and conflicts arise between the design and actual field conditions, you should always be supportive of the design and the products being installed. Explain in a positive manner what measures can be taken to produce, as closely as possible, the original design intent.

Maintain a professional attitude

Never degrade the design, the designer, the manufacturer, or other persons working on the project in front of the client, no matter how frustrated you might be. Just as importantly, never blame problems that you may be experiencing with the installation on any of the other members of the team. The client's confidence in the project will be lost. A healthy mutual respect for the skills and limitations of all of the team players is the key to any successful kitchen remodeling effort.

Job safety for workers

Anything that you can do to promote safety on the job is of great benefit to you; accidents are very costly to your business. Frequent accidents increase your workers' compensation rates, and you do not want to have to raise your prices to cover these higher rates. Injuries on the job also result in loss of time for the injured person and delays in your project.

You can prevent most accidents fairly easily if you recognize the high-risk situations that cause most injuries. Most of the key situations you encounter on the job include

1. Failure to use safety equipment such as face masks, goggles, or hard-soled shoes

2. Improper handling of large or heavy machinery or equipment

3. Careless or improper use of machinery or equipment

4. The collapse of ladders or other temporary platforms

5. Failure to leave safety guards on portable saws.

The wedging of safety guards has injured more workers than almost any other single factor. Take the time to anticipate safety risks, and plan to avoid accidents. The few minutes that it takes to avoid an accident can save hours or days of project time.

Since you cannot be on the job 24 hours a day, ensure that dangerous equipment and tools are not left out. Daily cleanup of the job site also eliminates many of the tripping and stumbling hazards common to construction projects. It is a good idea to post safety signs.

Perform routine inspections of your equipment to ensure that it is in proper working condition. If you are working alone, do not assume that because you are the only user of the equipment you are aware of its condition. Make it a habit to inspect your equipment before you begin work. If you are inspecting equipment for use by others, check to make sure that safety features have not been overridden and that electrical cords are not damaged.

Make certain that all workers are familiar with the locations and telephone numbers of nearby hospitals, ambulance services, and doctors. A professional-quality, well-stocked, first-aid kit should be kept on the job site at all times, and a second kit should be kept in your truck or car.

Clients may want to walk through the job site to view the progress. Discourage them from doing so unless you are available to accompany them. When clients are on the job site, they should also wear protective clothing. Children should not be allowed in the work area at any time. Again, post safety signs before you leave the job site.

OSHA regulations

Make sure that you are aware of all the Occupational Safety and Health Administration (OSHA) regulations that apply to your work, and be diligent in your compliance with these regulations. While OSHA regulations can seem unreasonable in some cases, they have been established for your safety and to reduce the injuries that lead to workers' compensation claims.

Due to increased injuries in residential construction, OSHA has increased inspections and imposed fines for violations such as lack of hard hats or failure to maintain a safe construction site.

Insurance

Business insurance is mandatory for today's small business. No one can afford to run a business without adequate coverage. Virtually all services that an installer offers are performed inside a client's home; therefore, the installer must be protected should anything happen to the home or to the client personally as a result of work being performed. One claim could ruin the entire business.

The most important source of coverage is general liability insurance, which protects against claims attributed to negligence or installer performance. This type of coverage would usually include uninsured or underinsured subcontractors. It is often available on a project-by-project basis. Some companies offer this coverage, and, if there are gaps between jobs, it may help lower insurance costs. If installers choose this type of insurance, they should verify that the insurance provisions remain active after a project is complete and the policy is canceled. Other types of insurance that may be needed by installers include automobile insurance for company vehicles, property insurance to cover tools and equipment, and health, life, and disability policies.

Installers must also ensure that subcontractors who work for them have sufficient coverage. Always ask for a certificate of insurance when a contract for services is executed. If subcontractors are uninsured, a claim against the installer's insurance can be placed for subcontractor negligence. Often installers can be listed as an additional name insured on the subcontractors' policies. Installers would then be notified if the subcontractor's insurance ceased to be in force.

Buying the correct insurance for an installation business is important. But even more important is working to ensure that claims against the company do not occur. Installers should review insurance requirements with an attorney or insurance agent knowledgeable in these areas. Nothing can drive the cost of business up more quickly than skyrocketing insurance rates due to claims.

Employee benefits

Employee benefits are a significant part of any compensation package and can account for nearly 40 percent of total payroll expenses. Employees are seeking health insurance coverage, vacation leave, sick or personal leave, paid holidays, and profit-sharing, pension, or retirement plans. As an employer, you must provide workers' compensation and Social Security benefits.

Health insurance. If you have only a few employees or are a self-employed installer, you may find it difficult to offer or obtain affordable health insurance coverage. Some companies pay the full insurance premium for an employee. However, it is becoming more common for the employee to be responsible for a portion of the premium.

Workers' compensation. According to recent industry surveys, the number-one business concern for remodelers is the high cost of workers' compensation. The cost of this mandatory benefit has tripled in just 10 years. You are required to provide and pay for workers' compensation insurance for all workers, whether full-time or part-time. Your rates are based on your record of safety on the job, so your attention to safety procedures is very important in controlling this cost.

Of course it is good to attract excellent installers to your company, and it is even more important to keep them. Offering good compensation and benefit packages builds a strong sense of job security. However, the costs for these fringe benefits must be included in your direct costs, and you need to add up the numbers to determine what you can realistically offer while still remaining competitive.

Codes and permits

As reported by the NKBA, there are three model codes which are widely used in the United States today for general construction purposes. The Building Officials and Code Administrators International (BOCA) code is used predominantly in the Northeast and the Midwest. Along the Gulf Coast and in the majority of the South, builders follow the Southern Building Code Conference International (SBCCI). In the West, the Uniform Building Code (UBC), published by the International Conference of Building Officials (ICBO), is widely used.

All three of these model codes must be adopted by states, cities, or counties before they can be enforced. In many cases, cities adopt different versions of the code. To make the matter even more complicated, local governments frequently amend model codes to include extras that they feel contribute to the health and safety of city residents. In addition to having a copy of the published code book, you need a list of the local amendments to the code. Local codes may change frequently, so be sure you have the latest amendments.

If your kitchen installation work is going to occur in one- or two-family residential dwellings, your code requirements come from yet another model code, known as the Council of American Building Officials (CABO) One and Two Family Dwelling Code. All three of the recognized model codes have agreed on this model code for residential use only. In most communities, the designer can choose either the model code adopted by the community or the CBO One and Two Family Dwelling Code.

Remember, building codes and regulations are minimum standards. They are designed to protect and enhance the life, health, and safety of occupants. By building to code, you are building to a minimum standard. Remember that as a professional kitchen installer, you should also be concerned about quality. Quality, in most cases, requires going beyond the minimum requirements of codes.

Obtaining permits

To begin work, you are required to obtain building permits from the local code enforcement officials. The requirements for filing for and obtaining these permits vary. In some cases, you or the contractor may be required to file for the permit; in other cases, the dealer, designer, or architect submits the required drawings for the permit. Some jurisdictions require the seal of a registered architect or engineer on the drawings to obtain a permit; however, this requirement is generally for larger projects. The fee for the permit is set by the locality and varies widely by the established rate and the value of the project. If the contract requires you to file for and pay for permits, make sure you have included this amount in your estimate.

In some jurisdictions, interior work that does not involve extending the house does not require a building permit. Thus many kitchen projects are done without a building permit, and many projects do not even pull an electrical or plumbing permit. It can be dangerous to do work that is not inspected for safety and code requirements. You should always follow the requirements of your local jurisdiction and obtain all required permits.

In addition, separate permits are required for mechanical trades, such as electrical or plumbing. In most instances, these permits can only be filed for and obtained by licensed plumbers and electricians. However, this requirement can also vary by jurisdiction.

Inspections

Once your project has been cleared for construction, you must post the required permits at the job site at all times. When you have completed certain portions of work, you request an inspection by the building inspector. If this portion of the work is found acceptable, the building inspector issues a certificate to this effect, and you are allowed to proceed to the next level of completion. It is important that you are fully aware of the required sequence of inspections and that you notify the building official when you are ready for a certain inspection. If you proceed beyond a required inspection, such as putting up the drywall without first obtaining a "close-in" inspection, the inspector can place a "stop work order" on your project and require you to take off all of the drywall. This can cost you time, materials, and money.

When you have passed all of the required inspections, the building inspector issues a certificate of occupancy for your project, giving your client permission to reinhabit the area of the home in which you were working. Working with codes can be very confusing. The most important thing is to make sure that you know which codes you are to follow and that you understand your responsibilities under the codes in force in your jurisdiction.

Licensing requirements

Because licensing requirements vary from state to state, and from locality to locality, licensing can only be discussed here in the most general terms. Installers must become familiar with the licensing regulations for their particular area. At minimum, most localities require all businesses to obtain a business license. There is an initial fee for the license and usually annual renewal fees.

In most major metropolitan areas, installers without a plumber's license are barred from handling any plumbing disconnects, rough-ins, or connections. The same is often true for electrical work in kitchen installation projects.

If installers are required to obtain building permits for kitchen installation projects, the building permit office requires proof of licensure for the regulated trades. Installers who do not have a plumber's or electrician's license must contract with subcontractors to accomplish this work. The subcontractors are responsible for pulling permits for their portion of the work. Building permits need to be on display at all times (Fig. 1.6).

Contracts and proposals

There are two agreements that involve a kitchen specialist:

1. A *contract* is a written and signed agreement between two parties (i.e., the installer and a dealer, designer, customer, or contractor) that describes in detail what work is to be accomplished, how and when it will be accomplished, what services are to be provided, and how much it will cost.

2. A *proposal* is a bid presented to the prospective customer to obtain work. Like a contract, the proposal spells out what services are to be provided for the proposed price. While it may not be necessary to detail all the specifications of the project, installers should be careful to qualify a proposal so that it is clear what is being provided. The customer may seek competitive proposals,

Figure 1.6 A building permit is one of the first steps needed for a kitchen addition. (Leon Korejwo)

and your most professional approach may be necessary to have the job awarded to you.

Contracts

When contracts are executed (or signed) by both parties, they become binding agreements both parties must observe. A party that does not live up to contractual obligations is considered to be in "breach of contract." Sometimes what an installer may think is perfectly clear may be understood quite differently by the customer. For this reason, installers should never work without a contract. Doing so leaves them open to loss of money and a variety of misunderstandings, with no legal means of setting things straight.

Installers running their own businesses should have standard written contract forms. The contract needs to be reviewed by an attorney who is familiar with construction law, to ensure the contract covers all business interests, while also checking it against federal and local laws and regulations to ensure compliance.

At a minimum, a contract should include the following:

- Name and date of the parties entering the contract.
- Legal description of the property where installation services are to be provided.
- Scope of work or description of the project.
- Any subcontractor's name, address, phone number, license number, proof of liability, and workmen's compensation insurance.
- Reference to the design drawings (the easiest and clearest way to describe the project, and limits the installer's responsibilities of the installation of the cabinets, fixtures, and equipment shown on the drawings).
- A schedule for completion of the project (with clauses so that the installer cannot be penalized if the project is not totally finished by the stipulated date, i.e., for delays in material deliveries, delays by subcontractors, acts of God, or if the customer fails to make scheduled payments).
- All financial terms clearly spelled out, including when payments are due and any penalties for late payments. Many contracts tie draws to the completion of a certain portion of work, like installation of cabinets or counter tops; however, it is wiser to schedule draws for the beginning of a portionof work rather than completion. Installers must take the time to becomefamiliar with state, provincial, or local regulations that govern payments and deposits received from customers. Ignorance of such regulations results in fines that penalize the contractor, therefore adding costs which were not figured into the original budget.
- A clause guaranteeing an installer's work and that of the subcontractors. It is customary to provide a one-year warranty on workmanship, but the warranty should specifically exclude any materials or equipment supplied

by the customer or by others. These warranties are typically provided by the manufacturers. At an additional cost, some installers offer extended warranties that cover materials, equipment, and workmanship for a longer period of time. For the additional fee, the installer fixes basically anything that goes wrong over the life of the warranty.

- Information about who is responsible for obtaining any necessary permits and inspections and who will pay for them.
- Change order procedures. If prepayment is required, or if there is an administrative fee for processing a change order, this must be stated.

At the time of signing, the installer should review the contract with the customer and ensure that all contract provisions are understood to avoid disputes caused by misunderstandings. In the United States, the 1974 Truth in Lending Act and Federal Trade Commission rules give customers the right to cancel a transaction. This right affects all agreements made in the customer's home or in any location other than the installer's place of business. The U.S. government guarantees a three-day right of cancellation or rescission, whether a job is paid in cash or financed.

When working directly for the homeowner, it is required by law to notify them of this right, usually stated in a contract clause. If you fail to notify the customer, that customer has the right to cancel the contract at any time during the project and get a full refund. Change orders under the original contract are not subject to this regulation. This forces the return of money for work that has already been completed should a customer invoke the right to rescind the contract.

Proposals

A proposal is simply a document that conveys a scope of work and a proposed price for installation of a project. The prospective dealer or designer or customer may solicit and review several proposals from several installers before deciding to enter into a contract.

Proposals are usually based on drawings and specifications that have been prepared by a dealer or designer for a customer. These drawings establish the services on which the contracted party is bidding. A standard proposal form can be used for each project. Like a contract, it should include the date, name of the customer, description of the project location, and the proposed price.

The proposal should always be hand-delivered to the prospective dealer or designer or customer, and, if possible, should be reviewed with them at the time of delivery. Proposals that are just dropped in the mail generally are not received as favorably. It should stipulate how long the prospective customer has to review and sign the document. Generally, proposals are good for 30 to 60 days. Beyond this point, the proposal may need to be revised to reflect current pricing and workload considerations. If left open-ended, the prospective customer could sign the proposal a year later and expect the installation at last year's proposed cost.

If the proposal has qualifications, installers should state them. Perhaps the drawings and specifications call for painting and wallpapering work but an installer may not handle this work. The proposal should simply state that the price does not include these services.

There should be a line at the end of the proposal requesting the contracting parties or customer's signature to accept the proposal, as well as the signing date. This is an expedient way for the customer to accept the proposal, indicating readiness to enter into a contract for the installation project.

Sizing up the job and the customer

As discussed in Chap. 1, keep in mind that while the client of the kitchen installer is often the home builder, the primary customer for the kitchen remodeler is the homeowner. In a project development, for instance, the installer works for the contractor or home builder. A builder's demands are usually different from those of an individual homeowner. When working for a contractor on a new custom house, you may have to work with the prospective owner as well as with the contractor, which may require considerable tactfulness. The new prospective homeowner and contractor are both your clients. It is vital to know from whom you take instructions. Do you have freedom to interview the owner?

When a homeowner wants a new kitchen, the professional kitchen remodeler's job may be to listen to the customer's complaints about the present kitchen and desires for a new kitchen and to make a personal inspection of the present kitchen (Fig. 2.1). After completing this procedure, a remodeler can make informed suggestions for the renovation of the space. In some instances, the kitchen installer is not going to have to help "sell" the kitchen; in most cases, that is the designer or salesperson's job. But in the case where an installer's responsibilities may include assisting a client with the initial stages (interview, inspections, etc.), the installer must take time to work with the homeowner to identify and propose solutions to existing kitchen problems.

What makes a good kitchen?

The kitchen in a home should be custom planned to provide for the needs and personal preferences of the individual family. Of course, how the family lives is most important in designing the kitchen. It is a good idea to have the customer make a list of the family's likes and dislikes about the kitchen. Good

Figure 2-1 Helping your clients choose what is best for their kitchen. *(Leon E. Korejwo)*

kitchens do not just happen. They start with careful planning to blend the ideal in attractiveness with practical working principles. The size of family, ages, activities, working and eating habits, availability of help, even the physical characteristics of the principal users of the room all have a bearing on planning a kitchen that is right for the individual family. Have your customer list the things that must be in the kitchen and the things it would be nice to have if the budget permits. It is a good idea to submit plans for approval before the work is done. It may be expensive to change things if the finished job does not pass inspection. The kitchen must reflect the individual needs, work habits, and lifestyle of the family, not those of anyone else.

Wearing a designer's hat

The project builder designs kitchen plans around an average homemaker and a typical life pattern. Builders, designers, and architects should keep up to date on all trends in kitchen styles and equipment. Although trend-setting surveys are good indicators as to what goes into home buying decisions, the best advice to follow is to use a good kitchen plan. A bad kitchen costs just as much money as a good one.

People who buy a house may not understand everything about electronics or thermodynamics; however, they should know at a glance whether they like the kitchen. They know whether it will be a pleasant and easy place in which to work, and whether it is conveniently located to watch the children and enjoy guests. An experienced homeowner usually rates design ahead of quality.

The remodeling of an existing kitchen does present problems. For instance, the physical limitations of the space available for a kitchen is perhaps one of the first considerations. A completely free plan usually occurs only in the planning stages of a house and very seldom in remodeling work. The kitchen remodeler is limited by the space already allotted for the kitchen. Even then, cost, limitations of size, and proportions of space with relation to the rest of the house frequently restrain the designer from complete freedom of design. Therefore, the size—the number of square feet within the space for the kitchen—is important.

Often, simple square footage does not permit the designer an ample layout of the necessary appliances, storage, and equipment. Just what is the basic plan? These considerations are a vital part in determining the kitchen installation. The location of the kitchen with reference to other rooms is often a determining factor in the design. Despite many physical limitations of an existing kitchen, you have a much greater chance of coming up with the perfect plan if you ask your customers a series of pertinent questions. It is then easier to design the kitchen around their lifestyles.

The answers to these and similar questions go a long way towards providing a solid basis as to what your customers' exact needs and desires are. Satisfactory design cannot be formulated without a thorough investigation of their needs, wants, and living habits. To ensure individuality, satisfy the functional requirements, and obtain the preferred appearance of a kitchen, you must be aware of the variables and work within the limitations. Each kitchen should be as different as each homeowner is different.

You should be able to guide the homeowner through the hundreds of decisions involved in kitchen remodeling, blending their tastes and desires into a practical, workable room. You need to ask the homeowners questions to help them to evaluate their needs.

After the client has shown interest in a new kitchen, the installer should visit the home to identify the needs and measure the space. At this point you can interview the customers and have them complete a survey form (Table 2.1). You should begin by measuring the space and examining the area for special conditions in the floors, walls, and so on—anything that may affect the installation project. There is no way to know if the new kitchen design will work until the clients have decided why the old design doesn't work. The questionnaire helps your client do just that. Have them write down things about their existing kitchen that bother them.

Table 2.1   Before you begin: a questionnaire

Name:
Residence Address:
Jobsite Address
Phone:
Work:
Date:

Your family

1. How many people are in your family? _____
   What are their ages? _____

2. Are you planning on enlarging your family while living
   in this house? _____

3. Who is the primary cook? _____
   Is the primary cook left-handed or right-handed?
   _____
   How tall is the primary cook? _____
   Does the primary cook have any physical limitations?
   _____

4. How many other household members cook? _____
   Will more than one person be cooking at one time?
   _____ If yes, how many? _____
   Is the secondary cook(s) left-handed or right-handed?
   _____
   How tall is the secondary cook(s)? _____
   Do any of these members have physical limitations?
   _____
   Is a specialized cooking center required for the
   secondary cook? _____

Kitchen logistics

5. How many hours is the kitchen in use during
   weekdays? _____ Weekends? _____

6. How often do you go grocery shopping
   (to determine amount of storage needed)?
   Daily   _____   Weekly   _____
   Bi-weekly   _____   Monthly _____

7. What types of products/materials do you purchase?
   ____ Predominantly fresh food purchased for a
        specific meal
   ____ Predominantly frozen foods purchased for stock
   ____ Pantry boxed/packaged/canned goods for stock
   ____ Cleaning products stocked in bulk
   ____ Paper products stocked in bulk
   ____ Other _____

8. What type of cooking do you normally do?
   ____ Daily heat & serve meals
   ____ Weekend quantity cooking
   ____ Daily, full-course, "from scratch" meals
   ____ Weekend family meals
   ____ Bulk cooking for freezing/leftovers
   ____ Other _____

9. What are your kitchen and dining area requests?
   ____ Separate table
        ____ New
        ____ Existing
        ____ Size
        ____ Leaf Extension
   ____ 30" table height dining counter
   ____ 36" counter height
   ____ Number of seated diners at one time
   ____ 42" elevated bar height dining center
   ____ Other _____

10. Do you have a separate dining room? _____

11. Do you like to entertain? _____
    Formally or informally? _____

12. When you entertain, do guests usually gather in
    the kitchen? _____

13. On what floor in the home is your kitchen located?
    _____

14. Is there a basement, crawl space, concrete slab,
    or finished ceiling below it? _____

15. What is above your kitchen? An attic or roof?
    The second floor? _____

Your present kitchen

16. What do you like and want to keep about
    your present kitchen? _____
    _____
    _____

17. Why do you want to change your kitchen?
    ____ Inefficient layout
    ____ It is too crowded
    ____ Needs more seating space
    ____ It looks dated
    ____ Needs more storage space
    ____ Needs more space
    ____ Too much maintenance required
    ____ Other _____

18. What type of activities generally take place in
    your kitchen?
    Eating   _____   Sewing   _____
    Baking   _____   Entertaining _____
    Growing plants   _____   TV/radio   _____
    Hobbies (arts and crafts) _____   Wet bar   _____
    Canning and freezing   _____   Laundry   _____
    ____ Other _____

Your dream kitchen

19. What elements/activity areas do you already know you want to incorporate into your kitchen design?
    ____ Dining area
    ____ Laundry center
    ____ Desk/planning center
    ____ Recycling center
    ____ Media/TV center
    ____ Home automations systems
    ____ Planning area
    ____ Breakfast nook
    ____ Snack bar
    ____ Peninsula
    ____ Storage shelves
    ____ Island
    ____ Open shelving
    ____ Pantry
    ____ Decorative end shelves
    ____ Computer
    ____ Telephone
    ____ Cookbook storage
    ____ Cutting/chopping surfaces
    ____ Waste disposal
    ____ Trash compactor
    ____ Large preparation area for baking, etc.
    ____ Other _____

20. What appliances will you keep or replace?

Appliance	Replace	Keep
Refrigerator	____	____
Dishwasher	____	____
Range, oven	____	____
Microwave	____	____
Sink	____	____
Separate freezer	____	____
Garbage disposal	____	____
Exhaust hood	____	____
Trash compactor	____	____
Washer/dryer	____	____
Other _____		

21. What color appliances would you like?
    White ____ Almond ____
    Paneled to match cabinets ____
    Other _____

22. What appliances require storage space?

Toaster	____	Mixer	____
Blender	____	Juicer	____
Coffee maker	____	Can opener	____
Food processor	____	Toaster oven	____
Crockpot	____		
Other _____			

23. For what items is specialized storage desired?

Food processor	____	Toaster oven	____
Bottles	____	Bread board	____
Bread box	____	Cookbooks	____
Cutlery	____	Dishes	____
Display items	____	Glassware	____
Lids	____	Linen	____
Soft drink cans	____	Plastics	____
Vegetables	____	Spices	____
Wine	____		
Other _____			

24. What type of cabinet interior storage are you interested in?

Lazy Susan	____	Tilt-out	____
Pantry	____	Drawer head	____
Vertical dividers	____	Drawer ironing board	____
Recycling/ waste bins	____	Toekick step stool	____
Roll-outs	____	Tilt bar	____
Other _____			

25. Where do you plan to sort recyclables?
    Kitchen ____ Laundry ____
    Garage ____ Basement ____
    Other _____

26. What else will you be replacing?

Wall coverings	____	Ceiling treatment	____
Countertops	____	Plumbing	____
Electrical	____	Lighting	____
Flooring	____	Windows	____
Sink(s)	____		

27. Is it possible to move doors, windows and/or plumbing? _____

Your personal taste

28. What type of style would you like your new kitchen space to have?

Traditional	____	Family retreat	____
Sleek/contemporary	____	Formal	____
Country	____	Open and airy	____
European	____	Southwestern	____
Strictly functional	____		
Personal design statement	____		
Other _____			

29. What color schemes do you like?

Earth tones	____	Black and white	____
Neutrals	____	Pastels	____
Bright colors	____	Current fashion colors	____
Other _____			

30. What type/color cabinets would you like?

Whitewashed	____	Light wood	____
White laminates	____	Medium wood	____
Colored laminates	____	Dark wood	____
Other _____			

31. Do you prefer framed or frameless cabinets?
    _____

32. What type of cabinet accessories interest you?
    _____

Table 2.1   Before you begin: a questionnaire (Continued)

The task ahead

33. What portion of the project, if any, will be your
responsibility? _____

34. Who will perform the following tasks if needed?
Obtain permits _____
Replace old cabinets _____
Remove walls and other structural changes

_____
Remove old floor _____
Replace plumbing and electrical _____
Paint or wallpaper _____
Install tiling or floor _____
Install new cabinets _____
Install new countertops _____

Add trim, moldings, etc. _____
Purchase appliances _____
Other _____

35. Realistically, by what date would you like to
have your new kitchen? _____

36. What is your preliminary budget? _____

37. What is your maximum budget? _____

38. Is financing a possibility? _____

Project management

Professional project management is an absolute must in today's kitchen installation market. You need to understand how to successfully market, schedule,
estimate, and construct kitchen installation projects. A well-managed project
starts with an open line of communication with the client, accurate field measurements, a detailed set of plans and specifications, a realistic cost estimate,
and a closely supervised installation schedule. The management of the job
requires you to handle specific project conditions as they arise.

As a kitchen installer, you need to understand your responsibilities, as well as
the responsibilities of the other professionals involved. In some cases, the dealer
or designer may sell the job, make the major decisions, order the materials, and
then look to you for the installation. You or the lead installer take full responsibility for the installation job from start to finish. Your project management
responsibilities include coordinating all subcontractors, ordering materials when
there are shortages, writing change orders, performing quality control, completing precompletion punch lists, and sometimes collecting payments. Your
management procedures will be continuously changing as you evaluate what is
most effective for your projects. The kitchen installation profession can be a
continuing learning process for you.

Many kitchen installers are in business for themselves because they want
to be independent; however, this independence does come with a price. A successful project installation requires someone to make the tough decisions and
direct the course of the project. You must be able to manage yourself as well
as others.

Scheduling the job

The most common questions asked by clients is when you can begin the
installation and how long it will take. This information is just as important
to you, too. To properly estimate the job start and end, you need an accurate

schedule to organize the project. A schedule is extremely important to make the best use of workers and subcontractors and avoid problems. If you do not prepare a realistic schedule, and the project takes longer than expected, you will have a very unhappy client.

You need to prepare a schedule that identifies the major categories of work to be done and the time required to accomplish the work. Remember that you will be disrupting your client's daily routine significantly. No client has ever been unhappy because a remodeling project was completed ahead of schedule. So, plan ahead and factor in extra time for delays, etc., which, unfortunately, are inevitable.

You must consider many factors when you prepare a construction schedule. The more experience you have and the more projects you have completed, the easier it is to identify all these factors and assign them a length of time for completion. Reviewing your cost estimate for the project is a good place to begin in preparing a schedule. Every item identified as a cost takes a certain amount of time to perform. Consider the time requirements for the delivery of materials and equipment, and plan accordingly. It is important not to begin until everything is in local stock.

Subcontractors must be scheduled so that they do not delay the project. Finding subcontractors who meet schedules and are disciplined is one of the hardest things in construction. Arriving on time is one of the most important qualities to look for in a subcontractor. Therefore, you must be sure to have the job ready for those subcontractors on schedule so they can do their work efficiently. If a delay is unavoidable, inform any subcontractors, and, as soon as possible, provide a revised schedule reflecting this delay.

When preparing the schedule, keep in mind that inspections are required by local code officials. Failure to schedule these inspections can cause a major interruption while you wait for the inspector to approve completed work and give you permission to continue.

Types of schedules

There are various visual aids that can be used as schedules. The first type is the bar graph schedule. The bar graph or bar chart schedule lists the categories of work in a column on the left side and dates for the duration of the project in a row across the top of the schedule. The projected start and completion of each category of work is plotted on the schedule.

To decide the best time for a task to begin, you need to know what categories of work must be completed. If your schedule shows overlaps for different categories of work, make sure that these tasks can physically occur within the kitchen space at the same time. A bar graph can be produced rather quickly with a computer. Revisions can be made easily when changes occur.

A critical path method, or CPM schedule, is useful for planning and management of all types of projects. It is a schematic diagram or network that shows the sequence and interrelation of all the categories of work for the project. It provides a more precise approach than conventional bar graphs. It clearly shows the difference between activities that are critical and those that are not.

Another method of scheduling is with a flowchart. It is simply a written, week-by-week list of the critical tasks to be done on a job. A flowchart is not as easy to quickly review as a CPM or bar graph; however, it is a good place to start. You may find that it will be a sufficient scheduling method. The more experience you acquire with installation projects, the better you will be at polishing your management and scheduling techniques.

A good way to keep everyone involved with the project up to date is to post the project schedule at the project site. Remember to update it as often as possible. Keep in mind that if your client sees the schedule on a daily basis, you must be prepared to keep to the schedule. A well-scheduled project with little or no wasted time can be very profitable for you.

Pricing, estimating, and quotations

The installer with the lowest price does not always get the job. Sales and marketing are at least as important, if not more important, than the price. An installer with a reputation for reliability, quality work, and effective people skills gets more business than a low bid.

Throughout the installation, many hours may be spent dealing with unforeseen problems on the job, meeting with clients, and coordinating with the architect or dealer or designer. These projects often have small budgets, which means that you may have several small projects running at the same time. A large amount of your time is spent traveling between jobs. If you want to be profitable, your estimating must take into account all of these inefficiencies.

Certain unexpected conditions need to be figured into the job estimate. It is difficult to put a reasonable cost on these problems. For example, in some homes, large appliances can be difficult to move through narrow hallways and doorways. Kitchen remodeling jobs in condominiums or other large buildings may present problems, for example, parking fees if free parking is unavailable for your service vehicle.

Figure in added costs for additional help if the project is not easily accessible. If the project is in an apartment building, for example, be certain to verify that the elevators are large enough to handle all of the materials and equipment needed. If you don't plan ahead for this potential problem, you will lose valuable time and money.

A few factors should be figured into all projected costs, including, as mentioned earlier, unusual job conditions such as difficult access; difficult customers—for example, a couple who can't agree on anything; unusual project requirements such as matching of existing materials; and company capability—do you have the workers and experience for the project?

The additional costs that come with any of the situations vary from project to project. Your ability to accurately identify the situations and figure in the costs can make the difference between making money and losing money on a project. Estimate one job at a time so that you do not confuse one job with another. The longer you wait to prepare the estimate, the harder it is to evaluate any of these mentioned situations.

Every element of the job must be included in the estimate. Leaving anything out seriously affects your net profit. If all of the items are included, the overall effect of misfiguring a minor item is insignificant compared to leaving something out.

To evaluate the actual cost of the job versus the estimated cost, keep a time sheet for each individual job. A comparison of these figures may help point out mistakes or shortcomings in the estimate. Even if you are installing highly specialized, custom kitchens, most of the job involves standard items and tasks. It is most efficient to use a computer for estimating. Add custom designs and finishes into the estimating database, and they then become standard items. When you are doing something for the first time, keep cost records so you can develop accurate unit costs for the next time that option is used.

Preconstruction conference

A preconstruction conference should be held after the client's three-day right of rescission (or cancellation) has passed, before any work begins. The contractor or production manager, dealer or designer, architect, installer, and homeowners are all generally included at this meeting. However, this may vary according to the size and type of project.

This meeting passes the chain of command from the dealer and designer to the installer, which must be made clear to the customers. They must understand that they will now be working with the installer rather than the salesperson. The installer will not be able to take full control of the project or gain the respect of the customers if the customers continue to call the dealer or designer during the course of the project.

The installer often presents the details of the project at the preconstruction conference, explaining the specifications in a way that is often more thorough than the dealer and designer may have done. While a good salesperson may mention some of the limits of a kitchen project, the sales emphasis is on exciting the customer to buy. The installer must point out potential misunderstandings in detail. Every time an installer undertakes an installation, this education process must take place.

Good communication skills are very important to the success of the project. The line of communication between installer and customer is established at the preconstruction conference. The more clearly an installer explains the limits of work and what the customer can expect, the happier the customer will be.

When there is more than one customer, one individual should be designated as key decision-maker and liaison with the installer. This person should also be responsible for handling payments so that the installer is not expected to communicate with both parties or agree to a change order with one but try to collect from the other.

As reported by Walter W. Stoeppelworth in his NKBA *Kitchen and Bathroom Installation Manual*, some of the most important issues to be covered at a pre-construction conference, which can last from one to two hours, include

- *Every specification in the contract is reviewed and explained.* This explanation is vital so that you know what is expected of you and the homeowners know what will be done to their house. For example, if an existing floor is not level, or if an existing wall has a bow in it, explain to the owner the impact on your work. Make certain that the client understands that these existing conditions will remain once the new work is completed, and that you will do your best to minimize any problems that may result from these conditions.

  Another common problem is the matching of new materials to existing materials. For example, if the client is not willing to repaint an entire room during the installation project, explain that you will do your best to match the color. They must understand that paint colors age over time, especially in kitchens, and repainting with the same paint color used previously does not guarantee a match.

  Clearly explain to the client what is and is not included in the project so that they have a good understanding of what the finished project will be like. Do not assume that they can read the blueprints or specifications, even if they continuously nod their heads and say that they understand. Always operate under the assumption that the client has a very limited knowledge of construction.

  Throughout the preconstruction conference, attempt to determine if the project that the client has contracted for is the project they believe they are getting. If they have unreal expectations for the project, this is the time to bring them back to reality. It is better to do this at the preconstruction conference than to deal with it later when all the materials and equipment have been ordered or installed.

- *Areas of the home affected by the work.* The homeowners must clearly understand which areas of the home will be affected, as well as what furniture and equipment must be moved or stored to permit the work to occur. They must be aware of the possible damage that can occur if they are not removed. Kitchen cabinets that are to be completely removed, will, obviously, have to be totally emptied of all dishes and food. Even if a refrigerator (as well as any other appliance) is to be reused, everything must be emptied from it.

- *Areas of the house off limits to workers.* If the client is going to allow workers to use a particular bathroom in the house, for example, it should be designated by the owner at the meeting.

- *Who will be at the job site on a regular basis and at what time.* Many people in the construction trades like to begin early in the morning, so everyone needs to be out of bed and out of the way by a mutually agreed-upon time. You may have to start a little later than you would like. Once the start time is set, you should have workers at the job site promptly.

- *The construction schedule.* The construction schedule should be presented at the conference. Review the basic steps that occur throughout the project. If the client must leave the home for an extended period of time, such as when the plumbing, heating, or electrical utilities will be shut down, schedule these times now.

- *Basic safety precautions.* Inform the client of all of the safety procedures that need to be followed during the remodeling process. If possible, the home-owners and the inhabitants of the house should stay out of the construction area. Explain to the client the possible dangers of walking around the construction area. The client should not attempt to turn on light switches or use outlets in the designated area until the project is complete.

- *Existing materials to be removed or salvaged.* A list of items to be salvaged should be completed by the client, along with your assistance. Items that look worthless to you, could be very important to the client. If items such as cabinets or plumbing fixtures are to be saved, the client needs to provide an area where these items can be stored. Also, decide who is responsible for the removal of salvaged items. Explain that if the client wants you to remove these items and put them in the basement or garage, you have to charge for this service. If the client decides they want to be responsible for removal of such items, give them a date by which the items must be removed. Inform the client that, while you will take every precaution to safely remove and store the items that are your responsibility to remove, you cannot be responsible for any damage that might occur during the removal or relocation process. In addition, if the plans call for the reuse of some or all of the existing appliances, you want to make sure that the client understands that you cannot ensure the working condition of the appliances at the end of the job. Obviously, you must see that the existing appliances are safeguarded during construction.

- *Verification that every important decision has been made before work starts.* Decisions such as paint color, tile selections, light fixtures, and so on should all be decided upon before work is started. Customers should have already made these decisions as well as been given samples to study, so they have not made decisions under pressure. Many contractors and kitchen specialists have showroom selection centers where customers can go to make all these choices at one time, which allows decisions to be made more quickly and efficiently. A customer can often select from predesigned ensembles that coordinate fixture selections with wallpaper, paint, and tile. If any decisions are not finalized at this time, a deadline should be set with the stipulation that the job will not start until the decisions are made.

- *Change orders.* Most installers do not like change orders because they seldom make any money on them. Inform customers that change orders can delay a project and can be quite expensive for both you and them. This conference is an important time to ask whether they have any additions to the project. If any such additions are brought up at this time, they can be priced according to normal policy and a change order written. If any changes come up later, the price might include, for example, a $50 administrative charge (to be decided upon by the company) added to the price of the work. This fee is charged to help pay for the cost of workers standing around while change orders are contemplated and figured, as well as cover the administrative costs of notifying subcontractors or suppliers of the change in plans.

- *Procedure for payment.* If you will be collecting payment from the homeowner, now is the time to discuss the procedures for doing so. If you are working for a lumberyard or home center or are a subcontractor, your arrangements for payments should be part of your contract with that company. In these cases, details of your payment schedule and contract should not be disclosed to the homeowner. You should not discuss the amount of money that the homeowner is being charged by the lumberyard, home center, or contractor.

  Payment procedures differ greatly throughout different parts of the country. The most common procedure is to collect 30 percent of the payment at the start of the job, another 30 percent midway, and 30 percent at substantial completion. The final 10 percent is due upon completion of the punch list. Some companies wait until the end of the job and collect all of the money at that time.

  If there is to be a down payment, explain when it is due (customarily at the time the contract is signed). Be familiar with the limits set by your locality or state on the amount of down payment you can legally charge. If you will be collecting payments at milestones during the progress of the project, it is better to tie the payment to the start rather than completion of a certain part of the work. For example, if one-third of the payment is due upon completion of cabinet installation and one cabinet is delivered damaged, you may have to wait to collect your money until the replacement cabinet arrives and is installed. If payment is due at the start of cabinet installation, this incident is not a problem.

  Final payment should be based on substantial completion. If there are punch list items to be corrected, a mutually agreed-upon amount can be withheld from the final payment. On a larger job, a percentage of your payment may be held in retainage (customarily 10 percent) until the punch list items are completed. Always collect your money in person. Never bill by mail. Going to the project site to collect money also allows you to check on the quality of the job. If something has gone wrong and the client wants to hold up payment, the sooner you find out and solve the problem, the better.

- *Items to be removed from the project area.* Working with the client to determine what furnishings, equipment, and personal belongings must be removed from the project area prior to the start of work is an important part of the preconstruction conference. Be realistic about how much of the home will be affected by the project, and make sure the client is aware of what will be affected. Rooms adjacent to the kitchen will also be affected. These spaces need to be cleared of furnishings and possessions as well to allow you enough room to work. When you need to have items moved, give the clients a couple days' notice. It is also important that they know you will not be responsible if damage occurs because the items were not previously moved.

If items must remain in the project work area, you need to protect them as much as possible. If there is a risk of damage, make sure that the client understands this from the beginning of the project. Drop cloths or polyethylene may be sufficient protection for most items, but larger items (i.e., china cabinets) may need a greater protection from accidental impact.

Before you start

Delivery and storage

For kitchen remodeling, many installers feel that storing materials in a warehouse is not cost-effective. This kind of storage is a waste of money when you take into account the cost for space, delivery time, and inventory. However, large kitchen companies might use a warehouse as a staging area. When cabinets are delivered, they are stored in the warehouse until all related items, such as appliances, light fixtures, and tile, are ready and are then transported to the job.

Kitchen installers have begun to form closer alliances with single-source suppliers. The suppliers provide storage and staging areas in their larger warehouses where all of the required materials and equipment for a project can be held. An initial delivery is made to the project at the beginning of the installation, and a second delivery is scheduled midway through the project. This trend toward the supplier-provided staging area is expected to continue to grow.

Scheduling delivery

If you do not use a warehouse or staging area, you need to schedule delivery of materials so they are on the job when you need them. It cannot be done so that you must store or work around materials and equipment that you are not ready to install. Some dealers and designers maintain a small storage area as a part of their showroom or at a separate location. If this is the case, cabinets should be delivered and held at this location until you are ready to install them. If your client has garage space available, it can sometimes be used.

Do not attempt to stock materials that you use for kitchen installation projects. Purchase what you need for each job when you need it. Even if you have materials left over at the end of the project, it is probably more cost-effective to offer them to the client than to take them with you and attempt to store them for later use. The cost of your labor to haul these materials quickly exceeds their value.

Protecting the area

Flooring finishes that are not being replaced must be protected, including any area that will be a path for workers. In most cases, drop cloths or polyethylene are sufficient to protect these surfaces, but in areas where you will be rolling or dragging heavy equipment, consider a more substantial material, such as plywood. Refinishing damaged hardwood flooring or replacing stained or torn carpet in a home can be very expensive and time-consuming. Take time to ensure all is protected.

In addition, rooms that will not be affected by the work should be separated from the work area by polyethylene or plywood barriers to reduce the passage of dust and dirt. The type of barrier selected depends on whether access is needed to the area closed off. Determine your required paths of travel through the home, as well as the client's, before making this decision.

You may sometimes need to suggest that some items be placed in an off-site storage facility for the duration of the project. This situation is most likely when the home has little or no storage space or if project will leave the home in an unsecured condition that would put the clients' possessions at risk of theft. If you do not feel that you can adequately protect the contents of the house, you must press the client to exercise this option.

Recording existing conditions

Some companies have instituted a policy of videotaping the kitchen prior to the start of work so that they have a permanent record of the existing conditions. This may not be necessary in all cases, but it does underline the necessity of making at least a written inventory of the existing materials and equipment and their condition before beginning your installation work.

Sequencing of new material delivery

Once the contract is signed and the clients have made themselves ready for their kitchen remodeling project, they want to see work begin immediately. However, you should not begin the tear-out or other preparation work until the new cabinetry, fixtures, and equipment are in local stock. The lead time for cabinetry is typically 4 to 12 weeks, but it may be as long as 20 weeks in some cases where the cabinets are totally built by hand. If you begin the tear-out process too soon, the client will be left with "dead time" between tear-out and installation. Because the clients cannot use the kitchen, this dead time can be very inconvenient.

Most installers have found it good practice to wait for everything to be in local stock. While this guarantees that the client will not experience dead time, the time between contract signing and the start of work can seem extremely long for the clients. Help them to understand that you are looking out for their best interests by compressing the time that their lives are disrupted by the project.

Prestart checklist

For planning purposes, you want to maintain a prestart checklist that allows you to keep track of items on order and their anticipated arrival date. An example of a typical prestart checklist is shown in Table 2.2. Obviously, the requirements of each individual job will dictate the contents of the prestart checklist. If you intend to delay the start of the installation until crucial materials for the project have arrived, track them on this checklist. Generally, special-order or custom items fall into the crucial category. This information allows you to gauge with some accuracy when to tell the client that the tear-out will begin.

Ordering materials

Before a job can begin, the installer must determine what materials are needed to construct the project and when they will be needed on the project site to keep the job running on schedule. If an insufficient quantity of materials is ordered, the job could be delayed while waiting for another delivery. On the other hand, if installers order too much, the materials are wasted and the profit on the project is reduced. Leftover materials are rarely usable on another job, and the cost of moving and storing them negates any savings.

In most cases, installers working directly for dealers do not order cabinets, fixtures, or appliances, or other major items, such as doors and windows. The order for these materials is placed by the salesperson once the design is accepted by the customer and the contract for the project is signed. Dealer-employed installers order other building materials for the project, such as framing lumber, drywall, and trim. Contract installers, on the other hand, when not working through a dealer/designer, order all materials required on a project.

Part of the process of turning over a project to the installer should be to establish what materials have already been ordered and what items are left for the installer to handle. Installers need to carefully review the paperwork for materials that have been ordered in advance so that they are clear as to what is and is not included in these orders. If they detect discrepancies or omissions in the orders, installers should notify the dealer or designer at once so that the problem can be corrected.

The prestart checklist is an excellent way to record items that have been ordered for the project, as well as the date ordered and the anticipated delivery date. Many of the materials used in kitchen projects are special-order products that have substantial lead times for delivery. The timing of orders for these materials and their anticipated delivery is extremely important to a project schedule. Special-order items should be ordered as soon as possible after contract signing and should be in local stock before a job is started. If precise measurements are required, they should be checked and double-checked.

Even when materials ordering is handled by others, installers have an important role to play. There should be an established line of communication between

Table 2.2   Kitchen installation prestart checklist

Name of client: _____

Home address: _____

Telephone (Home): _____

   (Work): _____

   (Job site): _____

Job site address: _____

Building permits

_____ Obtained and posted all required permits, including general building permit, plumbing
permit, electrical permit, etc.

Material/equipment Items	Rec'd in local stock	Date ordered	Estimated delivery
Cabinets	_____	_____	_____
Countertops	_____	_____	_____
Dishwasher	_____	_____	_____
Doors	_____	_____	_____
Dryer	_____	_____	_____
Flooring	_____	_____	_____
Food waste disposer	_____	_____	_____
Framing lumber	_____	_____	_____
Gypsum wallboard and accessories	_____	_____	_____
Hood	_____	_____	_____
Kitchen faucet(s)	_____	_____	_____
Kitchen sink(s)	_____	_____	_____
Lighting fixtures	_____	_____	_____
Microwave	_____	_____	_____
Oven	_____	_____	_____
Range/cooktop	_____	_____	_____
Refrigerator	_____	_____	_____
Wall finishes	_____	_____	_____
Washer	_____	_____	_____
Windows	_____	_____	_____
	_____	_____	_____
	_____	_____	_____

Subcontractors

_____ Scheduled electrician   Name: _____ Start date: _____

_____ Scheduled flooring installer   Name: _____ Start date: _____

_____ Scheduled plumber   Name: _____ Start date: _____

_____ Scheduled wallpaperer   Name: _____ Start date: _____

_____ Scheduled painter   Name: _____ Start date: _____

_____ Scheduled countertop fabricator   Name: _____ Start date: _____

_____ Others   Name: _____ Start date: _____

_____   Name: _____ Start date: _____

Color & material selections

_____ Client has made all color and material selections, and they have been checked against
the specifications for the project.

Miscellaneous

_____ The project site has been inspected. All field conditions have been confirmed.

_____ If necessary, contact has been made with the local utility companies to arrange for utility
changes or connections.

installers and those placing material orders. If installers have experienced problems with a certain product or supplier or know of products that may perform better for a given task, this information should be conveyed to the material orderer.

It is helpful for dealers or designers and installers to establish relationships with one or more suppliers that stock, or have available for special order, materials that are typically needed for kitchen projects. An established, ongoing relationship with a supplier that provides installers with priority service, quick turnaround on orders, and reliable delivery of material is an important factor in the on-time completion of a project. Installers who have established relationships with material suppliers know they can count on having their materials delivered when promised and that the deliveries will be complete and accurate.

Installers should carefully inspect all deliveries to the job site, especially special-order materials. Defects and deficiencies should be noted on the delivery ticket, and damaged items should be refused. Manufacturers are reluctant to accept responsibility for damage that is not noted at the time of delivery. While it does take time away from production for installers to do a complete and adequate check of a delivery of kitchen cabinets, for example, it relieves installers of responsibility for damages that occurred before the cabinets arrived at the site. Damage that is missed during this inspection can be very costly to the installer later in the project and can lead to a delay in completion.

Tools and equipment

Kitchen installations require a wide assortment of tools and equipment, as well as the knowledge to use and maintain them. If you are working for an installation company, some of the equipment may be provided for your use. In most cases, however, you need to provide your own tool box with a basic assortment of screw drivers, pliers, wrenches, hammers, tape measures, and other tools (Figs. 2.2 through 2.9). Having the right tools readily available makes your job go much more smoothly and therefore leads to a higher-quality installation.

Table 2.3 is a good base to get started for a kitchen installation. You won't need all of these tools for every project. These tools are considered your main basic tools, not a complete list. As you gain experience, you will find many more tools useful. Build upon this list. Plan ahead so that you know which tools are required for a project and have them on hand when you need them.

Equipment and tools brought into the home should be organized and well-maintained. No matter how carefully you use and maintain your tools, they eventually wear out. Invest in top-grade tools made by reputable manufacturers. A quality tool always carries a full warranty. Be sure to check for this feature when you purchase them.

Figure 2-2 Safety equipment includes a paper dust mask (a), a single filter dust-paint mask filter (b), safety goggles (c), ear protectors (d), a face shield (e), a fire extinguisher (f), and a double filter spray paint mask (g). (*Leon E. Korejwo*)

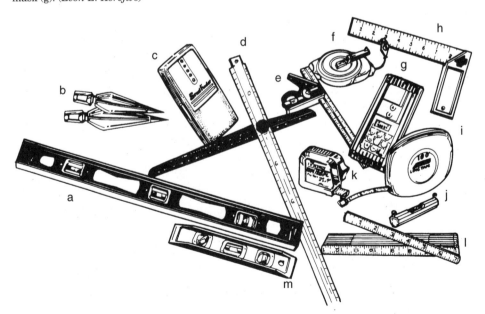

Figure 2-3 Measuring tools include a carpenter's level (a), plumb bobs (b), a stud finder (c), a dry-wall T-square (d), a combination square (e), a chalk line (f), a laser measuring device (g), a tri-square (h), a 100-foot retractable tape measure (i), a string level (j), an 8-meter (26-foot) tape measure (k), a folding ruler (l), and a torpedo level (m). (*Leon E. Korejwo*)

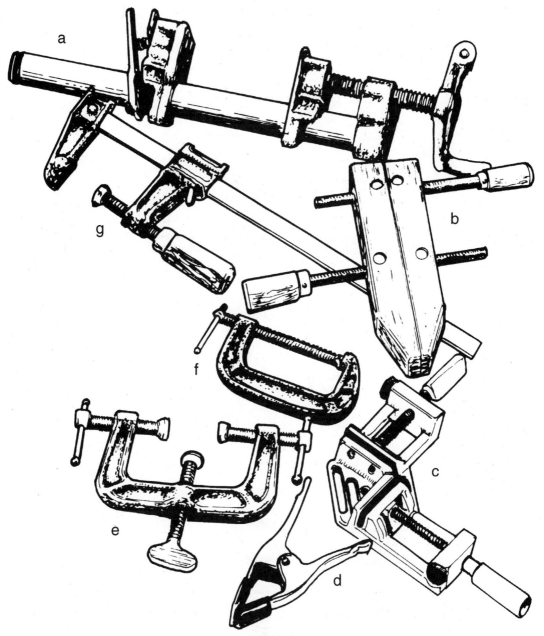

Figure 2-4 Various types of clamps: dual-action pipe clamp (a); hand-screw clamp (b); miter or right-angle clamp (c); spring clamp (d); three-way clamp (e); C-clamp (f); and bar clamp (g). (*Leon E. Korejwo*)

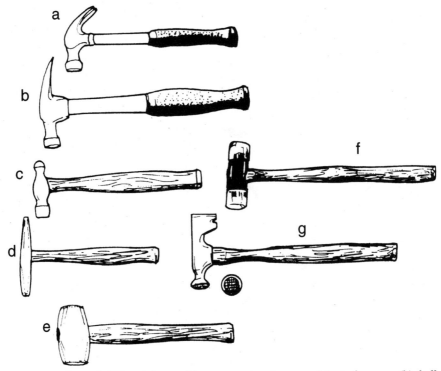

Figure 2-5 Several different types of hammers: claw hammer (a); rip hammer (b); ball peen hammer (c); magnetic tack hammer (d); mallet (e); soft-tip (plastic) hammer (f); and drywall hammer (g). (*Leon E. Korejwo*)

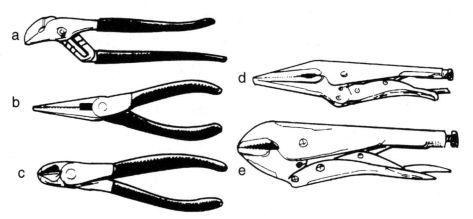

Figure 2-6 Types of pliers include channel-lock pliers (a), long- reach needlenose pliers (b), diagonal cutting pliers (c), long-nose locking pliers (d), and locking-jaw pliers (e). (*Leon E. Korejwo*)

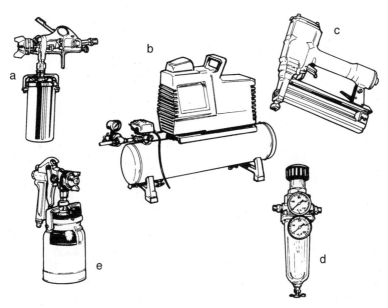

Figure 2-7 Various air tools useful to the kitchen installer include a touch-up spray gun (a); an air compressor (b); a 16-gauge finishing nailer (c); an air transformer that combines a pressure regulator and a filter (d); an external-mix, siphon-feed spray gun (e). (*Leon E. Korejwo*)

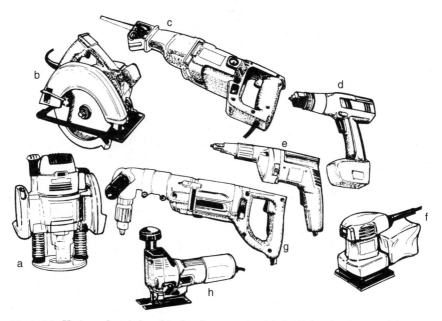

Figure 2-8 Various electric hand tools: plunge router (a); 2 1/4- hp circular saw (b); reciprocating saw (c); 3/8-inch cordless drill driver (d); 1/2-hp drywall screw gun (e); 1/4-sheet finishing sander with dust collector (f); angle drill (g); and variable-speed jigsaw (h). (*Leon E. Korejwo*)

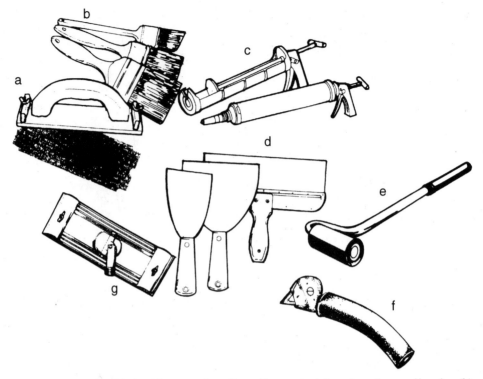

Figure 2-9 A sanding block with precut drywall sanding screen (a); various shapes of brushes (b); caulking and adhesive guns (c); scraper, putty knife, and spreader (d); J-roller (e); Douglas craft knife (f); and pole sander (g). (*Leon E. Korejwo*)

Table 2.3 Tools and equipment

Safety equipment	
1. Face Masks	8. Proper ear protection (ear plugs or muffs)
2. Filter mask	9. Safety glasses with side shields or wraparound safety goggles
3. Fire extinguisher	
4. First aid kit	
5. Hard hat	10. Safety guards on portable saws
6. Heavy-duty safety gloves	11. Safety signs
7. Lifting harness or back brace	12. Steel-tipped shoes with nonslip soles

Measuring equipment	
A. Squares	2. Line level
1. 12" combination square	3. Torpedo level
2. 4-foot drywall T-square	C. Measuring aids
3. Framers or carpenter's steel square	1. Folding rule
4. T-square	2. Metric tape measure
5. Tri-square	3. Retractable steel measuring tape (25', 30', 50')
6. Triangle	
B. Levels	4. Chalk line
1. Carpenter's levels, 2' and 4' (6' occasionally useful)	5. Plumb bobs
	6. Adjustable T-bevel

7. Contour gauge
8. Foot ruler
9. Laser alignment tool or laser measuring tool
10. Marking gauge
11. Scribing compass
12. Straightedge
13. Stud finder

Cutting tools

A. Chisels
  1. Masonry chisels
  2. Wood chisels
B. Drills
  1. Drill driver
  2. $\frac{1}{2}$"drill
  3. $\frac{3}{8}$" variable speed/reversible (corded or cordless)
  4. 90° right-angle drill (occasionally useful)
  5. Hammer drill (occasionally useful)
(NOTE: Drills with keyless chucks are time savers)
C. Drill bits
  1. Auger bits
  2. Carbide masonry bits
  3. Combination bits (used for hardwood floors)
  4. Complete assortment of standard twist bits
  5. Countersinks or counterbores
  6. Forstner bits (used for hardwood floors)
  7. Hole saw bits ($\frac{3}{4}$" to $2\frac{1}{2}$")
  8. Spade bits
D. Knives
  1. 4" to 6" finishing knife
  2. Drywall knife (taping)
  3. Hook-bill knife (linoleum knife)
  4. Utility knife (matte knife)
E. Planes
  1. Block plane (end grains)
  2. Jack plane (trimming)
  3. Power planer
  4. Smooth bench plane (general purpose)
F. Routers
  1. Laminate trimmer with offset base
  2. Standard router

G. Router bits
  1. Bevel cut
  2. Cove bits
  3. Dado bits
  4. Flush cut
  5. Rounded corner
  6. Variety of other profiles (carbide preferred)
(NOTE: Carbide preferred router)
H. Power saws
  1. $7\frac{1}{4}$"circular saw
  2. Bench-size table saw (10" preferred)
  3. Jig saw
  4. Power band saw
  5. Power miter saw
  6. Reciprocating saw
I. Hand saws
  1. Back saw with miter box
  2. Circle cutter
  3. Coping saw
  4. Cross cut
  5. Drywall saw
  6. Hack saw
  7. Keyhole saw
  8. Rip saw
J. Saw blades
  1. 40-tooth carbide circular saw blade
  2. Abrasive masonry-cutting blade
  3. Fine-tooth (80-tooth) miter box blade
  4. Jig saw, reciprocating saw blade assortment
  5. Metal-cutting blades
K. Miscellaneous
  1. Craft knife
  2. Floor scraper
  3. Metal snips (for straight or curved cut)
  4. Various files and rasps

Attaching tools

A. Clamps
  1. 3-way clamp
  2. Bar clamp
  3. C-clamps
  4. Hand-screw clamps
  5. Miter or right-angle clamps
  6. Pads (to protect finished surfaces)
  7. Pipe clamp
  8. Spring clamps
  9. Wood clamps (6", 8", 10")
B. Electric screwdrivers (corded or cordless)

C. Electric screwdriver tips
  1. Cabinet blades
  2. Extensions (up to 12")
  3. Flat
  4. Jeweler's screwdrivers
  5. Magnetic head
  6. Nut drivers
  7. Offset
  8. Phillips #1, 2, and 3
  9. Rachet driver
  10. Robertson (square drive)
  11. Screw holder equipped driver

Table 2.3 Tools and equipment (*Continued*)

12. Square heads—various lengths	3. Box wrenches (standard and metric)
13. Torx bit set	4. Open end wrenches (standard and metric)
E. Staple gun	
F. Stud driver (powder driven)	5. Pipe wrenches (12" to 18")
G. Wrenches	6. Ratchet and socket set: $\frac{3}{8}$" and $\frac{1}{2}$" drives (standard and metric)
1. Adjustable wrenches	
2. Allen wrench set (standard and metric)	7. Various extensions

Various hand tools

A. Hammers	3. Duckbill
1. 13-oz. claw hammer	4. Lineman's
2. 16-oz. claw hammer	5. Locking pliers (vise grip)
3. 20-oz. claw hammer	6. Needlenose
4. Ball peen hammer	7. Nippers
5. Drywall hammer	8. Various sizes of standard pliers
6. Heavy sledge	C. Other
7. Soft-faced hammer	1. Flat pry-bar
8. Tack hammer	2. Crow bars
B. Pliers	3. Cat's claw
1. Channel lock pliers (arc lock)	4. Nail puller
2. Diagonal cutters	

Air tools

A. Items required for basic air supply system	2. Finishing nailer
1. Air compressor	3. Spray gun (production or finish)
2. Air filter (in line)	4. Stapler
3. Air hose	5. Touch-up spray gun
4. Air regulator (in line)	C. Optional
5. Air tool lubricant	1. Caulker
6. Quick-disconnect couplings	2. Orbital sander
B. Tools	
1. Blow gun	

NOTE: Most hand tools available in electric are also available in air tools.

Additional equipment

1. Brushes	10. Ladders
2. Brooms	11. Mechanical taper
3. Doweling jig	12. Portable midget scaffold
4. Dust pans	13. Putty knives
5. Extension cords (proper gauge for length and tools)	14. Saw guides (edge guides)
	15. Sawhorses
6. Folding trestle	16. Tool pouch or holder
7. Hand truck	17. Vacuum cleaner(s)
8. Heat gun	18. Drop light
9. Hot-melt glue gun	19. Flashlight (rechargeable)

Optional but useful equipment

1. Automatic screw remover	3. Hardware drilling jigs
2. Cabinet installation jack(s)	4. Flooring nailer (hardwood floors)

Electrical

1. Circuit tester	4. Sniffer
2. GFI tester	5. Wire stripper
3. Lineman's pliers	6. Wire nuts

Plumbing tools

1. Basin wrench
2. Drain auger or "snake"
3. Flaring tool (copper tube)
4. Grade level
5. Pipe cutter
6. Plastic tubing cutter
7. Sink plunger
8. Spud wrench
9. Toiler plunger
10. Torch (for soldering copper pipe)
11. Tube bender
12. Valve seat dresser

Drywall tools

1. Abrasive mesh cloth
2. Adhesive spreader
3. Drywall dolly
4. Drywall panel lifter
5. Joint compound and tape
6. Mechanical cradle lifter
7. T-brace

Tools for laying tile flooring or tile countertops

1. Caulking gun
2. Float
3. Glass cutter
4. Grout float
5. Jointer
6. Notched trowels
7. Tile nipper pliers
8. Tile setting block (2" × 4") covered with towel or carpet
9. Tile spacers (plastic and wood)
10. Tile trimmer

Countertop tools

1. 3-way clamps
2. Bevel cut bit (router)
3. Cove bit (router)
4. Flush cut bit (router)
5. J-roller
6. Nail sets
7. Nail spotters
8. Rounding corner bit (router)
9. Router
10. Scribing compass
11. Straight bit (router)
12. Thin wood strip for laminate spacers

Paint

A. Supplies
  1. Angled trim brushes
  2. Brushes (various sizes and types)
  3. Cheese cloth
  4. Corner finishing pads
  5. Corner rollers
  6. Edge guides
  7. Paint mixers
  8. Paint scraper
  9. Putty knives
  10. Rollers
  11. Roller trays with disposable inserts
  12. Strainer cups
  13. Tack cloths
  14. Tape creasers
  15. Tape and dispensers
  16. Various grits of sandpaper
B. Sanders
  1. Belt sander with various paper grits
  2. Detail sander
  3. Drywall block handheld sander
  4. Electric hand sander (with dust collector)
  5. Long-handled pole sander
  6. Orbital sander
  7. Universal angle sander

Additional supplies

1. Carpenter's glue
2. Caulk
3. Construction adhesive
4. Contact cement
5. Degreasing agent
6. Drop cloths (preferably canvas)
7. Drywall screws
8. Duct tape
9. Electrical tape
10. Hanger wires
11. Lacquer thinner
12. Masking tape
13. Mild, nonabrasive cleaning products
14. Mineral spirits
15. Molly bolts and toggle bolts
16. Nails and brads

Table 2.3   Tools and equipment (*Continued*)

17. Polish and wax for wood finishes	22. Vapor barriers
18. Polyethylene sheets	23. Wax for screws
19. Rags	24. Wood shims
20. String	25. Wood screws
21. Touch-up paint and stain (from the cabinet and countertop manufacturer)	26. Wood filler putty (various colors)

Today, basic power tools can increase the speed and precision of your project. Many power tools are made in cordless versions. Cordless is fast becoming a good choice as more powerful, reliable cordless tools are available. They can be used anywhere and are not restricted by electrical power connections. When purchasing power tools, read specifications to compare features. More horsepower, faster motor speeds, and higher amperage ratings indicate a well-engineered tool. Better-quality tools also have roller or ball bearings instead of sleeve bearings. A good recommendation is to use or standardize the tools of one manufacturer so you are able to interchange rechargeable batteries. If you have several batteries on hand, you can keep one in the charger while operating another.

3

Basic principles of kitchen design

"Form follows function," great designers tell us. Certainly nothing could be more true when planning a new or remodeled kitchen. While clients want a kitchen to be beautiful, they also want it to be efficient. No matter how beautiful it is, if it is not functional, it is not good.

To be functional, the kitchen plan for either a new or remodeled house must take into consideration its location within the home, its size and shape, and the arrangement of its equipment and work centers (Fig. 3.1). Whether the kitchen installer is involved or not in the kitchen design process, it is important for all to be aware of the reasoning for the location of a kitchen within the home. This chapter gives the installer an overview of the basics for kitchen design.

Kitchen location

With regard to traffic in the home, there are several important connections between the kitchen and other areas of the house. First, there should be a direct connection from the kitchen to both indoor and outdoor eating and entertainment areas. Second, there should be a reasonably direct outside entry route to the garage or driveway. Delivery of groceries into the kitchen should be easy and quick and should not involve traipsing across carpeted dining or living areas. Third, the refuse disposal area should be directly accessible, without the need to pass through living areas or the main entrance.

While the kitchen should be situated to enable it to function as the home's center of activity and should be easily accessible from all parts of the house, it should not be a main thoroughfare for the rest of the house. Since passageways and access routes into the kitchen interrupt the continuous wall space for cabinets and appliances, a kitchen can be arranged to use floor space more efficiently. This arrangement can be accomplished best if there are only two, or at the most three, access ways, and if these access ways are located primarily in one part of the space.

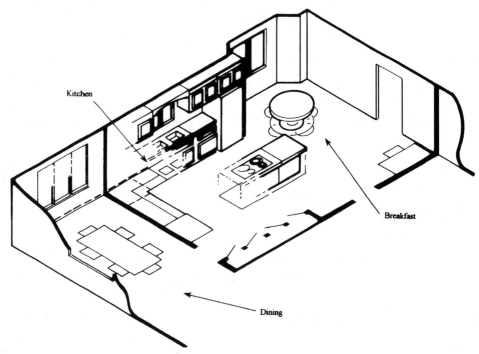

Kitchen

Breakfast

Dining

Figure 3.1 For serving convenience, a good plan locates a kitchen between the dining room and the breakfast nook. *(Leon E. Korejwo)*

An outside view can be another important factor in selecting the kitchen location, as well as the size and placement of the windows in the kitchen. Observation of community life and enjoyment of nature enhance the atmosphere, making routine chores more pleasant. If young children live in the house, the kitchen, if possible, should have a good view of and access to their play area.

Size and shape

When remodeling an old kitchen, of course, there is usually a limit to work within the existing area, although it is sometimes possible to expand the room by moving a wall or rearranging the existing space to obtain a more workable plan. With a new home—regardless of the size or price—the area allocated for kitchen work centers should not be less than 80 to 100 square feet and not more than 150 to 160 square feet. Areas in excess of these figures result in too great a distance between the work centers. Areas smaller than recommended create cramped, tight arrangements that are difficult to work in.

Frequently, many people make the mistake of assuming that more space is the solution to any problem of kitchen design. Sometimes it is. Other times, more space may simply mean that the client has farther to walk before he or she gets to other problems. Usually, inefficiency is the fault of poor planning (often in older homes) rather than of size. Thus, if more than 160 square feet is available, it is best to make it part of a separate dining or sitting area, a laundry room, or a playroom.

Functional kitchens must have adequate, organized storage at each work center. In addition, kitchens should have those "extras"—planning centers, pantries, and eating areas—so necessary for the easy, efficient functioning of the kitchen (see Chap. 4). However, while the location of the kitchen within the home is important, the efficiency of a well-planned kitchen depends on its size, shape, and the arrangement of its work centers.

Work centers

The basic kitchen is a place for storage of all the implements and ingredients necessary to the preparation of a meal. The activities of homemaking that are concentrated in the kitchen are complex, so they must be carefully planned in a sequence of events moving from storage, through preparation, to cooking and serving. To accomplish this sequence, there should be three main or basic activity areas, commonly referred to as work centers, in every kitchen:

1. The refrigeration or storage area
2. The sink or preparation and cleanup area
3. The food preparation and cooking area

The organization of these three main activity centers should be the first step in the formulation of any new kitchen plans, since they are the basis of any good design.

An ideal kitchen installation includes proper distribution of appliances, work counter surfaces, and cabinets within these three work centers. Each work center should include the major appliances, the food, the cooking equipment, and any other supplies used in that activity. In other words, each activity center should provide efficient working conditions for the particular functions that are performed at each appliance—refrigerator, sink, or range.

Refrigerator/storage work center

The refrigeration or storage area should be arranged to expedite the storage and removal of foods. Perishables must be moved to and from the refrigerator and the freezer. Canned goods, cereals, and staples are contained in the base, wall, and pantry cabinets that surround and are fitted to this center. In recent years, the refrigerator/storage center has become increasingly more important as food processing and prepackaging has changed. More ready-to-use foods are being used by families, and many of them require freezer storage. With families using more prepackaged foods, the pantry is considered an essential storage space in any well-planned kitchen.

To expedite the storage and handling of foods, the refrigerator/storage center should be located immediately adjacent to the kitchen's service entrance. A refrigerator/freezer combination or a separate freezer unit allows longer storage of many foods. As a separate freezer is not used as many times a day as the refrigerator, it may be located out of the main kitchen work area, for example, in a line of tall cabinets or another wall if it is an upright model, or in a garage, basement, or utility room. Allow at least 15 to 18 inches (38 to 46 cm) of countertop next to the refrigerator, with the refrigerator door opening on the side toward the counter. With a side-by-side refrigerator/freezer, remember it is easier to reach across the closed freezer door to load and unload the fresh food section than to reach around an open door to a counter on the wrong side. A combined total of at least 36 inches (91 cm) of counter space should be allowed for this center. Of course, with a side-by-side refrigerator/freezer, it is a good idea, when possible, to have an equal amount of space beside the freezer.

Some refrigeration models have doors that can be hinged to swing in either direction. Have the homeowner consider this idea carefully before ordering a new unit. Also, check the specifications of the intended model for the height of the box and recommended clearance for air circulation as well as for width. Avoid putting a refrigerator in a corner, beside a wall, or next to a line of cabinets. It is usually necessary to open the door more than 90 degrees to remove crispers and shelves.

When designing a kitchen, if the homeowners like pantries, the refrigerator/storage center is where the pantry belongs. If they prefer cabinet food storage, this is where the appropriate cabinets belong. Tailor everything to their habits and preferences, but be sure to group it all efficiently so they do not have to keep walking from one side of the kitchen to the other. When planning the storage space for the refrigerator/storage center, be sure to include several adjustable shelves to allow for odd-sized packages and bottles. The refrigerator/storage center should be placed next to the cleanup center, and there should be at least 42 inches (107 cm) of uninterrupted countertop between the two.

Food preparation/cook area

The main piece of equipment of the food preparation/cook area is the range (Fig. 3.2). Whether a free-standing, a built-in, a drop-in, or slide-in unit, the cooking equipment should be the latest design, offering all of the labor- and time-saving devices. Ideally, the range should be handy to the dinner area used more often by the family. However, do not place it next to a door that may open onto a constant parade of children. A cabinet or counter surface at least 15 inches (38 cm) wide between the door and the range cuts down the number of accidents. Do not forget that space should be provided on each side of the range for elbow room and pan handles, with at least 24 inches (61 cm) of counter on one side for serving and 12 inches (30 cm) on the other. Also, the range or built-in surface unit should not be installed under a window. Any curtains could catch fire, and the operation of such a window could be hazardous. Often local building codes have something to say about range placement. Check their requirements first.

Figure 3.2 A typical cook center contains a range (with oven) and vent hood. *(Fry Communications)*

While a built-in range top and separate oven or ovens are generally considered the most convenient of all range styles, they use up the most wall space. They also cost more to install, and replacement of the oven is difficult if the new model does not fit the cabinet for the old one. A separate oven should not be located directly next to the range; counter space is needed next to each. If necessary, locate the oven away from the busiest kitchen area. A wall oven should be installed so the inside top surface of the fully opened door is 1 to 7 inches (2.54 to 18 cm) below elbow height. If a double-oven unit is used, the bottom of the upper oven at counter height (36 inches, or 91 cm) is about right for most people. Never install a built-in oven too high. High mounting makes it difficult to remove pans and can result in burns. Allow 24 inches (61 cm) of counter on at least one side of the oven. A heat-resistant surface, such as stainless steel or ceramic glass, provides a place to place hot pots or utensils. A plastic laminate countertop should never be subjected to heat over 270 degrees F. Avoid having the wall oven at the end of a line of cabinets where the door opens into a traffic lane. Also avoid installation in a corner. Easier loading, unloading, and cleaning are possible with space all around the open oven door.

As for the range itself, several kinds are available. There are the familiar freestanding units of counter height as well as freestanding units of the "over and under" design (with an oven on the top, an oven on the bottom, and the cooktop in between). There are also drop-in or slide-in units with a separate built-in cooktop and wall oven or an eye-level oven and cooktop stacked on a base cabinet. The most common cooking facility for the least space is a 30-inch-wide (76 cm) range with an eye-level oven above the range top and another oven below it. However, a very short person often finds it difficult to handle a hot roasting pan in the high oven while a very tall person may find the view of the back burners on the range blocked by the upper oven. Have the home-owner stand in front of the model being considered and go through the motions of cooking to see if it can be worked with safely and comfortably. Handicapped, or seniors with restricted movement, must be kept in mind.

Whenever possible, the fumes from the food preparation/cook area should be vented to the outdoors. The most efficient arrangement is to have a hood and exhaust fan with a grease filter over the range, with the shortest possible duct to the outdoors. The duct may go through a wall, but avoid directing cooking odors toward an outdoor sitting area. The duct may also go through the roof, but it should never end in an attic. Check the local codes for installation standards. A ceiling or wall fan located away from the range draws the cooking fumes over the intervening surfaces. A nonducted hood and fan trap most of the grease and some of the odors but none of the heat and moisture from cooking. These hoods are rec-ommended only for interior locations from which outdoor venting is impossible.

Storage space is needed in the food preparation/cook area for small appliances, pots, pans, cooking utensils, and so on. For more on appliances for the food prepa-ration/cook area (ovens, vents, etc.), see Chapter 6. Wheelchair users benefit from roll-out storage, narrower doors on cabinets, and side-by-side refrigerators.

Sink/cleanup center

The sink and cleanup center involves the beginning and the end of the kitchen activities and consequently should be located where it is most convenient—usually between the other two centers. A minimum of 30 inches (76 cm) of counter surface should be on either side of the sink, and part of that should con-tain a chopping block or cutting board. The sink itself should be a double- or triple-bowl sink with one of the bowls deep enough and wide enough to allow the washing of pots and pans. In addition to the sink, the center should contain a disposer, a dishwasher usually adjacent to the sink, and a trash compactor. (If the dishwasher or trash compactor is not in the budget now, it is wise to plan ahead by installing proper-sized base cabinets that can later be replaced with an automatic dishwasher or trash compactor.) With today's increase in recy-cling the usage of compactors may have slightly declined. The sink and cleanup center must have good lighting, both natural and artificial. A light fixture should be located directly over the sink, and when possible, the sink should be placed close to a window. This is, however, a matter of personal preference. If there is only one window allotted to the area, some homeowners like it better in the eating area. Also, the installation of a sink along an interior plumbing partition usually costs less, if it is permitted by the local plumbing code.

As a general rule, store items in this center that require the use of water. Sufficient storage should be provided for fruits, vegetables, and other foods that do not require refrigeration and must be washed or soaked. In addition, space for small pots, a coffee maker, everyday dishes and glassware, dish towels, brushes, and dishwashing and cleaning supplies should be provided.

Supplementary work centers

Two supplementary work centers are also included in most well-designed kitchens. Although a considerable amount of preparatory work is done at the sink work center, studies of kitchen activities indicate that there is usually a need for a special preparation or *mix center*. Not as clearly defined, but nevertheless needed, is a *serve center*, which is often incorporated with the range center.

Mix center   For the sake of convenience, there is little difference whether the mix center is between the refrigerator and sink or between the sink and range. However, the latter arrangement involves more travel because the distance between mix center and refrigerator is usually longer. In recent years, the mix center has become unique in that the countertop height is lower than the normal countertop—32 inches (81 cm) instead of 36 inches (91 cm)— making it more comfortable for rolling dough and mixing. Plan on a counter 36 to 48 inches (91 cm to 122 cm) long with storage space to house such things as the mixer, blender, flour bin, baking pans, casseroles, larger canisters, mixing bowls of assorted sizes, and utensils for measuring and mixing. (Most packaged, canned, and bottled foods used in baking should also be stored near the mix center.) The mix center also requires a high intensity of illumination (at least the equivalent of two 100-watt incandescent lamps) mounted either under the cabinets or in hanging fixtures. The specialized storage can be supplied by most standard base cabinets and can be designed to roll out at the touch of a finger and spring up to counter height for easy use.

While separate wall ovens are most frequently a part of the cooking center, sometimes they are included in the mix area, and the complete unit is known as a *mix/bake center*. As such, it interrupts normal counter usage unless the oven is installed near the cooking surface. A single counter between the two may serve both, but if the oven and cooking surface are separated by a considerable distance, each should have an adjacent counter with cabinets. If both the built-in oven and refrigerator must be placed at one end of an assembly, a counter should be installed between them. The refrigerator should be closest to the sink/cleanup center. Wheelchair users require room for knees below work surfaces (for more on kitchens for the disabled, see chapter 4).

Serve center   The serve center is not used in conjunction with any specific appliance, but it is generally located between the range or cook center and the dining area. A counter area of at least 30 inches (76 cm) should be provided to facilitate food serving. In this center, keep in mind that moist and crisp warming drawers are a joy for party givers and those who dine in shifts. If space prohibits a separate small appliance center, it is a good idea to use small appliances near the range rather than close to the sink. A multi-outlet strip with a ground fault circuit accommodate a good number of appliances. Warming lights in this center are pleasant extras, too.

Storage for such items as a toaster, serving platters and dishes, table linens, cookies, cakes, and other foods that are purchased ready-to-eat should be included in the serve center. Of course, accessibility to the eating area is most important.

Combination centers   Frequently in small kitchens of studio apartments and similar areas, as well as in some second kitchens, the space is so limited that it is necessary to combine basic work centers. The cabinets and appliances of one center may be combined with those of the neighboring center to form a continuous assembly. It is usually best to combine at least two centers into one continuous assembly, with corresponding storage above and below.

One of the best ways to determine the proper counter space is to select the widest desirable counter of the two centers and then add 12 inches (30 cm) to it (Fig. 3.3). For instance, if combining the refrigerator/storage center with the sink/cleanup center, the counter required for the former is 18 inches (46 cm), while the minimum for the latter is 30 inches (76 cm). Thus the widest counter—30 inches (76 cm)—plus 12 inches (30 cm) would mean that the combined counter should be a minimum of 42 continuous inches (107 cm). The resulting counter permits both centers to operate simultaneously.

It is important to keep in mind when designing a kitchen with combination centers that it must contain at least 10 linear feet (305 cm) of full-use base cabinets plus 10 feet of full-use wall cabinets. Never cut down on counter space at the expense of base cabinets. Every kitchen needs at least 10 linear feet (305 cm) of base cabinets and 10 linear feet of wall cabinets. These are absolute minimums.

Specialty centers

While most kitchens could probably function adequately with only the three main work centers, there are some "extras," or specialty centers, that—like supplementary work centers—make the kitchen more workable and pleasant. Depending on the homeowner's specific needs and the available space, such items as eating areas, planning centers, pantries, barbecues, and laundry centers are no longer considered "extras." They have become essentials.

Eating area   Kitchen eating facilities can range from a counter-height eating bar to an almost-formal dining space. When planning for a kitchen eating space, a convenient location is of primary importance. The eating area should be easy to serve, yet away from the kitchen's main work areas. It should be large enough to contain a table and chairs and still allow easy circulation.

If there is a peninsula dividing the kitchen work area from the dining area, a snack bar built onto it solves the quick-meal problem (Fig. 3.4). Children love eating at counters, and innumerable steps can be saved by the homemaker. A dining bar should be about 29 inches (74 cm) high if chairs are used, about 40 inches (102 cm) with high stools. Allow 2 feet (61 cm) of counter or table space for each place setting. A counter should be at least 15 inches (38 cm) deep for meal service.

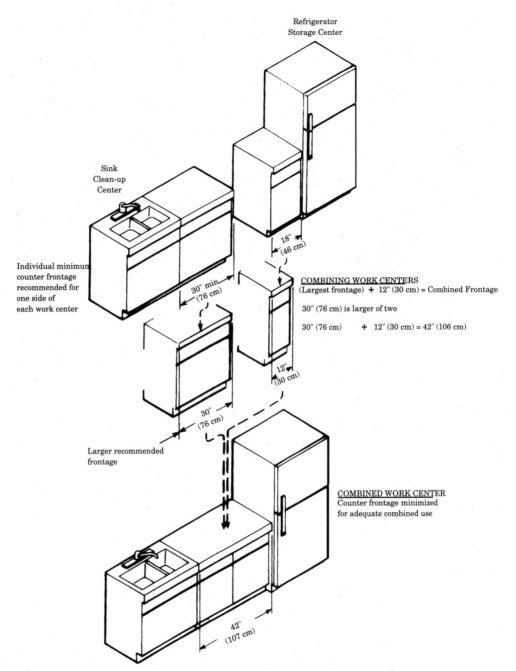

Refrigerator
Storage Center

Sink
Clean-up
Center

Individual minimum
counter frontage
recommended for
one side of
each work center

18"
(46 cm)

30" min.
(76 cm)

COMBINING WORK CENTERS
(Largest frontage) **+** 12" (30 cm) = Combined Frontage

30" (76 cm) is larger of two

30" (76 cm)   **+**   12" (30 cm) = 42" (106 cm)

12"
(30 cm)

30"
(76 cm)

Larger recommended
frontage

COMBINED WORK CENTER
Counter frontage minimized
for adequate combined use

42"
(107 cm)

Figure 3.3 Take the longest of the two required counter lengths and add 12" (30 cm) to determine the proper counter space. *(Leon E. Korejwo)*

Figure 3.4 An eating counter is ideal for snacks. *(Merillat Industries, Inc.)*

Planning center   Since the kitchen is where most household activities are likely to be organized, a planning desk, drop-leaf counter, or complete office space is a desired kitchen item. The work area of the planning or office center should be large enough to be functional (at least 30 inches, or 76 cm, long). It should have a telephone, a recipe file, cookbook shelves, and space for miscellaneous storage, with the work surface at desk height—29 to 30 inches (74 to 76 cm). A home office center could include a computer, telephone, mail organizer, and file drawers. The knee space should allow room for storing the chair, and the desk top must be well lit by natural or artificial light. Since the planning desk may be the homemaker's "office," every effort should be made to make it as pleasant as possible (Fig. 3.5).

Pantry   Before kitchens were "modern" in the sense of equipment and materials, the bulk of the kitchen's storage was in the pantry. Usually a closet-like space with a few shelves, it served as storage space for both food and cleaning equipment. Today there is a revival of the pantry, but its similarity to the pantry of yesteryear is minimal.

The contemporary pantry (every kitchen should have one) is a compact cabinet within easy reach of the work area. It should be handsomely designed. When wall space is limited, plot the pantry with pull-out shelves or drawers that use the full height and depth of the line of cabinets and bring all the stored items within easy reach. Pantry units can also be cleverly designed with hinged shelf sections on both sides, opening like leaves of a book to make everything accessible. Some manufacturers offer ready-made pantries. Slanting shelves, supermarket style, can also be used for canned goods storage with cans lying on their sides so that a replacement rolls into position as one is removed. The homeowner may specify pantry requirements not offered by cabinetmakers as part of their standard lines, although custom cabinetmakers can provide the builder with units directed more toward the client's needs. Removable adjustable shelving facilitates ease of cleaning and will be appreciated by the homemaker. Lazy Susan rotating units should be easily accessible for cleaning and removal of fallen items. As previously discussed, this pantry is correctly located as part of the refrigerator/storage center.

Laundry center If space can be found for a washer and dryer in another convenient location, it is better not to put them in the kitchen. Clothes and food just do not mix well together. It is convenient, though, to have the laundry center near the kitchen to help in dovetailing tasks.

If the laundry appliances must be in the kitchen, try to place them out of the refrigerator/sink/range area. Allow at least $4\frac{1}{2}$ feet (54 cm) of wall space for a

Figure 3.5 A well-designed planning center. *(Merillat Industries, Inc.)*

standard-size washer and dryer. Be sure to check the measurements of the models to be used. When space is very limited or a standard washer cannot be installed, a compact spinner washer 24 inches (61 cm) wide by 30 inches (76 cm) high by 15 inches (38 cm) deep can be stored out of the way and rolled to the sink for use. A matching compact dryer of the same size may be operated on 120-volt electricity. It can be hung on a wall, parked on a base cabinet, or placed standing on the floor. If there is room for a washer but not enough space or the required utility connections for a standard-size dryer, a compact dryer can be mounted on top of some washers with the use of a special kit.

More details on eating areas, planning centers, family rooms, and laundry centers, as well as other kitchen extras, are given in Chapter 4.

The work triangle

The three work centers should be placed in the kitchen so that movement between them is as efficient and direct as possible. The normal work sequence is from right to left. The ideal arrangement for a right-handed person, therefore, is to have the refrigeration/food-storage area on the right with a preparation area adjacent to it. The cleanup center, or sink center, with storage for dishes, sink, and dishwasher, should be next. The cooking and serving center, with storage for pots, pans and serving dishes, should be placed on the left.

A triangular arrangement creates a typical work pattern that occurs in the well-planned kitchen (Fig. 3.6). Food can be taken from the refrigerator and storage center and made ready for cooking (or eating) at the preparation center. Since water is likely to be needed in the process, the sink center will be used. Once prepared, the food may be moved directly to the serve center and then to the dining area, or it may go through a cooking process at the cooktop or oven.

Under certain conditions, a compromise may be needed and the centers may be placed out of sequence, reversed, or even separated. A work center can be isolated on an island or in a separate wall unit. When used in this fashion, it should be a complete center or unit. In any case, whether the sequence is left-to-right or right-to-left is not nearly important as arranging the kitchen to minimize the length of trips between the pairs of work centers. At the same time, ample room for counter and storage at each center should be allowed.

In arranging the three main work centers in a kitchen, a "triangle of convenience" (also known as the work triangle, the magic triangle, the step-saving triangle, etc.) is the prime consideration. This triangle represents the flow of work from one center to another. The sum of the three sides of the triangle should not exceed 26 feet (793 cm), to conserve the homeowner's energy. The sides of the triangle also should not be less than 13 feet (396 cm), to avoid an overcrowded work pattern. No single arm of the triangle should measure less than 4 feet (122 cm).

The most frequently trafficked leg of the triangle is that between the sink and the range. Ideally, this leg of the triangle would be the shortest. The next most heavily traveled route is between refrigerator and sink. This leg of the

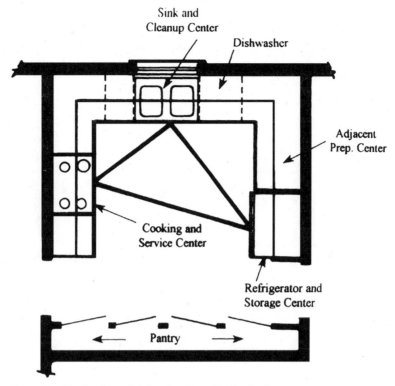

Figure 3.6 The basic work triangle. *(Leon E. Korejwo)*

triangle might be from 5 to 8 feet (152 to 244 cm). When measuring the distance between two appliances, measure from the center front of one to the center front of the other. By the way, since the separate-type oven is used relatively infrequently, it is often convenient to separate the oven center from the other centers of the kitchen; it need not be located on the triangle at all.

Before leaving the subject of kitchen traffic flow, it is worth remembering that common sense is most important (Fig. 3.7). Obviously, no doorways should funnel traffic directly into the middle of the work triangle. The number of doorways, their location, and the direction of the door swing affects the efficiency of the kitchen arrangement. Generally, doorways in corners should be avoided. It is also desirable to avoid door swings that conflict with the use of appliances or cabinets or with other doors. When remodeling a kitchen, rehang the door on the other side of the jamb or hinge it to swing out rather than in. In the latter case, make certain that the door does not swing into the traffic path in halls or other activity areas. A sliding or folding door avoids such a problem. But aside from such general thoughts on good traffic movement throughout the triangle, there are no hard and fast rules of traffic flow.

It is also important that adequate aisles between work centers be provided so that there is room to move about easily. Two people should be able to bend down

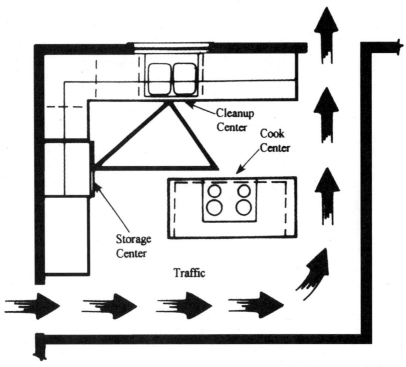

Figure 3.7   No major traffic patterns should cross through the work triangle. *(Leon E. Korejwo)*

back to back without touching. Between opposite work counters, allow at least 48 inches (122 cm). If two or more people are likely to be sharing the kitchen, allow 54 to 64 inches (137 to 163 cm). The same clearance is needed from a counter front to a table, a wall, or to the face of a storage wall if the space is a work area.

In recent years, the concept of the traditional work triangle has undergone some important changes. In today's world, many people may share kitchen responsibilities. Therefore, the concept of the double work triangle comes into play.

Double work triangles are based on the same idea as traditional work triangles, but they accommodate several cooks and allow for many different functions to take place within the kitchen at one time. In double work triangles, some appliances in work stations are shared, but since well-designed double triangles never cross each other, there is not a traffic problem. For example, as shown in Fig. 3.8, the addition of an island helps keep the work triangles separate. The cooktop is in one corner, the oven in the other. The refrigerator is shared, and both work spaces have their own sinks—the perfect arrangement for a cook and a baker.

Kitchen types

Well-designed kitchens take into account the special needs of a household. The following basic shapes of kitchens are explained with traditional and double work triangles. Help the homeowners decide which layout describes the best kitchen for them, and use the concept of the work triangle, or the double work triangle, in their kitchen design.

Although it is possible to have almost infinite numbers of arrangements and combinations of work centers, the most commonly used and generally the most efficient kitchens are U-shaped, L-shaped, corridor, one-wall, or island arrangements (Fig. 3.9). The arrangement used depends on the size and shape of the area allocated for the kitchen; the location of windows, doors, and services (plumbing, electricity, etc.); and the proximity of the kitchen to the home's other rooms and outdoor living spaces. It depends also on getting the most efficient arrangement of work centers arranged in a triangular pattern with the sequence of work centers being refrigerator/storage center, sink/cleanup center, and food preparation/cook area. This triangular arrangement and the sequence of the work centers along the triangle, as mentioned earlier, should not exceed 26 feet (292 cm) nor be less than 13 feet (396 cm). The distance between any two centers should not be more than 7 to 8 feet (213 to 244 cm). In the case of a U-shaped or L-shaped kitchen, the maximum distance between the two extreme centers should not be more than 10 feet (305 cm).

In planning a kitchen around the triangle, keep all other kitchen activities away from the three main work centers. Also, try to prevent the normal traffic lanes between other areas of the house from crossing the triangle, which

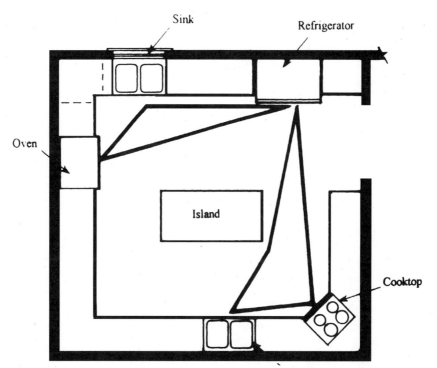

Figure 3.8 A double work triangle should be used to accommodate several cooks. *(Leon E. Korejwo)*

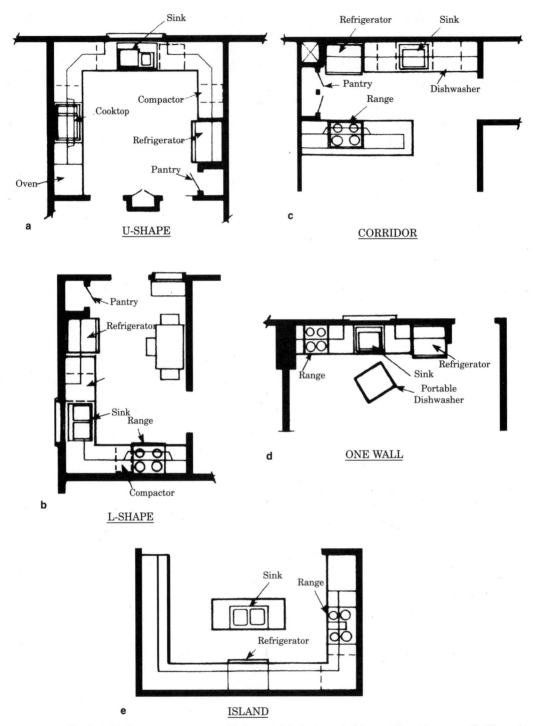

Figure 3.9 Typical kitchen arrangements: U-shaped (a); L-shaped (b); corridor (c); one-wall (d); and island (e). *(Leon E. Korejwo)*

would cut down on the plan's efficiency. Such through traffic may generally be avoided by U-shaped, L-shaped, and one-wall kitchen layouts. With some arrangements, such as the corridor and island types, cross traffic often can pass through the work triangle. However, it need not interfere with kitchen activities if the plan provides an alternate route of travel.

Architectural design dictates to a great extent the type of kitchen plan. However, imagination and basic knowledge of traffic patterns in the kitchen can help the planner overcome most any obstacle in planning a beautiful new kitchen. By becoming acquainted with the five basic shapes, one can soon know what to look for in arrangements. Of course, all five basic kitchen types have many variations. The U-shaped, for example, can become round or octagonal. But, it should be remembered that each one is planned for a specific type of use and each has its unique advantages and disadvantages. Select the plan that best suits the needs of the cook or cooks and the available space.

The U-shaped kitchen

In spite of the fact that experts seldom agree on anything, most kitchen design experts seem to agree that the U-shaped arrangement is the most desirable kitchen plan from the standpoint of efficiency. This arrangement divides the work centers and their appliances among three walls that are placed in a "U" configuration (Fig. 3.10a). The sink/cleanup center is usually placed at the base of the U, with the refrigerator and cook centers installed on the facing legs, creating a tight work triangle that eliminates wasted effort. In fact, the U-shaped kitchen's greatest advantage is the short distance between work centers and the privacy it ensures for the cook. The U-shaped kitchen also allows more than one cook to work comfortably and efficiently in a reasonable amount of space.

Because of its shape, traffic patterns form naturally outside the work areas and do not affect the U-shaped kitchen's efficiency. Counter space is continuous, and ample storage is generally made available. The U-shaped plan is also adaptable to both large and small kitchens (Fig. 3.10b). However, if the U is less than 6 feet (183 cm) between base cabinets, the work area gets too cramped. More than 10 feet (305 cm) across the work area results in too many steps between work centers. A 5-foot (152 cm) aisle and another wall storage and counter space is better than more aisle and no counters.

Frequently, one leg of the U may extend into the kitchen without wall support. Such an arrangement is generally used in connection with an informal dining space, with one leg of the U serving as a room-dividing peninsula to separate the kitchen and dining area or kitchen and family room.

The minor advantages of the U-shaped kitchen are the extensive countertops and the need for special cabinets to fit the corners. In the U-shaped plan, it is important to locate the dishwasher so that the door can open without hitting the adjacent corner cabinet. In a tight U plan, this can be accomplished with a filler at the corner.

Figure 3.11 shows a variation of the U-shaped plan, the so-called G-shaped kitchen, which has the addition of an elongated partial wall. Again, there is

Figure 3.10a The U-shaped kitchen allows great flexibility in arrangement. *(Leon E. Korejwo)*

plenty of counter space and the arms can be used for a multitude of purposes. This layout is ideal for larger families that need extra storage. It also offers the same accessible work area as the U-shaped kitchen. The peninsula works as a partition between kitchen and family areas.

The L-shaped kitchen

The L-shaped kitchen divides the work centers and appliances between two walls that meet at right angles. Another variation of this layout extends the base of the L into the kitchen area. The L shape is a good plan for a large kitchen and generally provides sufficient room for more than one cook to work at a given time. Ideally, the two work spans should not be broken by openings other than windows.

While the L-shaped kitchen permits less space for counters and storage, it has the advantage of creating an eating area without sacrificing space from the work area. This free space on the two unoccupied walls can also be used for a laundry area, child's play space, or other activity center. The L-shaped kitchen also keeps the work triangle free of through-room traffic and allows an efficient arrangement of the work centers. In any L-shaped kitchen, the sequence of work centers should provide a work flow from refrigerator to sink to range (Fig. 3.12). Any other arrangement would be awkward and difficult to work with. But keep

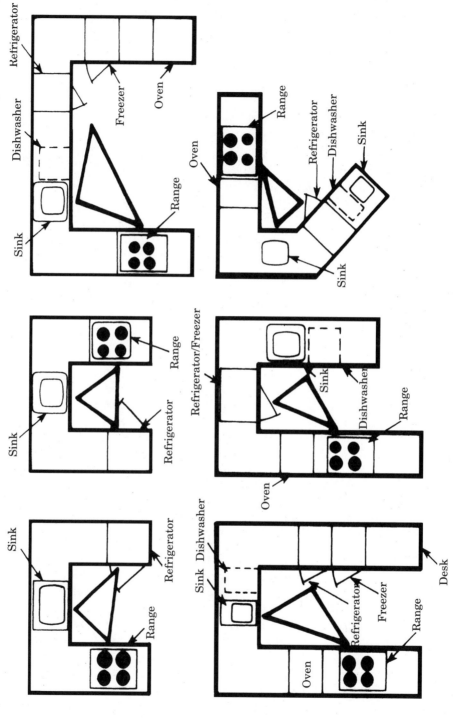

Figure 3.10b Six arrangements for a U-shaped kitchen. *(Leon E. Korejwo)*

71

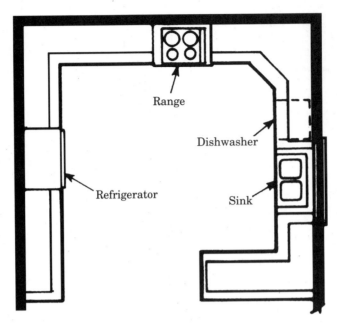

Figure 3.11 A typical G-shaped kitchen plan. *(Leon E. Korejwo)*

in mind that if the refrigerator, sink, and range are too far apart, the work triangle would be such that serving a meal could be a very exhausting task.

In most L-shaped kitchens, the sink/cleanup center is placed on the window wall. Sometimes the short leg of the L can be widened to create a peninsula between the eating space and the work area, with the food preparation/cook area placed on the kitchen side of the peninsula and an eating bar on the other side.

It may be necessary, because of doors, windows, or offsets, to adapt one of the five basic kitchen types. For example, the most efficient utilization of space may result in a broken U or a broken L where doors interrupt the wall pattern. As a rule, however, a broken U or L layout is a sign of failure in the house's design.

The corridor kitchen

The corridor type of kitchen, sometimes called a *galley kitchen*, divides the work centers and their appliances between two parallel walls, with two work areas on one wall and one on the other (Fig. 3.13). Because this arrangement is well-suited for long narrow spaces, it is one of the most popular kitchen types used in apartment construction. The corridor kitchen is also particularly useful in homes where there is a premium on space. Often its compact dimensions limit it to one or two cooks, but when well-planned, this design can squeeze the maximum number of work surfaces, cabinets, and appliances into the smallest space. The aisle in the corridor should be at least 4 feet (122 cm) wide and, if possible, should reach a dead end to prevent casual traffic from passing through the work triangle. When this happens, the corridor kitchen resembles

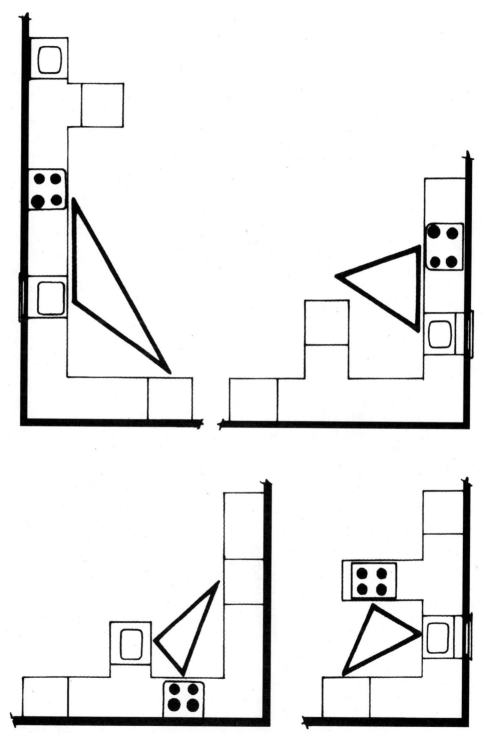

Figure 3.12 Four arrangements for an L-shaped kitchen. *(Leon E. Korejwo)*

a small U, and contains many of its advantages. It is also perhaps the most economical type since there are no corners to turn. However, if the corridor is open at both ends, it is a passageway from one part of the house to another, which could drive the cook crazy.

The one-wall kitchen

The one-wall type kitchen, frequently called a *pullman*, *line-a-wall*, or *strip kitchen*, positions all work centers and appliances along one wall (Fig. 3.14).

Figure 3.13   The layout of a corridor kitchen is useful when space is limited. *(Leon E. Korejwo)*

Figure 3.14 A one-wall kitchen has all work centers and appliances along one wall. (*Leon E. Korejwo*)

Where space is very limited—mini-kitchens in studio apartments, second kitchens in a family room, or a vacation home kitchen—this plan is frequently employed. It also finds favor in some contemporary open-plan houses where the kitchen is part of the social space. But, because of the cramped space, there is no possible way of forming the step-saving triangle of convenience. Thus the challenge in planning a one-wall kitchen is to provide adequate storage and work surfaces without the distances between the work centers becoming too great for efficiency.

To overcome some of the disadvantages of the one-wall kitchen, consider converting a nearby closet to a pantry for canned foods or for cleaning supplies, or let a mobile cart or two give some extra counter work space. A portable dishwasher also adds to counter space by serving as a traveling work surface. An eating area will probably have to fit into an adjoining room where space is available—either the family room or a dining room. In this plan, as in all others, route traffic so it bypasses the important kitchen work areas.

The island kitchen

All layouts except the one-wall kitchen can benefit from the addition of an island. In recent years, this type is usually a modification of either the U-shaped kitchen or the L-shaped kitchen and should not be confused with what is normally called a broken U-shaped plan. The island kitchen retains much of the efficiency of both the U-shaped and the L-shaped kitchens while providing a focal point and design feature (Fig. 3.15a). Island kitchens should be used when space is not a problem or as part of a combined kitchen/family room; the island serves as an excellent room divider. Sometimes, one of the work centers may be moved from the normal sequence and placed on the island (Fig. 3.15b). When this is done, care must be taken not to increase or decrease the distance between work centers more than recommended. The island should be a minimum of 25 by 48 inches (64 by 122 cm) and permit an aisle between island and nearby work counters of at least 3 feet (92 cm). If the range and sink are on opposite walls, never place the island between them unless the homemaker is out to set some sort of a hiking record.

The island counter immediately adds storage space. Base cabinets can fit on all sides, or shelving may be fitted into one or both ends. The counter may be used to house many of the items that cannot fit onto counters along the kitchen wall—a bar or vegetable-washing sink, a barbecue grill, a hard maple cutting board, or a small appliance center. If space permits, the island might be widened to add a food-serving extension or even a drop-leaf for quick snacks.

Where there is not enough floor space for a permanent island, the so-called *floating* type, which is an island counter on heavy-duty casters, may be an ideal solution. In addition to providing an extra work surface in the kitchen where it is needed, the floating island can be rolled to other rooms in the house or outside to the terrace.

The peninsula feature in a kitchen is closely akin to the island. A peninsula can bring the three kitchen work centers closer together and reduce the size of the work triangle. Often, the center floor space in a kitchen is not put to good use. A large, old, square kitchen or an open, new kitchen can benefit from the addition of a peninsula. Even a short lineup of cabinets extending into the center of the room can divide the working and eating areas and give additional storage and work tops. A peninsula is often designed so cabinets open

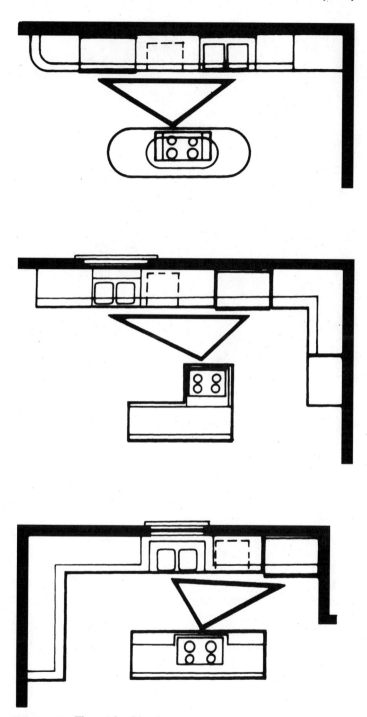

Figure 3.15a   Three island kitchen arrangements. *(Leon E. Korejwo)*

Figure 3.15b Islands can be functional, as well as quite elaborate. *(KraftMaid Cabinetry, Inc.)*

from both sides. This feature is particularly good if the table-setting supplies are housed in the peninsula. Wall cabinets can be hung over the top of the peninsula, or the view can be left open. The use of a peninsula also permits the so-called *four-wall* type of kitchen for spacious, gracious freedom of design.

The base cabinet at the end of the peninsula should have a toe or kick space on the end side, as well as on the kitchen side, so a person can work in a normal manner. Also, depending on how the peninsula is employed, it may be necessary to have toe space on both sides of all base cabinets. Most island units have a toe space all the way around them.

Peninsula wall cabinets are usually laid out so that their run is shorter than the base cabinet units. Besides reducing head-bumping incidents, this layout contributes to a more open appearance for the entire kitchen. To cut down on bruised arms and hips, it is wise to figure on curved corners rather than square ones at the end of the peninsula.

The family kitchen

The *family* kitchen is just an open version of any of the five basic plans (Fig. 3.16). Its function is to provide a gathering place for the entire family in addition to providing space for everyday kitchen functions. Because of its dual function, a family kitchen is normally divided into two sections. One section includes the three work centers, while the other contains the dining area and family-room facilities.

Common mistakes

Frequently, a little replanning can make a kitchen easier to work in and cut the builders' costs in the bargain. In the two plans shown here, the "befores" were taken from actual builders' plans and the "afters" were taken after a little replanning.

Kitchen No. 1 began as a shallow U in which the counter range could not be correctly positioned (Fig. 3.17a). A room-divider counter changed it to a straight-line plan that easily accommodated the range and, as a bonus, provided an eating bar (Fig. 3.17b).

Kitchen No. 2 suffered from badly allocated work surfaces, a poorly positioned refrigerator, and lack of room for both a washer and a dryer (Fig. 3.18a). The first revision provided work space around all centers and room for a separate dryer, but left the refrigerator blocking part of the window (Fig. 3.18b). The second revision cleared the window and put the dryer on the outside wall, where it is easier to vent (Fig. 3.18c).

All these mistakes were taken from actual builders' plans. For example, in Fig. 3.19a and b, the dishwasher's location is the problem. Since people often leave the machine open while it accumulates dirty dishes, it must be positioned so its door does not block a work area.

The mistake in Fig. 3.20 stems from the fact that a refrigerator is deeper than other appliances and may block a window. The mistake in Fig. 3.21, a surprisingly common one, results from just plain forgetting that a refrigerator has to be moved in and out when the kitchen is finished or needs to be cleaned. Appliance-door clearances, easy to overlook on paper, create the mistakes in Fig. 3.22 and 3.23. The mistake in Fig. 3.24 is usually the result of trying to jam appliances into inadequate space. The mistake in Fig. 3.25 occurs when a freestanding appliance is set into a counter that also serves as a room divider.

Guidelines for kitchen planning

To summarize the important points of design, the National Kitchen and Bath Association and the University of Illinois Small Homes Council developed standards based on extensive research conducted in conjunction with the University of Minnesota, known as *The 40 Guidelines for Kitchen Planning.* Our society is growing more diverse, which requires that the guidelines be constantly revised. It also requires that they remain under scrutiny for further changes in our population and the living habits of the various population segments. These revisions more fully incorporate universal design as clarified by the Uniform Federal Accessibility Standards (UFAS) and the American National Standard for Accessible and Usable Buildings and Facilities (ANSI A117.1-1992). The dimensions included in these NKBA guidelines are based on ANSI and UFAS, but they are not intended to replace them.

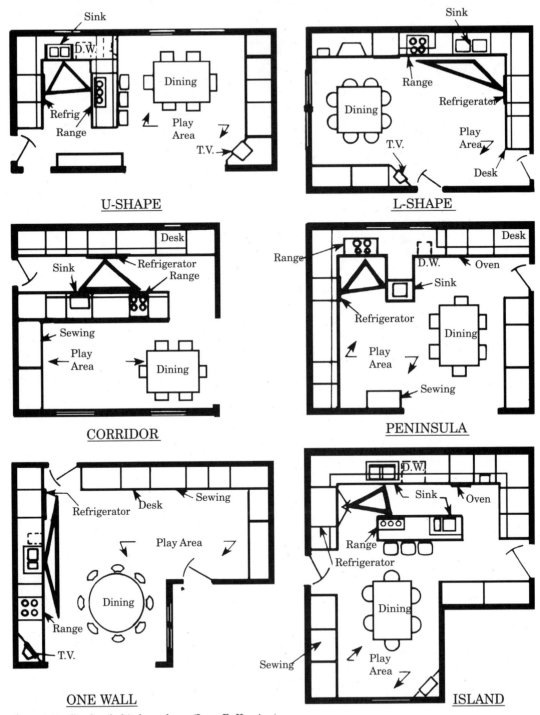

Figure 3.16 Six-family kitchen plans. *(Leon E. Korejwo)*

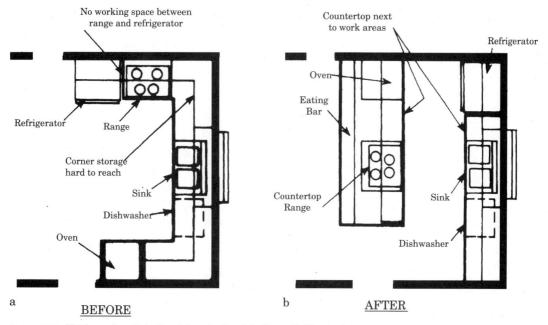

Figure 3.17 Kitchen plan #1 before (a) and after (b). *(Leon E. Korejwo)*

To date, most single-family residential projects do not fall under any standard for accessibility, but this fact is changing. If a particular project is subject to local, state, or national laws or codes, the designer must comply with those requirements. These guidelines, and the space planning provided here, are intended to be useful design standards, supplemental to the applicable codes, not hard-and-fast rules, but guidelines that will help in planning kitchens that are functional and flexible, or universal, to better meet the needs of today's varied lifestyles.

Traffic and workflow

1. (a) Doorways should be at least 32" (81 cm) wide and not more than 24" (61 cm) deep in the direction of travel.

   (b) Walkways (passages between vertical objects greater than 24" (61 cm) deep in the direction of travel, where not more than one is a work counter or appliance) should be at least 36" (91 cm) wide. The 36" (91 cm) width of a walkway allows a person using a wheelchair to pass. However, in order to turn when two walkways intersect at right angles, one of the walkways must be a minimum of 42" (107 cm) wide.

   (c) Work aisles (passages between vertical objects, both of which are work counters or appliances) should be at least 42" (107 cm) wide in one-cook kitchens, at least 48" (122 cm) wide in multiple-cook kitchens.

2. The work triangle should total 26' (792 cm) or less, with no single leg of the triangle shorter than 4' (122 cm) nor longer than 9' (274 cm). The work triangle

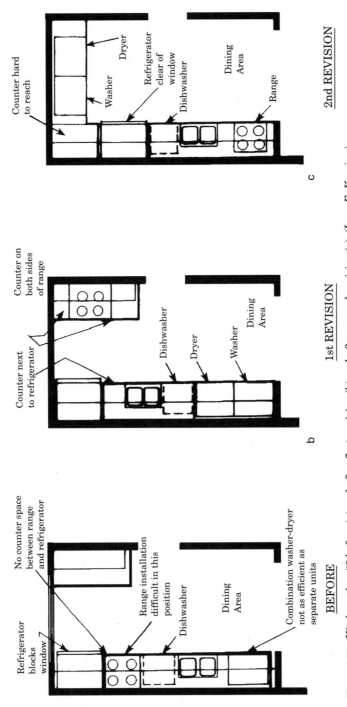

Figure 3.18 Kitchen plan #2 before (a) and after first revision (b) and after second revision (c). (*Leon E. Korejivo*)

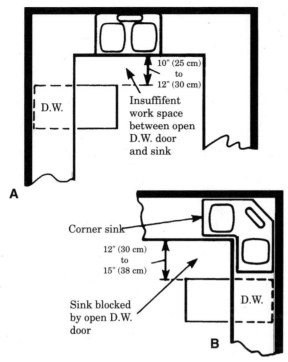

10" (25 cm)
to
12" (30 cm)

Insuffifent work space between open D.W. door and sink

D.W.

A

Corner sink

12" (30 cm)
to
15" (38 cm)

Sink blocked by open D.W. door

D.W.

B

Figure 3.19   Kitchen mistake #1; Dishwasher door interferes with work space. *(Leon E. Korejwo)*

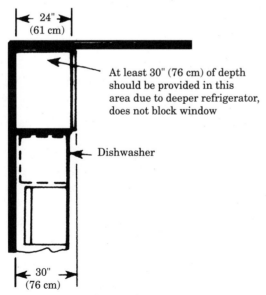

24"
(61 cm)

At least 30" (76 cm) of depth should be provided in this area due to deeper refrigerator, does not block window

Dishwasher

30"
(76 cm)

Figure 3.20   Kitchen mistake #2; Refrigerator too wide, blocks window. *(Leon E. Korejwo)*

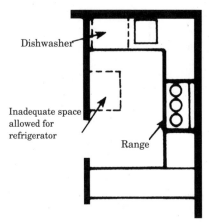

Figure 3.21 Kitchen mistake #3; Refrigerator in poor location. *(Leon E. Korejwo)*

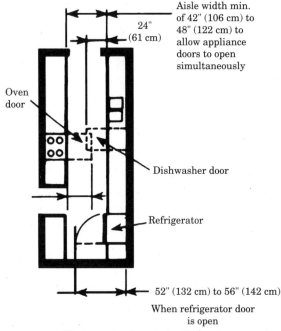

Figure 3.22 Kitchen mistake #4; Appliance door openings. *(Leon E. Korejwo)*

should not intersect an island or peninsula by more than 12" (30 cm). (The triangle is the shortest walking distance between the refrigerator, primary food preparation sink, and primary cooking surface, measured from the center front of each appliance.)

3. No major traffic patterns should cross through the work triangle.

4. No entry, appliance, or cabinet doors should interfere with one another.

5. In a seating area, 36" (91 cm) of clearance should be allowed from the counter/table edge to any wall/obstruction behind it if no traffic will pass behind a seated diner. If there is a walkway behind the seating area, 65" (165 cm) of clearance, total, including the walkway, should be allowed between the seating area and any wall or obstruction. The 65" (165 cm) walkway clearance called for behind a seating area allows room for passage by or behind the person using a wheelchair.

Cabinets and storage

6. Wall cabinet frontage, small kitchens—under 150 sq. ft. (14M²) (14 sq. meters) —allow at least 144" (366 cm) of wall cabinet frontage, with cabinets at least 12" (30 cm) deep, and a minimum of 30" (76 cm) high (or equivalent) which feature adjustable shelving. Difficult-to-reach cabinets above the hood, oven or

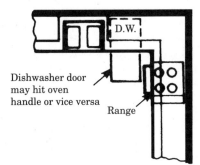

Figure 3.23 Kitchen mistake #5; Appliance doors interfere with each other. *(Leon E. Korejwo)*

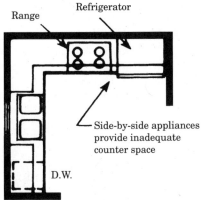

Figure 3.24 Kitchen mistake #6; Appliances are jammed too close together. *(Leon E. Korejwo)*

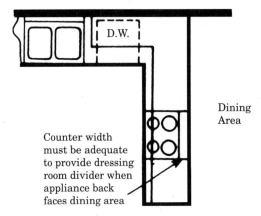

Dining
Area

Counter width
must be adequate
to provide dressing
room divider when
appliance back
faces dining area

Figure 3.25 Kitchen mistake #7; Freestanding appliance fits poorly in counter/room divider. *(Leon E. Korejwo)*

refrigerator do not count unless devices are installed within the case to improve accessibility.

Wall cabinet frontage, large kitchens—over 150 sq. ft. (14M²)— allow at least 186" (472 cm) of wall cabinet frontage, with cabinets at least 12" (30 cm) deep, and a minimum of 30" (76 cm) high (or equivalent), which feature adjustable shelving. Difficult-to-reach cabinets above the hood, oven, or refrigerator do not count unless devices are installed within the case to improve accessibility.

7. At least 60" (152 cm) of wall cabinet frontage, with cabinets at least 12" (30 cm) deep, a minimum of 30" (76 cm) high (or equivalent), should be included within 72" (183 cm) of the primary sink centerline.

8. Base cabinet frontage, small kitchens—under 150 sq. ft. (14M²)—allow at least 156" (393 cm) of base cabinet frontage, with cabinets at least 21" (53 cm) deep (or equivalent). The blind portion of a blind corner box does not count.

    Base cabinet frontage, large kitchens—over 150 sq. ft. (14M²)—require at least 192" (488 cm) of base cabinet frontage, with cabinets at least 21" (53 cm) deep (or equivalent). The blind portion of a blind corner box does not count.

9. Drawer/roll-out shelf frontage, small kitchens—under 150 sq. ft. (14M²)— allow at least 120" (305 cm) of drawer or roll- out shelf frontage.

    Large kitchens—over 150 sq. ft. (14M²)—allow at least 165" (419 cm) of drawer or rollout shelf frontage. Multiply cabinet width by number of drawers/rollouts to determine frontage. Drawer/roll-out cabinets must be at least 15" (38 cm) wide and 21" (53 cm) deep to be counted.

10. At least five storage/organizing items, located between 15" to 48" (38 cm to 122 cm) above the finished floor (or extending into that area), should be included in the kitchen to improve functionality and accessibility. These items may include, but are not limited to, lowered wall cabinets, raised base cabinets, tall cabinets, appliance garages, bins/racks, swing-out pantries, interior vertical dividers, specialized drawers/shelves, etc. Full-extension

drawers/rollout shelves greater than the 120" (305 cm) minimum for small kitchens or 165" (419 cm) for larger kitchens may also be included.

11. For a kitchen with usable corner areas in the plan, at least one functional corner storage unit should be included.

12. At least two waste receptacles should be included in the plan; one for garbage and one for recyclables, or other recycling facilities should be planned.

Appliance placement and use/clearance space

13. Knee space (which may be open or adaptable) should be planned below or adjacent to sinks, cooktops, ranges, dishwashers, refrigerators, and ovens whenever possible. Knee space should be a minimum of 30" (76 cm) wide by 27" (69 cm) high by 19" (48 cm) deep under the counter. The 27" (69 cm) height at the front of the knee space may decrease progressively as depth increases.

14. A clear floor space of 30 by 48" (76 by 122 cm) should be provided at the sink, dishwasher, cooktop, oven, and refrigerator. (Measure from face of cabinet or appliance if toekick is less than 9"—23 cm—high.)

15. A minimum of 21" (53 cm) clear floor space should be allowed between the edge of the dishwasher and counters, appliances and/or cabinets [that] are placed at a right angle to the dishwasher.

16. The edge of the primary dishwasher should be within 36" (91 cm) of the edge of one sink.

17. If the kitchen has only one sink, it should be located between or across from the cooking surface, preparation area, or refrigerator.

18. There should be at least 24" (61 cm) of clearance between the cooking surface and a protected surface above, or at least 30" (76 cm) of clearance between the cooking surface and an unprotected surface above. (If the protected surface is a microwave hood combination, manufacturer's specifications may dictate a clearance less than 24"—61 cm.)

19. All major appliances used for surface cooking should have a ventilation system, with a fan rated at 150 CFM (cubic feet per minute) minimum.

20. The cooking surface should not be placed below an operable window unless the window is 3" (8 cm) or more behind the appliance and more than 24" (61 cm) above it. Windows, operable or inoperable, above a cooking surface should not be dressed with flammable window treatments.

21. Microwave ovens should be placed so that the bottom of the appliance is 24" (61 cm) to 48" (122 cm) above the floor.

Counter surface and landing space

22. At least two work counter heights should be offered in the kitchen, with one 28 to 36" (71 to 91 cm) above the finished floor and the other 36 to 45" (91 to 114 cm) above the finished floor. Varying counter heights will create work spaces for various tasks and for cooks of varying stature, including seated cooks.

23. Countertop frontage, small kitchens—under 150 sq. ft. (14M²)—allow at least 132" (335 cm) of usable countertop frontage.

 Countertop frontage, large kitchens—over 150 sq. ft. (14M²)— allow at least 198" of usable countertop frontage.

 Countertop frontage—Counters must be a minimum of 16" (41 cm) deep, and wall cabinets must be at least 15" (38 cm) above their surface for counter to be included in total frontage measurement. (Measure only countertop frontage; do not count corner space.)

24. There should be at least 24" (61 cm) of countertop frontage to one side of the primary sink, and 18" (46 cm) on the other side (including corner sink applications) with the 24" (61 cm) counter frontage at the same counter height as the sink. The countertop frontage may be a continuous surface or the total of two angled countertop sections. (Measure only countertop frontage, do not count corner space.) For further instruction on these requirements see Guideline 31.

25. At least 3" (8 cm) of countertop frontage should be provided on one side of secondary sinks, and 18" (46 cm) on the other side (including corner sink applications) with the 18" (46 cm) counter frontage at the same counter height as the sink. The countertop frontage may be a continuous surface or the total of two angled countertop sections. (Measure only countertop frontage, do not count corner space.) For further instruction on these requirements see Guideline 31.

26. At least 15" (38 cm) of landing space, a minimum of 16" (41 cm) deep, should be planned above, below, or adjacent to a microwave oven. For further instruction on these requirements see Guideline 31.

27. In an open-ended kitchen configuration, at least 9" (23 cm) of counter space should be allowed on one side of the cooking surface and 15" (38 cm) on the other, at the same counter height as the appliance. For an enclosed configuration, at least 3" (8 cm) of clearance space should be planned at an end wall protected by flame-retardant surfacing material and 15" (38 cm) should be allowed on the other side of the appliance, at the same counter height as the appliance. For further instruction on these requirements see Guideline 31.

28. The plan should allow at least 15" (38 cm) of counter space on the handle side of the refrigerator or on either side of a side-by-side refrigerator or at least 15" (38 cm) of landing space [that] is no more than 48" (122 cm) across from the refrigerator. (Measure the 48" [122 cm] distance from the center front of the refrigerator to the countertop opposite it.) For further instruction on these requirements see Guideline 31.

29. There should be at least 15" (38 cm) of landing space [that] is at least 16" (41 cm) deep next to or above the oven if the appliance door opens into a primary traffic pattern. At least 15" (38 cm) by 16" (41 cm) of landing space [that] is no more than 48" (122 cm) across from the oven is acceptable if the appliance does not open into a traffic area. (Measure the 48"—122 cm— distance from the center front of the oven to the countertop opposite it.) For further instruction on these requirements see Guideline 31.

30. At least 36" (91 cm) of continuous countertop [that] is at least 16" (41 cm) deep should be planned for the preparation center. The preparation center should

be immediately adjacent to a water source. For further instruction on these requirements see Guideline 31.

31. If two work centers are adjacent to one another, determine new minimum counter frontage requirements for the two adjoining spaces by taking the longest of the two required counter lengths and adding 12" (30 cm).

32. No two primary work centers (the primary sink, refrigerator, preparation or cooktop/range center) should be separated by a full-height, full-depth tall tower, such as an oven cabinet, pantry cabinet, or refrigerator.

33. Kitchen seating areas require the following minimum clearances:

    30" (76 cm) high tables/counters: Allow a 30" (76 cm) wide by 19" (48 cm) deep counter/table space for each seated diner, and at least 19" (23 cm) (48 cm) of clear knee space

    36" (91 cm) high counters: Allow a 24" (61 cm) wide by 15" (38 cm) deep counter space for each seated diner, and at least 15" (38 cm) of clear knee space

    42" (107 cm) high counters: Allow a 24" (61 cm) wide by 12" (30 cm) deep counter space for each seated diner, and 12" (30 cm) of clear knee space

34. (Open) countertop should be clipped or radiused; counter edges should be eased to eliminate sharp edges.

Room, appliance, and equipment controls

35. Controls, handles, and door/drawer pulls should be operable with one hand, require only a minimal amount of strength for operation, and should not require tight grasping, pinching, or twisting of the wrist (includes handles/knobs/pulls on entry and exit doors, appliances, cabinets, drawers and plumbing fixtures, as well as light and thermostat controls/switches, intercoms, and other room controls).

36. Wall-mounted room controls (i.e., wall receptacles, switches, thermostats, telephones, intercoms, etc.) should be 15" (38 cm) to 48" (122 cm) above the finished floor. The switch plate can extend beyond that dimension, but the control itself should be within it.

37. Ground fault circuit interrupters should be specified on all receptacles within the kitchen.

38. A fire extinguisher should be visibly located in the kitchen, away from cooking equipment and 15" (38 cm) to 48" (122 cm) above the floor. Smoke alarms should be included near the kitchen.

39. Window/skylight area should equal at least 10 percent of the total square footage of the separate kitchen or a total living space [that] includes a kitchen.

40. Every work surface in the kitchen should be well-illuminated by appropriate task and/or general lighting.

4

Kitchen extras

In many cases your client will not require you, as the installer, to assist in choosing kitchen extras, but it is a good idea to be aware of the different options that are available. For the installer who may need to handle this part of the business, this chapter provides many unique ideas that you or your client may have never thought about installing in the kitchen.

As stated in Chap. 3, a kitchen should always be designed to give some "extras." Sometimes it is necessary to do a great deal of planning to obtain the necessary space in which to put these extras. Remodeling a roomy old kitchen and condensing the work area may allow extra space for other kitchen activities. Adding on a new kitchen automatically leaves the old kitchen area to be transformed into a family room, playroom, laundry room, or dining area.

A kitchen for the disabled

In today's world, the aging population and people with a wide range of physical and mental abilities and impairments need to be recognized. A person in a wheelchair should be able to peel potatoes comfortably at a sink that is height adjustable for the spouse who stands. The sink is smoothly raised by simply pushing a button (Fig. 4.1), making it easier for a tall person to work at the sink. The mechanized sink is just as easily lowered for a seated user. The knee space under the sink covers the pipes and drain and provides a place to sit while working. The 9-inch (23-cm) toe kick, when combined with 36-inch (91-cm) base cabinets, allows additional toe space for mobility aids and provides taller people with a 45-inch (114 cm) countertop.

Planning a barrier-free arrangement for a disabled person depends, to some extent, on the desires of the persons involved. Some persons desire a kitchen that would be functional for both disabled and for those who are not, while others want a kitchen that is designed and completely arranged for only a wheelchair-bound cook.

As seen in Fig. 4.2, a kitchen provides three counter heights: 30 inches (76 cm), a good height for a person who is seated; the traditional 36 inches (91 cm); and

Figure 4.1 With the push of a button, raising the sink makes it easier for a tall person to use. (*GE Appliances*)

45 inches (114 cm), a height that is appreciated by a person who is taller. Varying the height of the cabinets makes their contents more accessible to a wide range of people. In the snack area, the drawers, roll-out shelves, island storage, mug rack, and the bottom of the open plate rack can be designed to be reached by a cook who is seated, as well as by a child. When planning, be sure to check the turning diameter of the particular wheelchair, motorized cart, etc.

Accessibility to appliances and storage is of great importance to such cooks, with safety as a major consideration. For example, a cooktop is a better choice than a freestanding range because the controls are located at the front of the unit, with the heating elements staggered, so the cook does not have to reach across a front burner to place a pot on a burner in the back. Also, it is best that a disabled person have wall-mounted ovens and microwave units, most importantly those that open from the side rather than from the top. The side-opening position is typically at a better height for a disabled person to reach from a wheelchair.

The refrigerator and freezer drawers should be the low-mounted, slide-out type. Make sure the refrigerator/freezer doors, when open, do not present a barrier to the wheelchair user. Refrigerators can be purchased to provide with side-by-side storage, as well as with frozen and fresh food at varied

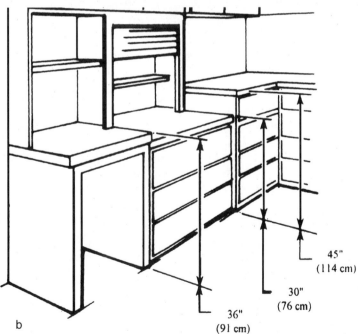

Figure 4.2 This kitchen provides three varying counter heights: 30" (76 cm), for a person who is seated (a); 36" (91 cm), the traditional countertop height; and 45" (114 cm), comfortable for the taller person (b). (*GE Appliances*)

heights for all users. Today, items are not only easy to reach, they are also easy to see because refrigerators have additional lighting, plus clear shelves and see-through storage bins. Additional modern conveniences are ice and water service on the outside of the door, as well as slide-out, spill-proof shelves, modular door bins, and adjustable shelves, which all help in making a disabled person's life a little easier. There are many ways of color coding floors and surface textures on countertop edges to aid the visually handicapped. For example, a contrasting color in a row of tile in front of the range.

A rolling table can be conveniently placed beneath the built-in convection microwave, giving a seated cook plenty of working space. An additional rolling cart fits neatly beneath the rolling table and is easily moved around the kitchen to assist in food preparation, serving, or cleanup.

Conventional floor or wall-mounted kitchen storage cabinet arrangements are not suitable for most wheelchair users, since, for the most part, they are located out of a seated person's reach. To alleviate this problem, use drawers and turntables (lazy Susans) instead of the usually stationary shelves.

Small electric appliances are extremely handy for disabled cooks. Therefore, in planning, make sure that there are plenty of outlets and that the electrical circuits have the capacity to handle the necessary electrical load. Of course, the outlets should be within the cook's reach.

Other design standards that should be considered when designing a kitchen for a disabled person are:

1. Keep the work triangle between the three major work areas as short as possible. To make it bigger would mean extra cross-kitchen travel for even the simplest of tasks.

2. If feasible, keep all work areas and countertops around the kitchen at a height of 30 inches (76 cm) and depth of at least 24 inches (61 cm).

3. Providing a free space of 36 inches (91 cm) wide and at least 27 inches (69 cm) high under the work area will enable a wheelchair user to pull his or her knees under the countertop. To accommodate the arm rests of the wheelchair under the countertop as well, the free space under the work area should measure 36 inches (91 cm) wide, 30 inches (76 cm) high, and 24 inches (61 cm) deep. This extends one's forward reach to 24 inches (61 cm)—the full depth of standard kitchen countertops.

4. The sink should be located no higher than 34 inches (86 cm) above the floor and have a single-lever fitting no farther than 21 inches (53 cm) from the front edge of the counter. The sink itself should be no more than 5 inches (13 cm) deep and recessed about 3 inches (8 cm) from the front of the counter to provide free space for elbow support.

Dining areas

Dining areas are the most popular kitchen extra—whether a table and chairs, a snack bar or counter, or a complete home party center. The size of the table or counter, as well as the space for seats, determines the dining capacity. This informal dining space should be of sufficient size to accommodate the family.

Ideally, the informal area should be separated from the working kitchen by a decorative divider that allows the transfer of light but screens unavoidable kitchen clutter (Fig. 4.3). This divider may be a freestanding serving unit, with two-way doors for easy access to dishes and glassware from either side. Where there is no room for a table and chairs, careful planning can often find space for a snack bar or dining counter.

A snack bar or counter can be many things to the careful planner. It can be a homemaker's dream of convenience, combining snacks, buffet dining, and a pass-through to the dining room, or it can even include the family message center. Plan enough space to serve the family adequately, from breakfast through school lunches to a quiet candlelight supper when all the day's work is done. A dining counter, as mentioned in Chap. 3, should usually be 29 inches (74 cm) high for chairs and no higher than 40 inches (102 cm) for stools (Fig. 4.4). The minimum counter depth of 15 inches (38 cm), 30 inches (76 cm) if both sides of the counter are used, is adequate for family breakfasts or for quick snacks. Typically, it is better to allow a 24-inch (61 cm) depth or more if the counter is extended for serving dinner or for a lineup of buffet foods.

The dining area—either table or counter—should be located for convenient serving. The amount of space available is important when planning the family's dining area. For comfortable dining, the desirable minimum spacing for persons seated side by side at a table or counter is 24 inches (61 cm). This

Figure 4.3 The working kitchen is separated by see-through cabinets and a peninsula eating counter from the informal dining area. (*Leon E. Korejwo*)

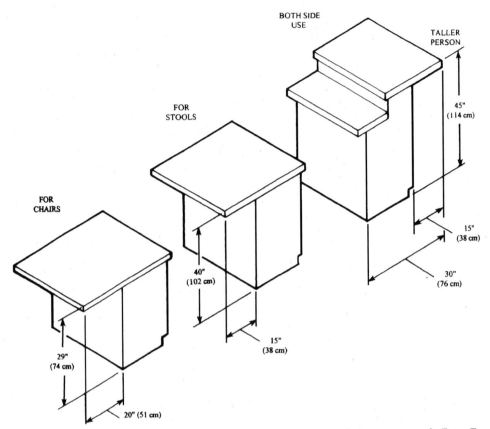

Figure 4.4 Dining counter levels for chairs, stools, and if both sides of counter are used. *(Leon E. Korejwo)*

spacing usually provides ample elbow room for dining and accommodates the usual armless dining chair. Chairs with arms are sometimes wide enough that the 24 inch (61 cm) spacing must be increased. In this case the spacing should be equivalent to the width of the chair plus 2 inches (5 cm) (Fig. 4.5a).

The recommended clearances, from table to wall, for access to the dining table or counter are as follows: to allow room to pull out and return the chair when sitting or rising—32 to 36 inches (81 to 91 cm) (Fig. 4.5b); for additional space for a second person to walk behind one who is seated—44 to 65 inches (91 to 165 cm) (Fig. 4.5c). If the table faces a closet, the clearance should be at least 2 inches (5 cm) more than the door swing. A greater clearance is required if the chairs at the table cannot be pushed under the table and out of the way.

If the space is to serve four people, a 36-inch-wide-by-48-inch-long (91-cm-by-122-cm) table is sufficient (Fig. 4.6a). If six people will use it, then a 36-inch-by-72-inch (91-cm-by-183-cm) table is required. A 42-inch-by-96-inch (107-cm-by-244-cm) table seats eight (Fig. 4.6b). There should be, as already mentioned, 36 inches (91 cm) minimum of clearance on all sides of the table.

The amount of space necessary to accommodate dining can be reduced considerably by using a longer table, by eliminating seating at the ends of the table, or by placing the table in a nook. The minimum serving clearance of 36 inches (91 cm) allows the server to edge behind the people seated in their chairs (Fig. 4.7a). The liberal serving clearance permits the server to walk behind the chairs. If the serving clearance in back of the seating is not needed, a booth may be used (Fig. 4.7b). The width of the booth can be reduced, in comparison to a nook, an additional 20 inches (51 cm). In this case, the diners may have to bend their knees and enter the booth in a slightly crouched position.

When planning built-in seating and a table in the kitchen, be sure to set aside an area no less than 48 by 66 inches (122 by 168 cm). This space is designed to serve four people in comfort.

Dining facilities are often provided in the same space with the kitchen working area. In this case, the table and chairs should be located so that when they are occupied, the kitchen working area can be used by at least one person. The table should be placed a minimum of 50 inches (127 cm) from the counter or preferably 58 inches (147 cm) from an appliance such as a dishwasher or range. These dimensions include 20 inches (51 cm) of space for an occupied chair (Fig. 4.8a).

Different clearances are recommended for the space beside a dining table that does not face a counter or appliance. If there is no passageway, the clearance from the table to a wall or counter back should be 36 inches (91 cm) for

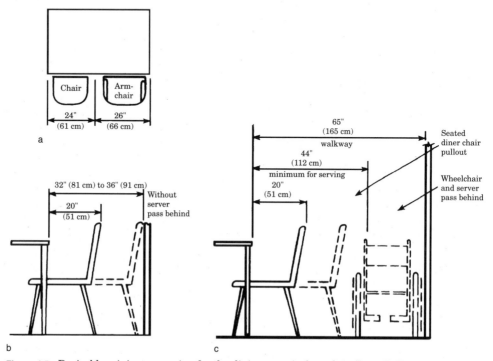

Figure 4.5 Desirable minimum spacing for the dining area (a, b, and c). *(Leon E. Korejwo)*

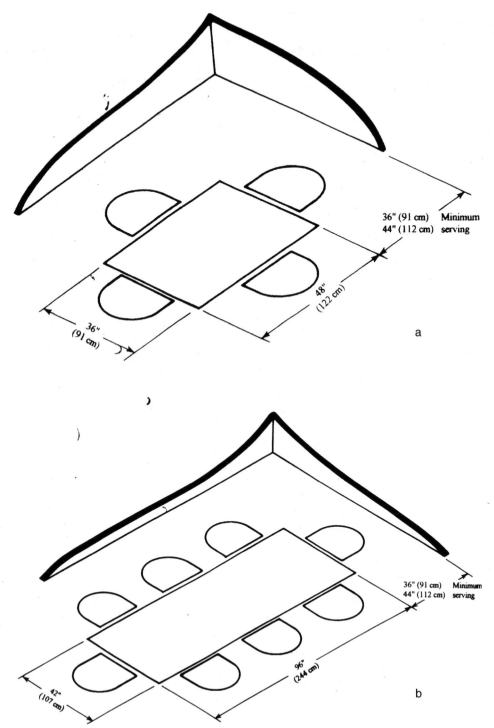

36" (91 cm)   Minimum
44" (112 cm)   serving

48"
(122 cm)

36"
(91 cm)

a

36" (91 cm)   Minimum
44" (112 cm)   serving

96"
(244 cm)

42"
(107 cm)

b

Figure 4.6 Desirable minimum spacing for the dining area to serve four people (a) and eight people (b). *(Leon E. Korejwo)*

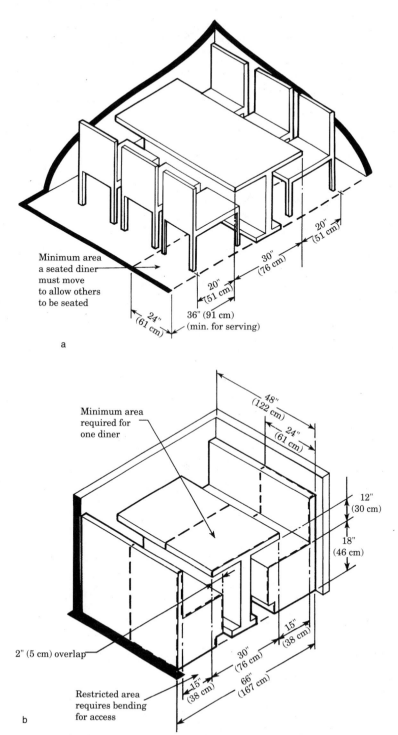

Minimum area
a seated diner
must move
to allow others
to be seated

20"
(51 cm)

30"
(76 cm)

20"
(51 cm)

20"
(51 cm)

24"
(61 cm)

36" (91 cm)
(min. for serving)

a

Minimum area
required for
one diner

48"
(122 cm)

24"
(61 cm)

12"
(30 cm)

18"
(46 cm)

15"
(38 cm)

2" (5 cm) overlap

15"
(38 cm)

30"
(76 cm)

66"
(167 cm)

Restricted area
requires bending
for access

b

Figure 4.7 A typical nook arrangement (a) and a typical booth arrangement
(b). (*Leon E. Korejwo*)

liberal space, 30 inches for medium space (76 cm), or 26 inches (66 cm) for minimum. Liberal clearance allows a person to leave the table without disturbing others (Fig. 4.8b).

If there is a passageway along one side of the table, the clearance to a wall or counter back should be 44 inches (112 cm) for liberal space, 36 inches (91 cm) for medium space, or 30 inches (76 cm) for minimum space. The liberal clearance allows room for a person to walk past a seated person.

It is frequently possible to get the necessary space for a dining area by changing the floor plan only slightly. In the L-shaped kitchen in Fig. 4.9a, for example, the relocation of the rear door and the sink allowed for the desired eating space (Fig. 4.9b). In Fig. 4.10a, a two-wall corridor kitchen, the relocation of the utility closet changed the shape of kitchen to an L and provided the necessary dining area (Fig. 4.10b).

Before leaving the subject of eating, it would be wise to mention pass-through counters between kitchen and dining areas. The judicious use of such openings between adjoining rooms significantly increases the available effective space. In addition, pass-throughs have the virtue of opening up small, cluttered rooms, making them feel larger and more airy. The step savings gained from a pass-through are a more concrete benefit.

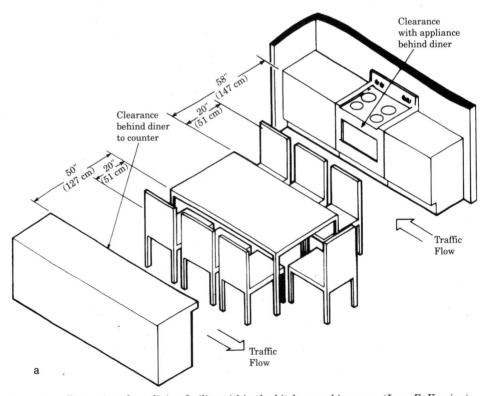

Figure 4.8a Dimensions for a dining facility within the kitchen working area. (*Leon E. Korejwo*)

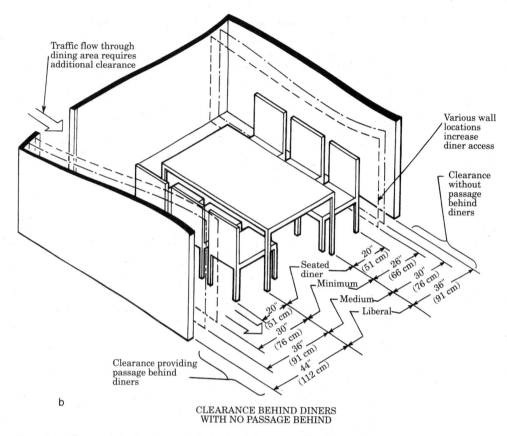

Figure 4.8b Dimensions for clearance from the table to a wall or counter. *(Leon E. Korejwo)*

Laundry center

Another suggestion for a "kitchen plus" is a laundry center. If the current laundry facilities are located in the basement, you can save the homeowner time and trouble by installing a washer and dryer in or near the kitchen—close to the action. So dirty clothes do not end up on kitchen counters, there should be some type of separation from the kitchen. Place laundry equipment around a corner or on the other side of a divider or in any area of the home that allows easy access to water and drain connections and permits the dryer to exhaust to the outside. Bifold doors provide a quick access and visual barrier, allowing sorting chores to be done out of sight.

A laundry area must be planned as logically as the kitchen. Activity centers in the laundry include

1. soiled clothes storage

2. sorting and preparation area

3. washing and drying centers

4. ironing center and clean clothes storage

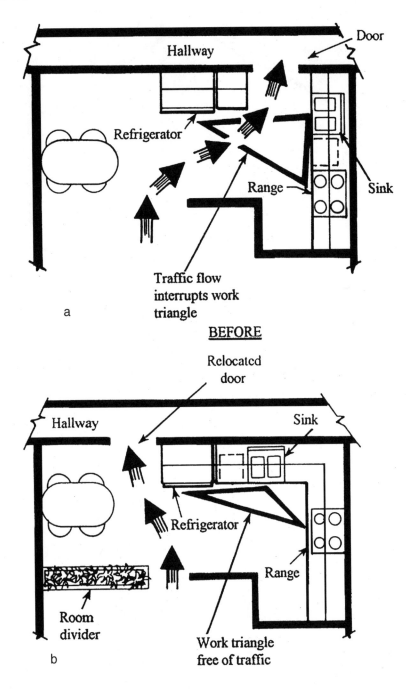

Figure 4.9 A typical L-shaped kitchen before (a) and after (b) remodeling to allow for desired eating space. (*Leon E. Korejwo*)

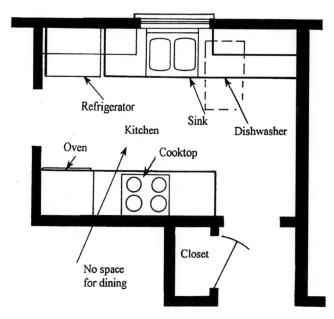

a

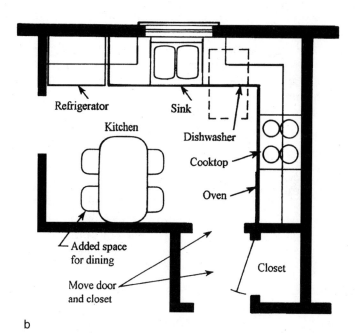

b

Figure 4.10 A typical two-wall corridor kitchen before (a) it is reconstructed into an L-shaped kitchen and (b) to provide for the necessary dining area (*Leon E. Korejwo*)

Each center should contain the appliances, storage space, and work surfaces needed for that task. The basic kitchen can also be used to organize laundry activity centers into a work triangle. Even if space for the laundry center is extremely limited, there are certain basics necessary for a minimum installation: appliances (washer and dryer); storage for soiled clothes (preferably at least three bins); counter space for sorting, pretreating (nearby water supply necessary), and folding; and storage for laundry aids.

An optimum laundry center would include the following:

1. *Laundry equipment consisting of an automatic washer and dryer.* Because of floor space or preference, a combination washer-dryer may be chosen. Remind the homeowner to inquire about load capacities of the appliance selected. Frequently, in the case of a big family, a large-capacity washer and dryer saves time and money. Washers and dryers in these rooms need special features, however. Look for extra insulation and heavy-duty stabilizing springs, which reduce vibration and subsequent noise.

2. *Space to presort and store soiled clothes.* Providing storage for each of the categories into which the laundry is sorted makes it easy to know when a washer load of each has accumulated. A minimum of three storage units is required for adequate presorting by laundry procedure. Types of storage containers can vary. Tilt bins or large roll-out drawers built in under a counter are convenient (Fig. 4.11). The bins or drawers can be labeled, and the family can put laundry in the proper bin as it is soiled.

3. *Storage for laundry aids and stain-removal supplies.* Adequate space must be provided near the washer for all detergents, bleaches, and other laundry aids used. To eliminate stooping and to keep these items out of reach of small children, overhead storage is best. However, it may be necessary to set aside some under-the-counter storage for extra-large boxes of detergent and heavy bleach bottles. The most frequently used items should be easiest to reach.

4. *Sink for pretreating and other laundering needs.* This sink should be located between the sorting bins and the washer. However, if the kitchen sink is located nearby, it can be used instead.

5. *Space to fold clothes and store those that require ironing.* A counter is most desirable for folding, but if the space is inadequate, the tops of the appliances can be used. A shelf over the appliances is useful for the temporary storage of folded items. However, if the washer is the top-loading type, the shelf should be high enough to allow the top to open without interference. A drop-down or pull-out table or ironing board can provide this utility.

6. *Place to hang permanent-press items.* A full-length hanging closet next to the dryer is especially desirable for hanging permanent press items as they are removed from the dryer. A clothes rack can also be used for this purpose, or, if no room is available, a wall hook can be installed. Retracting clothes line reels may be used in unique situations.

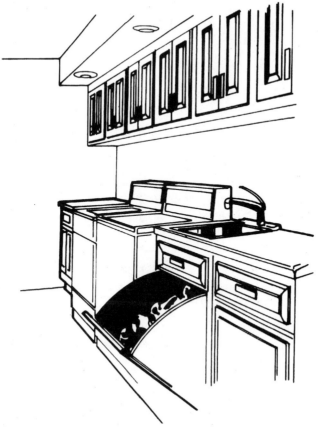

Figure 4.11 Built-in tilt bins provide convenience for the laundry center. (*Leon E. Korejwo*)

7. *Provision for sewing supplies.* Mending supplies—needles, thread, and scissors may be all that are needed—can be stored in a small drawer or in one of the cupboards. However, if much garment-making is done, a full sewing center will be needed. A shelf that pulls up and out from under the counter (similar to a typewriter shelf in a desk) is especially handy for a portable sewing machine, because it eliminates the need to lift the machine. Perforated board on the inside of a cupboard door is convenient for hanging scissors, thread, and other sewing accessories.

Laundry center space requirements

The space required for the laundry center varies with the type of appliances selected and the other activities planned for the area. A washer and dryer typically require a space from 54 inches (137 cm) up to 63 inches (160 cm) wide and 30 inches (76 cm) deep, depending upon the widths of the appliances chosen (Fig. 4.12a). The amount of space needed for a washer and dryer depends on

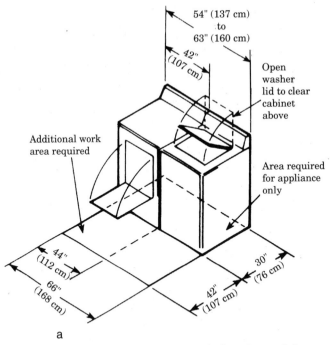

Figure 4.12a Dimensions for storing a typical washer and dryer. (*Leon E. Korejwo*)

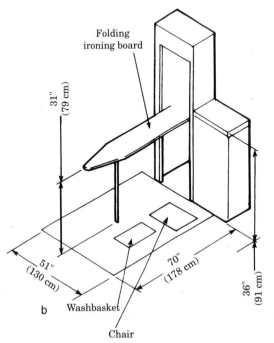

Figure 4.12b Required dimensions for a typical ironing board, laundry cart/basket, and chair. (*Leon E. Korejwo*)

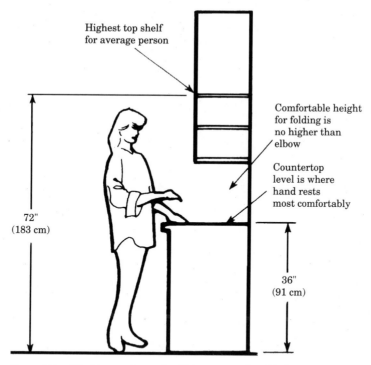

Highest top shelf
for average person

Comfortable height
for folding is
no higher than
elbow

Countertop
level is where
hand rests
most comfortably

72"
(183 cm)

36"
(91 cm)

Figure 4.12c Working surfaces should be at a comfortable level. (*Leon E. Korejwo*)

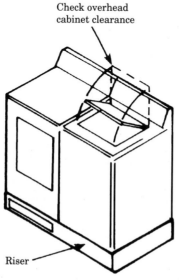

Check overhead
cabinet clearance

Riser

Figure 4.12d A typical dryer placed on a riser. (*Leon E. Korejwo*)

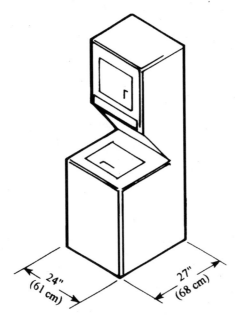

Figure 4.12e Dimensions for a typical washer
and dryer stack-on unit. (*Leon E. Korejwo*)

whether the dryer is wall-hung or equipped with casters and rolled out of
storage only when needed. If the dryer is wall-hung over a narrow sorting
counter next to the washer, for example, only 42 inches (107 cm) would be
required. But adequate work space should either be provided or be accessible
nearby.

A University of Illinois study of home laundry operations shows that the
amount of work space needed for most people is relatively constant. The following recommendations are the minimum to permit freedom of action. These
measurements are in addition to space for the appliances.

- *Washer and dryer*: 66 inches (168 cm) wide by 42 inches (107 cm) deep in
  front of the appliance's floor space.

- *Washer or dryer alone*: 44 inches (112 cm) wide by 42 inches (107 cm) deep
  in front of the appliance. If the appliances are located in a traffic area, or if
  the washer and dryer are opposite each other, the work space should be
  increased to at least 48 inches (122 cm) deep.

- *Ironing*: A space at least 70 inches (178 cm) wide by 51 inches (130 cm)
  deep is required for an ironing board, a chair, and a laundry cart or basket
  (Fig. 4.12b). If a clothes rack is used, 28 inches (71 cm) of working space
  should be allowed in addition to the rack measurements.

To help justify the space needed for an adequate laundry center, consider
additional uses for the space. Many of the facilities required for the laundry

can be used for other purposes, depending on the location in the home and the amount of space available. Ironing is a part of sewing as well as of laundering. Thus, the laundry center may be combined with a sewing room. The laundry sink can also be used for arranging flowers or as a place to wash up. Using the space as a family hobby center is another possibility.

Planning a laundry center

When planning the laundry center arrangement, consider the homemaker's height and build. Working surfaces should be at a comfortable level. A counter is at the correct work height when the homemaker can stand and rest his or her hands on the counter with arms comfortably relaxed from the shoulder. Hands should not have to be raised above the level of the elbows while folding clothes. People of average height maintain good posture and avoid fatigue when working at a counter 36 inches (91 cm) above the floor (Fig. 4.12c). The depth of the counter depends on the length of the homemaker's arms and his or her build and physical agility.

Wall cabinets should be low enough so that the homemaker can easily reach the top shelf. For the average person, this is about 72 inches (183 cm) from the floor. Shelves should be adjustable. When a wall cabinet is to be directly above the washer, allow for clearance of the washer lid when open. Hamper-type cabinets are very popular in a laundry center, especially when located near the washer.

Placing the dryer on a riser to prevent stooping is often convenient (Fig. 4.12d), especially for the older homemaker. The riser can be a solid base, or it can contain a drawer.

For the washer and dryer, the use of stack-on units is ideal. By using stack-on units, a floor space of only 27 inches (69 cm) by 24 inches (61 cm) is required, depending on the model selected. The two stack-on units (washer on the bottom, dryer on top) (Fig. 4.12e) also have the advantage that a standard wall cabinet can be placed over them to serve as a storage area.

Because one of the purposes of a laundry center is to ease the homemaker's workload, it should also be easy to keep clean. Wall cabinets built to the ceiling or soffit and flush drawers and doors without paneling prevent dust from accumulating. Wall coverings or paint should be washable. Durable, stain-resistant countertops make cleaning simpler.

A pleasant decor can do a great deal to make work more pleasant. Therefore, choose cheerful colors and patterns that agree or complement those of the kitchen. Follow suggestions for kitchen countertops, floors, walls, and ceilings, since the problems these surfaces pose are about the same for both areas (see Chap. 11). The laundry area also needs good general lighting, plus specific illumination for the pretreating, mending, and ironing centers.

Before deciding to locate the laundry in the kitchen, the homeowner should consider where soiled clothing accumulates and the desire to hang wash on outside lines. The lugging of wash baskets in either case may affect the laundry location.

Kitchen/family room

A family room and kitchen are a perfect combination for several reasons. A parent can keep an eye on the children and be readily available in a crisis. In addition, children, especially young ones, like to know that a parent is nearby. Being close to the kitchen also simplifies the serving of food and refreshments that would otherwise require a trip through the house for delivery. A location on the opposite side of the kitchen from the living or dining room can be particularly effective. The kitchen serves as a buffer between adults in the living or dining room and the noise of children in the family room (Fig. 4.13). At the same time, the kitchen can serve both areas.

Figure 4.13 (a)   A kitchen and family room combination. (*Leon E. Korejwo*)

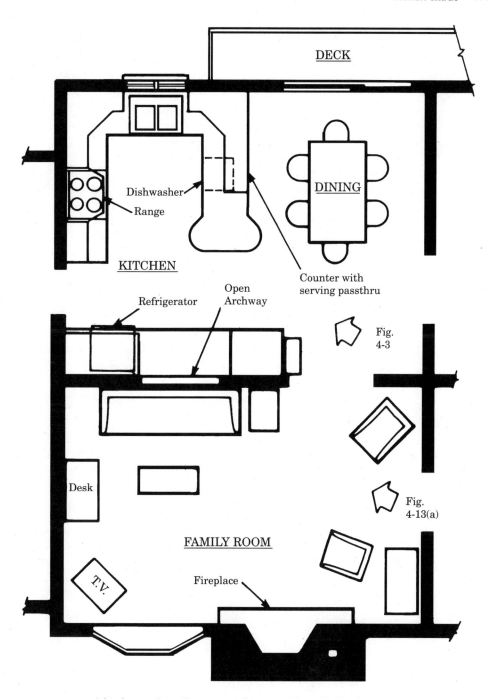

Figure 4.13 (b) A kitchen and family room combination. (*Leon E. Korejwo*)

The decorative theme of the kitchen is usually carried over into the family room. Most of the materials—floor, walls, and ceiling—may be used in the family room. There should be plenty of storage for the activities that will be carried out there. It should be kept in mind that the family room is for informal everyday living and entertaining, for television, reading, and children's play. In short, it neatly handles those overflow activities that often occur by default in a crowded kitchen or formal living room, neither of these rooms being satisfactory as a genuine family room.

U-shaped and L-shaped kitchens adapt particularly well to kitchen/family room plans. If the kitchen is not close to the family room, a small, one-wall snack and entertainment kitchen may be built at one end of the family room.

Adding a wet bar to the family room or kitchen is a good way to make the area more operational, as well as the social center of every party (Fig. 4.14). To create a wet bar, consider any or all of these options:

- *Sink.* A plumbing hookup and counter space is needed to accommodate the sink. Hospitality sinks range in size from 15 inches (38 cm) square to 15 by 25 inches (38 by 64 cm). (See Chap. 3). A hospitality sink with pass-through above it on the wall adjacent to the family room is another idea. A welcome option is a lined receptacle for bottles.

- *Refrigerator.* An undercounter refrigerator is a convenient addition to the hospitality bar. It requires a power outlet and a space 17 to 18½ inches (43 to 47 cm) wide. Some compact refrigerators are square and can be built into a base cabinet unit; others are countertop height with laminated tops that give a built-in look.

- *Storage.* Consider the items used with the bar and plan space to house them. For example, glasses may be stored on shelves, in cabinets, or hung from racks. Racks to store stemmed glasses can be built or purchased and installed on the underside of shelves or wall cabinets. Beverage storage requires a closed-door cabinet. Depending on family security measures, the homeowner may want to provide a lock for the liquor cabinet. Bar accessories, such as the blender, ice crusher, mixing and stirring tools, an ice bucket, and any other miscellany, require storage in drawers or cabinets. Wine racks are an attractive way to store more pleasing and exotic bottles.

Desk/menu-planning area

A planning center need not take up much of the kitchen's area, is a real convenience, and, if standard furniturelike kitchen cabinet components are used, need not include costly built-ins or special cabinetry. A kitchen desk is a perfect place for paying the household bills, writing a letter while a meal is cooking, or planning menus for the week. If it is built around the telephone and intercom service, it is possible to create a small business office right in the kitchen. Ideally, the location of a planning center should be convenient for kitchen activities but removed from the basic work area.

Figure 4.14 An elaborate wet bar added to a family room. (*Wellborn Cabinet, Inc.*)

The desk/planning center may be a computer/home office area. For the kitchen-management functions, the computer can help with shopping lists, meal and menu planning, recipe displays, food inventory, and other similar data. In home management, the computer can assist in balancing the checkbook, keeping the family budget in line, monitoring various household activities, and reminding one of upcoming events. When establishing a computer/home planning center, keep the following in mind:

- Bookshelves, files, telephones, and storage space for office supplies.
- A 120-volt power outlet with a "surge" device to prevent overloading.
- A sturdy, nonvibrating desk or table. The typical computer weighs approximately 35 pounds. There should be enough table space around the computer to accommodate all necessary work papers and materials. Remember that frequently upgrading the computer may increase the size or shape of a unit.
- Consider a separate phone line for a fax/modem.
- Keep heat-producing appliances, such as ranges or ovens, or heat- and moisture-producing appliances, such as dishwashers, away from the computer.

Potential household office spaces in and around the kitchen might include an end wall, an alcove, the end of a counter in a corner of the kitchen, or a closet that is ventilated.

In a U-shaped or L-shaped kitchen, the planning desk can be placed at either end of the principal work areas. The desk can also be placed along the free wall that is opposite the cabinet-appliance lineup. In the one-wall and the corridor kitchens, try to fit the desk along the end of the kitchen, just outside the traffic path. That is, locate the desk, which should be about 29 or 30 inches (74 cm or 76 cm) high, out of a traffic lane, so that someone can work at it without interruption and a passerby won't be dropping clutter on it. Be sure it is not near the spatter of a sink or range. It may be wise to provide space above the desk for recipe or reference books. A file drawer can be used for business supplies and materials if the space is used as an office, or it can keep children's medical and school records, appliance instruction books, and the family's business and social records organized. A typewriter or personal computer (PC) keyboard can be located on a spring-up platform that lifts out and up to the correct working height.

Kitchen greenhouse

Plants are a vital part of everyone's lifestyle. They are found not only outdoors but in most every room of the house. Kitchens are natural habitats for plants and the growing of herbs—and even vegetables—since this room usually includes plenty of light and humidity. Such plant-growing areas help make a "kitchen plus." Potting centers are also becoming a current trend.

In new construction, the growing center can be figured in the kitchen layout (Fig. 4.15). In the remodeling of a kitchen, adding a plant center usually involves either bumping out a wall or a window. The latter simply involves extending the window out from the wall into a bay-window type frame. There are also prefab, three-sided window greenhouses manufactured that fit most window-size openings. These window units usually come complete with glass shelves and hanging basket attachments.

It is often possible to stretch a portion of the kitchen out to the very edge of the roof overhang. In most cases a foundation is not usually necessary, nor is a change in the heating or plumbing system. If the floor cannot be cantilevered by tying the extension into existing floor joists, pour a couple of concrete piers, one for each end of the new area. The actual extension should be made following conventional construction methods. Once completed, it offers a functional and atmospheric greenhouse center for the kitchen.

A kitchen garden center might include a separate utility sink, watering hose, grow lights, and abundant storage for pots, sacks of soil, and other supplies. The center could be as simple as a deep window sill surfaced with tile to withstand watering mishaps.

Craft center

It is possible in a "plus" kitchen for arts and crafts to share floor space with the culinary arts. If the homeowner's creativity goes beyond cooking, a kitchen

Figure 4.15   An elaborate, spacious kitchen greenhouse. (*GE Appliances*)

craft center may be desirable. For example, if the homeowner's interest involves a craft that involves a large item such as a loom, a needlework frame, an easel, or a quilting frame, select and arrange the necessary furniture to the space available.

It is possible to use the materials for some crafts as part of the room decoration. For example, bolts of fabric and colorful hanks of yarn make interesting accents that can become an integral part of a room's decorative scheme. A wall-mounted shelf or cubbyhole unit keeps supplies neat and organized and shows off their bright colors or interesting textures to advantage.

When planning a kitchen craft center, some important considerations are

- Easy-to-clean surfaces on walls and floor if the craft tends to be messy.

- A sink or close proximity to the kitchen sink if the craft requires water or water cleanup.

- Power outlets near the work area if the craft calls for electrical tools or accessories.

- Good lighting. Plan the intensity and amount of light according to the detail required by the craft; try a floor lamp or wall-mounted draftsman's lamp.

- Counter space or folding tables for family snacks if the craft temporarily occupies the dining table.
- Wall or shelf space on which to display the craft items. The whole cooking and eating area takes on the character of the craft when adopting finished products as one-of-a-kind accessories for the kitchen craft center.
- Aroma-producing products used in some crafts may be more acceptable with added ventilation fans.

Mini entertainment center

Many people lead very busy schedules, running through the kitchen just to grab a quick dinner before moving on to the next activity or meeting. Rather than traipsing their dinner into the family room to watch the news, having a television in the kitchen provides a nice plus. There are many TV/VCR combinations designed expressly for the kitchen, and many of the 13-inch sets fit perfectly between most kitchen cabinets. They come with their own turntable that allows family members to watch from the stove, the counter, the breakfast room, or anywhere in between. The mini entertainment center can provide all the performance of a full-featured television with the convenience of a built-in, easy-to-use VCR. It is also convenient for the avid chef who likes to watch cooking videos.

A safe and wise feature for homes, apartments, and condominiums is a security system with cameras to view who is coming and going. This type of security can be viewed right on the television in the family room or kitchen. The mini entertainment center in the kitchen can also be used for a stereo or intercom system. Speakers can be wired throughout the home, with this center as the central location. The latest home automation systems deserve consideration. All of these tools will need room to expand.

Mud room

A mud room is currently an attractive feature for any type of home (Fig. 4.16). It is typically an area with access from the outside leading into the kitchen. It provides the family with an area to come in from the outside to unload boots, outside gear, dirty clothing, school books, coats, etc., and not drag them across the new carpet or clean kitchen floor. It can be equipped with large closets and a storage bench to sit down, often with cabinet work matching the kitchen. In addition, it is a very convenient place to have a large pantry to store canned goods or extra household supplies, as well as a good location for a recycling center. The mud room provides space away from the major work areas for the family pet to eat or sleep. Additional pet needs, such as grooming or drying off can also be attended to in the mud room.

Additional kitchen extras

A few additional inspired touches to consider to give any kitchen that "plus" feeling are

Figure 4.16 A mud room can be equipped with large cabinets, a pantry, and even a storage bench for outside gear. (*Leon E. Korejwo*)

- Snack and sandwich center near the refrigerator with chopping block insert, bread box, sandwich bags, canisters for chips, and so forth
- Pull-out racks for soft drinks near the refrigerator
- Turntables in the cabinet over refrigerator
- Children's corner for a high chair and drawer or cabinet for toys
- Cabinet for baby foods and utensils used in preparing bottles and feeding the baby

- Undersink heater to serve a tap with instant, near-boiling water for coffee, dehydrated soups, gelatin desserts, and the like
- Vertical dividers in a deep drawer or cabinet for baking pans
- A pull-out lid rack in a cabinet near the range
- Tall utility cabinet for cleaning supplies to be used in and near the kitchen
- A chimney divider between the kitchen and family room, with a fireplace on the family-room side and a charcoal-cooking hearth on kitchen side.

5

Kitchen cabinets, storage, and countertops

Regardless of the kitchen's size, the arrangement of its work centers, and the variety of its equipment, it won't function properly unless it has an adequate amount of storage space. Depending on the kind of kitchen being built or remodeled and the location of the work centers, there are various standards and recommendations for the amount and kind of kitchen storage required.

Kitchen cabinets

Now it is time to actually decide on the element that sets the kitchen's tone more than anything—the cabinets. Before the clients or homeowners select cabinetry, they should determine how they will use the kitchen and what their daily and special storage needs are (these considerations should have been addressed in the questionnaire in Chap. 2). This chapter provides the knowledge necessary for assisting your clients in choosing their cabinets. Cabinets are the key element in kitchen storage, as well as one of the most visible features in the kitchen—they determine how a kitchen looks and functions (Fig. 5.1). Because they come in many sizes or can be made to order, they provide the flexibility that enables the designer to fit a kitchen to their customers' needs and to the room space.

Remember that cabinets are the most permanent part of any kitchen; therefore, they should be chosen with care for their quality and lasting value. When a portion or all of the cabinets become obsolete in usefulness or appearance, they cannot be replaced without also disturbing countertops, flooring, wallcoverings, and perhaps the plumbing, wiring, and appliances.

Kitchen storage is usually in the form of built-in wood cabinets—base cabinets and wall cabinets. Kitchen cabinets can be built on the job, built in a local cabinet shop, built as finished prefabricated units by a manufacturer, or purchased completely assembled in a knockdown state or ready-made.

Figure 5.1   The cabinets set a warm atmosphere for the entire kitchen. *(Wellborn Cabinet, Inc.)*

Preassembled and prefinished cabinets are the quickest way to finish kitchen cabinets; however, they are also the most expensive way to go. Although there are some well-made cabinets available, beware of budget cabinets held together partly with wire staples in knockdown pieces to be assembled, shelves supported on pins instead of being securely recessed into dadoes, and a coverup coating of thick stain in place of a lasting finish.

Since the number of skilled cabinetmakers is decreasing, and because it is difficult to get a fine finish on job-built cabinets, most builders, designers, and homeowners are relying on prefabricated wood kitchen cabinets. Prefabricated cabinets are manufactured to standard sizes and dimensions, and most cabinet manufacturers do not offer special sizes or shapes. There are cabinet components designed for use with each of the kitchen's work centers and for special requirements.

It is important when using ready-made cabinets to look for the round blue and white certification seal of the Kitchen Cabinet Manufacturers Association Certification Program. This seal assures the user of kitchen cabinets that the cabinet complies with the rigorous standards set by the American National Standards Institute (ANSI) and sponsored by the Kitchen Cabinet Manufacturers Association (KCMA) (Fig. 5.2). It ensures that the cabinets meet certain national standards for quality, including structural testing, finishes, and laboratory tests to determine durability under stress.

Cabinets come in a variety of styles, from classic, country, or contemporary, and in a variety of colors and finishes to suit any decorative scheme. Different styles of doors, drawer fronts, and hardware give cabinets their individual character and personality. Although cabinets are available in a wide array of shapes and finishes, their overall basic construction is similar.

As is the case in the purchase of all furniture, good design coupled with expert workmanship and a durable finish should be the first criteria in selecting kitchen cabinets. Styles and colors should be ones that can be lived with for many years. The colors that are "in" today can and should be used in the kitchen but should be limited to wall areas and decorative accessories that can be replaced easily and relatively inexpensively tomorrow.

Method of cabinet construction

There are two basic construction styles for cabinets: framed and frameless.

Framed construction. Framed cabinets provide kitchens with a traditional look. With framed cabinets, the doors are mounted on a frame, which is then mounted to the front of the cabinet box (Fig. 5.3). They have openings that are completely surrounded by face frames made of stiles (vertical side members) and rails (cross pieces). Better cabinets have adjustable hinges that allow door realignment. With the exception of full overlay doors, the frame is visible from the front of the cabinet and can be seen as edges around doors and drawers. Full overlay cabinetry features oversized doors and drawer fronts built around the frame.

The frame takes up space; it reduces the size of the door opening, so drawers or slide-out accessories must be significantly smaller than the width of the cabinet. The interior size of drawers or roll-out accessories is usually 2 to 3 inches

Figure 5.2 Look for the Kitchen Cabinet Manufacturer Association's round blue-and-white seal when using ready-made cabinets. (*KCMA*)

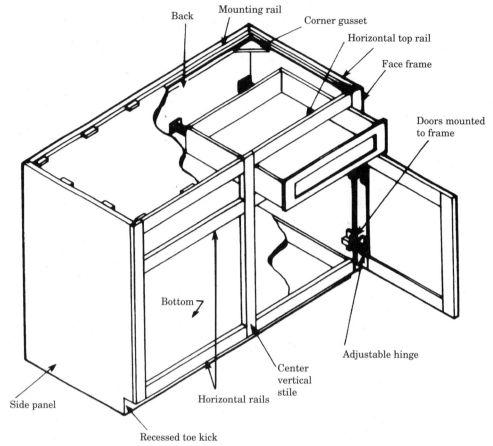

Figure 5.3 Typical framed cabinet construction. *(Leon E. Korejwo)*

(5 to 8 cm) smaller than the overall width of the cabinet. Although some framed cabinets may not have quite the interior storage space of frameless designs, they are much easier to install, therefore saving you time and your customer installation costs.

Unlike frameless cabinets, framed cabinets have some base cabinet units that are only cabinet fronts. These are most commonly used as sink bases where the back of the cabinet would interfere with plumbing rough-ins at the back wall. They offer somewhat more flexibility in irregular spaces than frameless; the outer edges of the frame can be planed and shaped, which is called *scribing*, to conform to unique discrepancies.

Frameless construction. Frameless cabinets, sometimes referred to as European style, provide a more sleek, contemporary styling. With frameless cabinets, the doors are mounted directly to the sides of the cabinet box (Fig. 5.4). Core material sides $\frac{5}{8}$ to $\frac{3}{4}$ inch (1.59 to 1.9 cm) thick are connected with either a mechanical fastening system or a dowel method of construction. Because of their thickness, these case parts form a box that does not need a front frame for stability or squareness. The hinges on frameless cabinets are screwed directly to the inside of the cabinet, eliminating the need for face frames (hinges are hidden, providing a cleaner look). Because they have no face frames, frameless cabinets offer slightly more storage space than framed cabinets—for example, they permit easier access for storing larger items, such as platters—and feature a continuous surface of drawers and door fronts. A simple narrow trim strip covers raw edges; doors and drawers usually fit to within $\frac{1}{4}$ inch (0.6 cm) of each other, revealing a thin sliver of the trim. Interior components, such as drawers, can be sized larger, practically to the full dimension of the box.

Door hinges are mortised into the sides and the doors usually fit over the entire front of the case flush with each other and with drawer fronts. This method dictates a very tight reveal, usually $\frac{1}{8}$ inch (.3 cm) or less.

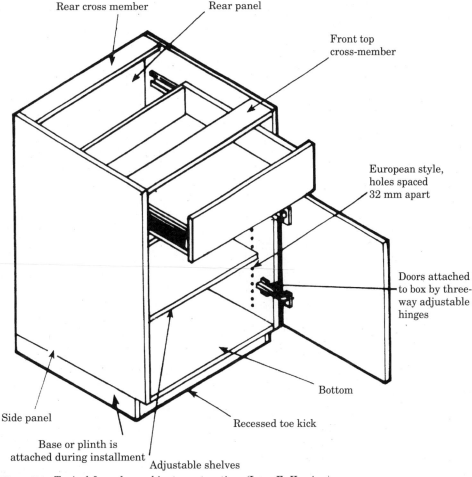

Figure 5.4 Typical frameless cabinet construction. *(Leon E. Korejwo)*

Frameless cabinets typically have a separate toe space pedestal, or *plinth*, which allows counter heights to be set specifically to the homeowner's liking, stacking base units, or making use of space at floor level. Because of absolute standardization, every component (shelf supports and connecting hardware) is inserted into standard 32-mm (1¼-inch) on-center predrilled holes. The terms *System 32* and *32-millimeter* refer to the basic matrix of all these cabinets: all the holes, hinge fittings, cabinet joints, and mountings are set 32 millimeters apart.

The sizes of European styles differ slightly from those of standard cabinets. The depth of the wall cabinet may vary from the 12 to 13 inches (30 to 33 cm) of the standard units.

Cabinet manufacturing options

The type of cabinets your customers choose affects the cost, overall appearance, and workability of their kitchen. Cabinets are manufactured and sold in the following different ways.

Stock. Stock cabinets are literally in stock wherever they are sold. They are made in quantity, in advance, to go into distributor warehouses for quick delivery. They are made in a wide variety of standard sizes that can be assembled to suit the kitchen space. The quality of standard cabinets may be fair, good, or excellent, depending on the manufacturer and price, with a limited number of styles and colors. Most stock systems do have cabinets that can be ordered for peninsulas or islands, with doors or drawers on both sides and appropriate toespaces, trim, and finishes. Stock cabinets are sold mainly to distributors and to builders.

Custom. The word custom can be used to describe any cabinetry or manufacturer that builds product to the measurements of a specific project. Custom cabinet manufacturers make cabinets kitchen by kitchen after a kitchen has been designed and sold. Generally, they are made in the same 3-inch (8-cm) modules as stock cabinets, but special sizes are also made for a perfect fit in the destination kitchen. Custom producers offer a wide range of wood species, finishes, and special units.

Because custom cabinets are made to order, delivery may take from 4 to 16 weeks. Place the order well in advance of the date you have scheduled to install the cabinets. Custom cabinets almost always are delivered completely finished, like fine furniture. Prices run from moderate to very expensive. Though it is generally a more expensive approach, custom shops can match old cabinets, build to odd configurations, and accommodate details that can't be handled by stock cabinets.

Semicustom/built-to-order. Semicustom cabinets are produced by both stock and custom manufacturers. These cabinets are usually produced on a stock basis, but with many more standard interior fittings and accessories than regular stock units, although not as many as are available on custom units. They are available in a variety of standard sizes, finishes, and styles. Each kitchen is built to order based on the kitchen design and preference. Custom offers many more choices, special shapes, special sizes and, usually, better joinery and finishes.

Specialty cabinets. Specialty cabinet manufacturers are those that are set up to supply specific housing tracts where hundreds of houses might have only three or four kitchen floor plans, and they make cabinets only to fit those floor plans. For example, the specialty manufacturer might take 100 identical single cabinets to fit a wall 96 inches (244 cm) wide that would be the same in 100 houses. A similar wall with stock or custom cabinets would have combinations of cabinets in different sizes.

Cabinet types

There are three classes or types of kitchen cabinets: base cabinets, wall cabinets, and tall cabinets.

Base cabinets. Base cabinets are usually installed under all kitchen counters, except in those places where the sink and dishwasher are located. Base cabinets do double duty, combining storage space with working surface. They rest on the floor and against the wall and form both the lower storage areas and a base for the countertops.

Though usually equipped with only one top drawer, some base cabinets have 3 or 4 drawers, making them particularly useful near a sink, range, or refrigerator. Sink units have a false drawer front at the top or a very narrow drawer that tilts out for storage of soaps or brushes. The standard base cabinet has a single drawer over a half shelf and a full shelf. The half-depth shelf is above a full-depth shelf at the bottom. Typically, the front-to-back dimensions are 23 to 24½ inches (58 to 62 cm), with 24 inches (61 cm) being the most common (Fig. 5.5). The units are 34½ inches (88 cm) high and support a 1½-inch-thick (4-cm) counter, making the counter surface 36 inches (91 cm) above the floor (4 inches (10 cm) for toe space and base, 30½ inches (77 cm) for the cabinet itself, and 1½ inches (4 cm) for the countertop). Some homeowners complain that this height is too low. It can be adjusted upward by increasing the cabinet base or adding another layer to the countertop base material (usually plywood). For people taller than 5 feet 4 inches (163 cm), 37½ inches (95 cm) is a better height. Most base cabinets are equipped with built-in toe space. Storage in standard base cabinets is primarily for cooking utensils and other larger or heavier kitchen items, small appliances, and food supplies.

Base cabinets range from 9 to 48 inches (23 to 122 cm) in width, increasing in increments of 3 inches (8 cm) from 9 to 36 inches (23 to 91 cm) and in increments of 6 inches (15 cm) after that. Single-door base cabinets are generally available in widths of 9, 12, 15, 18, 21, and 24 inches (23, 30, 38, 46, 53, and 61 cm). Double-door cabinets are usually available in widths of 24 to 48 inches

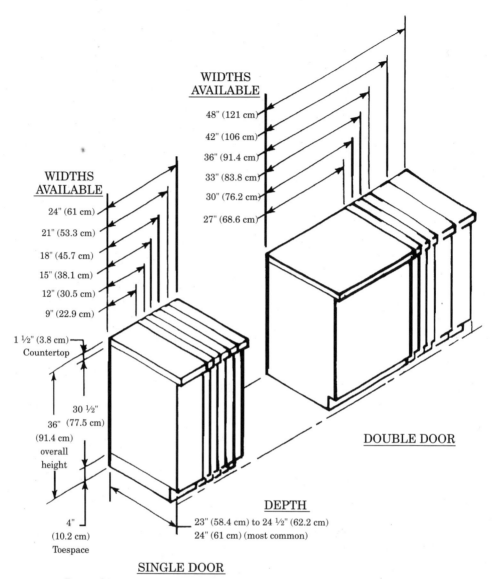

WIDTHS
AVAILABLE

48" (121 cm)
42" (106 cm)
36" (91.4 cm)
33" (83.8 cm)
30" (76.2 cm)
27" (68.6 cm)

WIDTHS
AVAILABLE

24" (61 cm)
21" (53.3 cm)
18" (45.7 cm)
15" (38.1 cm)
12" (30.5 cm)
9" (22.9 cm)

1 ½" (3.8 cm)
Countertop

30 ½"
36" (77.5 cm)
(91.4 cm)
overall
height

4"
(10.2 cm)
Toespace

DOUBLE DOOR

DEPTH
23" (58.4 cm) to 24 ½" (62.2 cm)
24" (61 cm) (most common)

SINGLE DOOR

Figure 5.5   Base cabinets come in a variety of sizes. *(Leon E. Korejwo)*

(61 to 122 cm). They are also available in 3-inch (8 cm) increments, but some manufacturers do not provide 39-inch (99 cm) and 45-inch (114 cm) units. Remember that using fewer wider cabinets rather than more narrower-width cabinets reduces the cost of the installation.

Base cabinets provide a major portion of the storage needed in the average kitchen. In planning kitchens, it is convenient to assess the amount of available base-cabinet storage in terms of accessible frontage. Accessible frontage for the typical base cabinet is equal to the actual width measured across the front of the cabinet. Accordingly, a cabinet that is 24 inches (61 cm) wide has an accessible frontage of 24 inches (61 cm). Since a cabinet installed below a single built-in oven has a reduced volume, it should be credited with an amount of accessible storage equivalent to only two-thirds of the actual width of the unit.

No accessible frontage credit is given for storage cabinets located under the sink, because plumbing lines and traps prevent the installation of drawers and shelves. Furthermore, the space behind the front may be further restricted if a garbage disposer unit is installed. No storage credit can be given for appliances that are 36 inches (91 cm) high or higher and occupy base-cabinet space, even though in some instances a single drawer may be provided in the appliance.

Early studies of storage space in kitchens recommended certain amounts of base-cabinet frontage to accompany each kitchen activity center. However, in actual practice, base-cabinet storage is usually distributed by providing base cabinets in all under-counter space, and then adding additional frontage if necessary to meet the standards selected. When this practice is followed, the distribution of the base-cabinet storage is usually adequate. The recommended standards for total base-cabinet frontage are still as follows: liberal, 10 feet (305 cm); medium, 8 feet (244 cm); and minimum, 6 feet (183 cm).

Other forms of base cabinets. Other types of base cabinets have several drawers, or pull-out trays, in place of the shelves. For example, the base drawer units may have no doors or shelves; they may contain only drawers—usually three. While they are used in all kitchens, regardless of size, they are found most frequently in small kitchens to comply with HUD/FHA minimum drawer requirements. They are available in widths of 12, 15, 18, 21 and 24 inches (30, 38, 46, 53, 61 cm); the most popular sizes are 15 and 18 inches (38 and 46 cm).

Base tray cabinets generally are available in various sizes from 9 to 24 inches (23 to 61 cm). The 12- and 15-inch (30 and 38 cm) widths are most popular. Tray cabinets provide convenient storage space. There is no shelf below the drawer, and the entire height of the storage area is left free for large, flat items (trays, broiler pans, etc.)

Base corner cabinets are employed to turn a corner where another run of cabinets join at a right angle (Fig. 5.6). They are blank (no drawer or door) in the area where the other cabinets must butt up against them. Base corner cabinets eliminate the use of base fillers since the cabinets can be pulled away to adjust to odd dimensions along the wall. Base corner cabinets are usually blind corner (reach-in) units or rotating shelf (lazy Susan) units. If equipped with three shelves at least 26 inches (66 cm) in diameter, or two shelves at least 32 inches (81 cm) in diameter, an extra 6 inches (15 cm) of accessible frontage can be credited. Lazy Susans, while more expensive than other corner cabinets, provide the most efficient use of corner space. However, a blind corner cabinet gives you the adjustability to fill a larger range of widths. These units must be installed while the corner is open, usually during initial construction or a major remodeling.

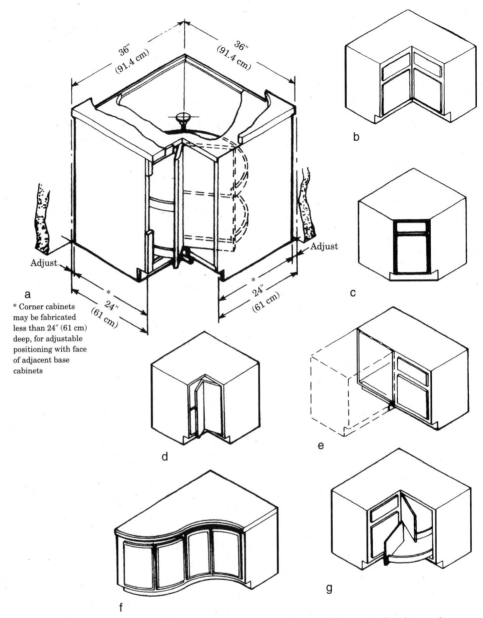

a

* Corner cabinets may be fabricated less than 24" (61 cm) deep, for adjustable positioning with face of adjacent base cabinets

Figure 5.6 Corner cabinets: lazy Susan (a); base corner cabinet with drawers (b); diagonal corner cabinet (c); base corner cabinet with hinged door (d); blind corner cabinet (e); curved corner cabinets (f); and rotating shelf (g). *(Leon E. Korejwo)*

Peninsula base cabinets (Fig. 5.7) are used in the same manner as standard base units, but they are generally positioned as dividers between rooms—usually between the kitchen and dining area. These cabinets are also used in island installations. Several widths are available, from 18 to 48 inches (46 to 122 cm), with finished (paneled) backs or with door access from both sides. Blind island base cabinets are available for "turning corners" into a peninsula.

The primary function of the *sink/range base cabinet* is to support the sink or range. Widths are available from 24 to 48 inches (61 to 122 cm) and most with 3-inch (8 cm) increments in between. Drawer fronts are fixed ("dummies") as there are no functional drawers, but ample storage space is provided behind the door area.

Sink fronts, which are basically employed to reduce the overall cost of the kitchen cabinets, are used in conjunction with sink bowls and countertop range installations in much the same way as the sink/range cabinet except they have no floor and back. (Some manufacturers have sink-front floors available.) These units can be had in one- or two-door patterns, in widths similar to base cabinets (Fig. 5.8). Some fronts are designed to permit trimming so that the cabinet assembly frontage may be adjusted in width. A variation of the sink base, the sink front is the cabinet front only. It can be attached between two adjacent cabinets to make up a sink base. Sink fronts are especially common in corner sink installations. Fit with a plywood floor, use of the sink front allows access to the entire corner with no loss of space.

Figure 5.7 A typical peninsula base cabinet. *(Merillat Industries)*

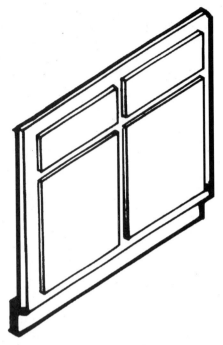

Figure 5.8 A two-door pattern sink front.
(Leon E. Korejwo)

Some manufacturers build other special base units, such as the pantry base, sliding-shelves base, buffet base, and bread-board base cabinets. Desk units 30 inches (76 cm) high are also made by some firms. For available sizes of these specialty units, check the manufacturer's literature.

Wall cabinets.   Wall cabinets are the upper cabinets that are mounted on the wall or from the ceiling, above the countertops (Fig. 5.9). Wall cabinets come in singles, doubles, and various specialty configurations. They are available in several heights and widths. The heights chosen depend on the ceiling and whether or not the kitchen design includes a soffit. Wall cabinets range in width from 9 to 48 inches (23 to 122 cm), increasing in 3-inch (8-cm) increments. Framed single-door wall cabinets range from 9 to 24 inches (23 to 61 cm) wide; wall cabinets from 27 to 48 inches (69 to 122 cm) wide use two doors, with the center stile included. Frameless single doors range from 9 to 21 inches (23 to 53 cm); double doors begin at 24 inches (61 cm) up to 36 inches (91 cm) without a center stile, but 36 to 48 inches (91 to 122 cm) with a stile included.

Wall cabinets at each work center are important to supplement the base-cabinet storage. They are generally used for storage of dishes, glassware, staples, canned goods, spices, and other small items used in the preparation of meals. To be classified as net accessible frontage, however, a wall cabinet should be at least 30 inches (76 cm) high and have two adjustable shelves in addition to the bottom shelf. The top shelf should be within 72 inches (183 cm) of the floor. Wall cabinets that extend higher may be desirable for storage of infrequently used items, but no use can be made of shelves that are higher than 72 inches (183 cm) above the floor. Except in unusual instances, wall cabinets over the refrigerator, oven, sink, or cooking surface should not be included in the readily accessible cabinet frontage for evaluating the kitchen, because their shelves are limited and not easily accessible. However, if wall cabinets are pulled forward and positioned over the front of the refrigerator, one-third of the shelf that is 72 inches (183 cm) or less above the floor can b0e used for storage.

The minimum clearance between the countertop and the bottom of the wall cabinet should be 15 inches (38 cm). Actually, with a clearance of 15 to 18 inches (38 to 46 cm) above a 36-inch-high (91 cm) counter, the top shelf of a 30-inch (76 cm) wall cabinet with fixed shelves will not be more than 72 inches (183 cm) above the floor. With adjustable shelves in the wall cabinet, the top shelf can be set at an accessible height even if the cabinet is 18 inches (46 cm) above the counter.

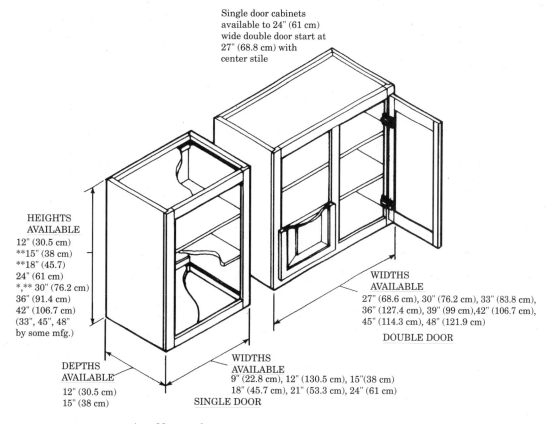

Single door cabinets available to 24" (61 cm) wide double door start at 27" (68.8 cm) with center stile

HEIGHTS AVAILABLE
12" (30.5 cm)
**15" (38 cm)
**18" (45.7)
24" (61 cm)
*,** 30" (76.2 cm)
36" (91.4 cm)
42" (106.7 cm)
(33", 45", 48" by some mfg.)

DEPTHS AVAILABLE
12" (30.5 cm)
15" (38 cm)

WIDTHS AVAILABLE
9" (22.8 cm), 12" (130.5 cm), 15"(38 cm)
18" (45.7 cm), 21" (53.3 cm), 24" (61 cm)
SINGLE DOOR

WIDTHS AVAILABLE
27" (68.6 cm), 30" (76.2 cm), 33" (83.8 cm),
36" (127.4 cm), 39" (99 cm),42" (106.7 cm),
45" (114.3 cm), 48" (121.9 cm)
DOUBLE DOOR

*   Most popular
**   Most frequently used

Figure 5.9a Typical framed construction wall cabinets. *(Leon E. Korejwo)*

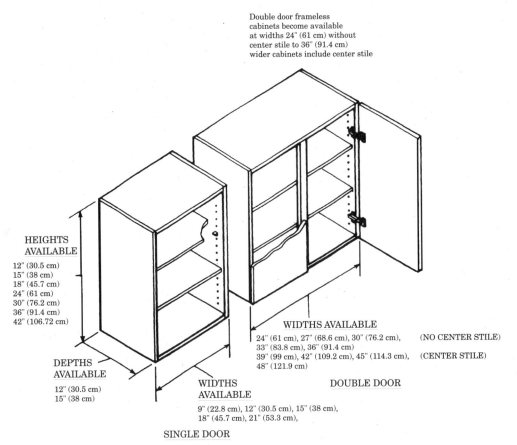

Double door frameless
cabinets become available
at widths 24" (61 cm) without
center stile to 36" (91.4 cm)
wider cabinets include center stile

HEIGHTS
AVAILABLE

12" (30.5 cm)
15" (38 cm)
18" (45.7 cm)
24" (61 cm)
30" (76.2 cm)
36" (91.4 cm)
42" (106.72 cm)

DEPTHS
AVAILABLE

12" (30.5 cm)
15" (38 cm)

WIDTHS
AVAILABLE

9" (22.8 cm), 12" (30.5 cm), 15" (38 cm),
18" (45.7 cm), 21" (53.3 cm),

SINGLE DOOR

WIDTHS AVAILABLE

24" (61 cm), 27" (68.6 cm), 30" (76.2 cm),   (NO CENTER STILE)
33" (83.8 cm), 36" (91.4 cm)
39" (99 cm), 42" (109.2 cm), 45" (114.3 cm),   (CENTER STILE)
48" (121.9 cm)

DOUBLE DOOR

Figure 5.9b Typical frameless construction wall cabinets. *(Leon E. Korejwo)*

For wall cabinets, the criteria for minimum accessible frontage is 72 inches (183 cm) and for liberal accessible frontage, 120 inches (305 cm). The amount of dinnerware is the key to wall-cabinet storage requirements. The recommended criteria provide ample space to accommodate a single set of dinnerware for four people. For many families who have dinnerware for 12, an additional 48 inches (122 cm) of wall-cabinet frontage may be recommended.

Wall cabinets should be located over counters, and, whenever possible, some wall cabinets should be located at each center. To provide a convenient and adequate location for everyday dishes, at least 42 inches (107 cm) of the wall-cabinet frontage should be placed within 72 inches (183 cm) of the center front of the sink. Extra wall-cabinet storage may be provided in tall cabinets if they are located in a direct path from a work center or in a sideboard located in the dining room.

The most commonly used wall units have a front-to-back depth of 12 to 15 inches (30 to 38 cm). The most popular height is 30 inches (76 cm). Heights of 12 to 15 inches (30 to 38 cm) are commonly used over refrigerator installations. Heights of 15 to 24 inches (30 to 61 cm) are sometimes used over ranges. Though the most frequently used heights are 15, 18, and 30 inches (38, 46, 76 cm), units range from 12 to 36 inches (30 to 91 cm) high or more. In addition, shorter cabinets are sometimes mounted over sinks. Some manufacturers provide cabinets with heights of 18, 20, and 24 inches (46, 51, 61 cm). Also, a few companies make cabinets 33, 36, 45, or 48 inches (84, 91, 114, or 122 cm) high. The selection of the height depends largely on the space between the counter surface and the bottom of the cabinet (15 or 18 inches, or 38 or 46 cm) and on the height of the top of the door trim, as well as ceiling height.

Single-door wall cabinets are usually available in widths from 9 to 24 inches, in 3-inch increments (23 to 61 cm). Double-door cabinets run from 24 to 48 inches (61 and 122 cm) in width, in 3-inch (8-cm) increments. Cabinets can vary in width from 9 to 60 inches (23 to 152 cm) in 3-inch (8-cm) increments.

The accessibility of small items often can be improved by equipping some wall cabinets with narrow shelves on the back of the cabinet door. The regular shelves must then be reduced to an 8-inch (20 cm) depth. These 8-inch (20 cm) shelves accommodate most food packages stored with the narrow end out, and the $1\frac{1}{2}$- to 3-inch-deep (4- to 8-cm) shelves on the back of the door hold small items, such as spice containers.

Wall cabinets usually come with reversible doors, or the entire cabinet can be inverted so that the door is a left- or right-hand swing, as needed. Some have fixed shelves, and others have adjustable shelves, depending on the manufacturer.

Other forms of wall cabinets.   Wall cabinets are available in the blind corner configuration and in corner cabinets with diagonal front. A few manufacturers make diagonal-front cabinets with rotating shelves (24 by 24 inches, or 61 by 61 cm) or with 90-degree lazy Susan arrangements.

Peninsula wall cabinets are used in a peninsula arrangement and are finished with either paneling on the back side or access doors on both sides. They are usually available in standard wall-cabinet widths and in heights of 24 and 36 inches (61 by 91 cm). Peninsula blind corner wall and peninsula diagonal wall cabinets are also available.

Tall cabinets.   Tall cabinets are generally constant at 84 or 96 inches (213 or 244 cm) high—the height accepted by the kitchen cabinet industry as the standard hanging height of cabinets. Tall cabinets are most commonly made in 12-, 15-, and 24-inch (30-, 38-, and 61-cm) depths (front to back). Common frontage widths are 15, 18, and 24 inches (38, 46, and 61 cm). Larger widths are available, particularly for oven cabinets. Designed to span the entire distance from the floor to the top of the wall cabinet, oven cabinets range from 18 to 36 inches (46 to 91 cm) wide.

There are several designs for tall cabinets. They can be designed to house ovens (Fig. 5.10a), broom closets, or pantries. For instance, the *utility style cabinet* (Fig. 5.10b) has one tall door with no shelves in the lower portion of its storage area. Above that, the upper portion has a fixed flat shelf behind a smaller door. Storage, therefore, is intended for brooms, mops, and other tall utility items below and small utility items above.

The utility cabinet should be accessible to the rest of the house without interfering with access to any kitchen cabinets, counters, or appliances that are used daily.

Pantry-type tall cabinets usually combine a stack of sturdy adjustable shelves with some type of swing-out storage shelves (Fig. 5.11). In deeper models, inner as well as outer swing-out shelves may further increase shelf storage capacity. There is much to be said for the old-style walk-in pantry for storage of canned and packaged foods, small appliances, dishes, and glassware. When wall space is limited, plot the pantry with a pull-out unit that uses the full height and depth of the line of cabinets and brings all the stored items within easy reach. Pantry units can also be cleverly designed with hinged shelf sections on both sides, opening like leaves of a book to make everything accessible. Slanting shelves, super-

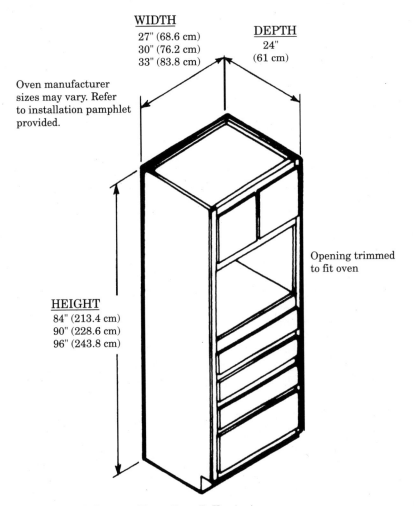

Figure 5.10a   Tall oven cabinet. *(Leon E. Korejwo)*

WIDTH
24" (61 cm)
27" (68.6 cm)
30" (76.2 cm)
36" (91.4 cm)

DEPTH
12" (30.5 cm)
24" 64 cm)

HEIGHT
84" (213.4 cm)
90" (228.6 cm)
96" (243.8 cm)

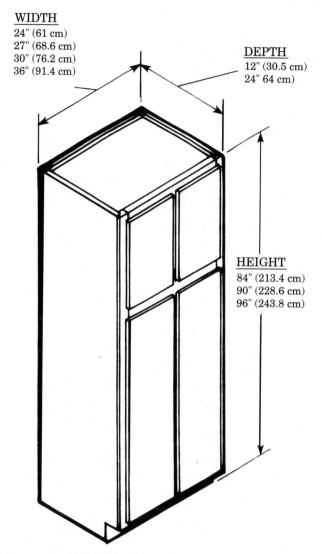

Figure 5.10b   Tall utility cabinet. *(Leon E. Korejwo)*

market style, can also be used for storage of canned goods, with cans lying on their sides so that a replacement slides into position as a can is removed.

Tall oven cabinets are intended for housing built-in or slide-in ovens. They are available with openings designed to stack one, two, or even three ovens as required. Available in several widths, these cabinets usually come so that the cutouts (openings) can be trimmed to meet the exact requirements of a particular oven. Frequently, oven cabinets are designed for using oven-hood vent fans in conjunction with the built-in ovens. As a money-saving item, some manufacturers make tall oven and utility fronts.

Gas ovens must be vented from the front of an oven cabinet. Each manufacturer supplies dimensions for their models. Allow adequate side clearance for easy installation. Also, check for position of utility input lines, whether gas or electric. Wall ovens may also be built into a cabinet.

Accessories

Different manufacturers provide a wide variety of accessories that may be added to cabinets. These items include metal bread boxes (Fig. 5.12a), ventilated slide-out fruit and vegetable bins, cutlery drawer dividers (Fig. 5.12b), sturdy and versatile pull-out counters (Fig. 5.12c), built-in wine racks, drawers for pots and pans (Fig. 5.12d), wall-cabinet spice racks (Fig. 5.12e), tambour door cabinets for appliances, multistorage pantries (Fig. 5.12f), a microwave shelf, easy-reach triple waste baskets for recycling (Fig. 5.12g), pull-out spice racks (Fig. 5.12h), sliding under-sink storage, tilt-out utility trays, pull-out chopping blocks, canned goods storage units, and so on.

Figure 5.11   Multistorage pantry-type tall cabinet. *(KraftMaid)*

Refer back to the worksheet in Chapter 2 to review what the family's needs are. Serious cooks may opt for cutlery trays and knife organizers. Frequent entertainers who want a wet bar in the kitchen might like a wine rack (Fig. 5.13a) and mounted stem glass holder (Fig. 5.13b) to display glasses. Busy families will appreciate the convenience of a swing-out pantry. Not only can the right accessories add that special touch, but they can save a lot of time, effort, and space.

In addition to the cabinets, some manufacturers have decorative wall and base shelves, valances (Fig. 5.13c) that can be used as decorative trim across windows between two cabinets, molding for trim work around cabinet installations (Fig. 5.13d), end panels to finish runs of wall and base cabinets, and filler strips (Fig. 5.13e). Batten strips are used to cover joints or seams (Fig. 5.13f). Crown molding is used to trim along the top corner where cabinets extend to the ceiling (Fig. 5.13g). Dishrails or galley rails attach to shelves such as those at valances (Fig. 5.13h). The strips should not be used except to take up or fill in odd dimensions, at the end of a given run or in conjunction with corner base cabinets to provide drawer clearance in L-shaped or U-shaped kitchens (Fig. 5.13i). A few companies also produce refrigerator and freezer fronts paneled to match the cabinetry. Optional extras allow you to help customize your clients' new kitchen cabinets in any manner that suits them.

Cabinet materials

Cabinet manufacturers use a variety of materials for the construction of cabinets. The major classifications predominantly used in the kitchen are solid wood (hard or soft), decorative laminates (usually on industrial-grade particleboard), wood products, and furniture-grade steel.

Wood. Both hardwoods and softwoods are used in more than 70 percent of U.S. kitchen cabinet manufacture. Wood is still the most popular kitchen cabinet material. Of course, wood cabinet construction varies among manufacturers to some extent, but it is fairly well standardized for stock cabinets. Softwood species commonly used are ponderosa and sugar pine, Idaho (western) white pine, Douglas fir, and western hemlock. Knotty pine and western red cedar are used for some rustic cabinet styles and for units that are sold as unfinished products. Mixing grains, or wood species, using one type for the door stile and another for the center panel, is a current trend which adds unique contrast.

Birch, oak, maple, ash, walnut, and cherry are used in cabinets that are prefinished with the natural look. The cabinets are stained and finished with clear lacquer or other plastic finishes that are sprayed on.

Plywood has always had a prominent place as a basic material for cabinets. In recent years the use of particleboard has expanded greatly. Particleboard has proven particularly useful as a base for cabinets finished with wood or plastic laminates.

Laminates. The use of laminated panels for cabinets has become popular in recent years. They are available in a wide variety of colors, patterns, performance characteristics, thicknesses, and prices. Particleboard is the predominant core stock for this kind of cabinet. However, some plywood is still used as a core material in cabinet construction.

Today, with new technology and improved resin and glue methods, the best interior surface for many cabinet applications is some type of artificial board that has been covered with either a laminate solid-colored surface or a laminated wood-grain surface.

Figure 5.12 Typical cabinet storage accessories: knife drawer/utensil tray/bread box (a); cutting board/cutlery drawer (b); pull-out counters (c); pots and pans drawer. (d) *(A, D, E, F, G: KraftMaid) (B & H: Wellborn Cabinet, Inc.) (C: Aristokraft)*

Figure 5.12 Typical cabinet storage accessories *(Continued)*: wall cabinet spice rack (e); base pantry (f); triple wastebasket (g); and pull-out spice rack (h). *(A, D, E, F, G: KraftMaid)* *(B & H: Wellborn Cabinet, Inc.)* *(C: Aristokraft)*

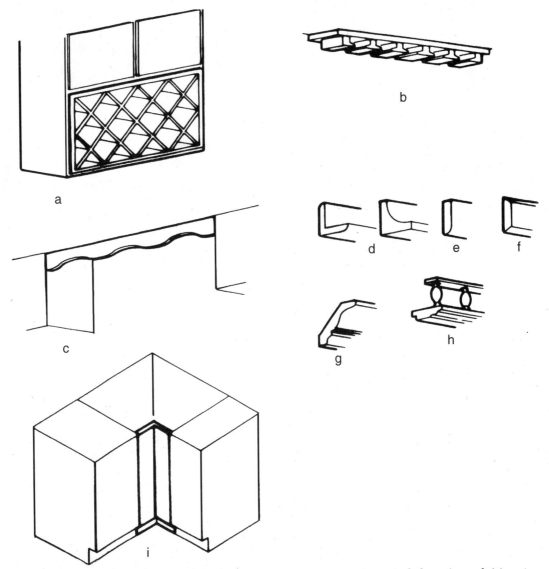

Figure 5.13 Optional accessories available to complete the cabinet installation include a wine rack (a); a stem glass holder (b); a valance (c); outside corner and inside corner cove moldings (d); scribe filler (e); batten (f); crown (g); a galley rail (h), corner filler (i). *(Leon E. Korejwo)*

Steel. With proper finishing, furniture-grade steel is a desirable cabinet material. Steel is used as the basic material for the case work in some cabinets. The doors, drawer fronts, and other exposed elements of the cabinets may be of steel, but steel cabinets may also be finished with wood and plastic laminates. Steel cabinets are often constructed so that doors and drawer fronts occupy the full front of the cabinet and no stiles or rails are visible on the front of the cabinet. With this design, fillers are always needed to turn corners in an assembly of cabinets. The major benefits of steel are durability, strength without bulk, and maintenance-free service.

Cabinet construction

The major components of typical cabinets, regardless of the type of construction, are the front frames, end panels, backs, bottoms, shelves, doors, drawers, and hardware.

Case parts. The front frames usually are made of hardwood, $\frac{1}{2}$ to $\frac{3}{4}$ inches (1 to 1.9 cm) thick. Rails, stiles, and mullions are doweled (or mortise-and-tenoned), as well as glued and stapled for rigidity. Lap joints and screw joints are also used.

End or side panels typically consist of $\frac{1}{8}$- or $\frac{1}{4}$-inch (0.3 or 0.6 cm) plywood or hardboard, glued to $\frac{3}{4}$-inch (1.9 cm) frames, or $\frac{1}{2}$-inch (1-cm) and thicker plywood or particleboard without frames. The end panels frequently are tongue-and-grooved (Fig. 5.14a) or dovetailed (Fig. 5.14b) into the front frames. The mortise-and-tenon (Fig. 5.14c), butt joints, rabbet (Fig. 5.14d), dado (Fig. 5.14e), and dowel (Fig. 5.14f) are also used to affix the end panels to the front frames. Furniture-grade plywood and particleboard with veneer facings are the most common materials for end panels.

The backs of cabinets range in thickness from $\frac{1}{8}$ (0.3 cm) to $\frac{1}{4}$ inch (0.6 cm) and are of plywood or hardboard construction. The backs are fastened to the side panels by insertion and with glue blocks that are pinned with staples. The ledger at the back of the cabinet provides a solid surface through which screws are driven to anchor the unit to the wall.

The tops and bottoms of cabinets also vary from $\frac{1}{8}$ to $\frac{1}{2}$ inch (0.3 to 1 cm) in thickness, and they are let into the sides (dadoed) and fastened with glue blocks. Base cabinets generally do not incorporate tops; however, some manufacturers still use the dust caps as cabinet tops on the base units.

Shelving in both base and wall units vary in material composition from plywood to particleboard with or without wood or plastic-banded edges. Thickness varies from $\frac{3}{8}$ to $\frac{3}{4}$ inches (0.95 to 1.9 cm). Shelving in wall cabinets is either fixed or adjustable with the trend toward adjustability, using either plastic or metal adjusting hardware.

Doors. The doors and drawer fronts are the most visible parts of cabinets; therefore, they determine the style of the cabinets. There is a wide array of cabinet door and drawer styles from which to select. Actually, hundreds of variations exist; however, the construction of doors is of two types: flush and panel. The flush doors may be hollow or may have a particleboard, wood, high-density fiber, or foam-plastic core. Veneer overlays may be wood or plastic.

Cabinet door panels may also be decorated with vinyls, brass grilles, cane, spindles, reinforced plastic, colored glass, or glass mullion doors. The panel doors consist of stiles, rails, and panel fillers of various types. These flush panels or

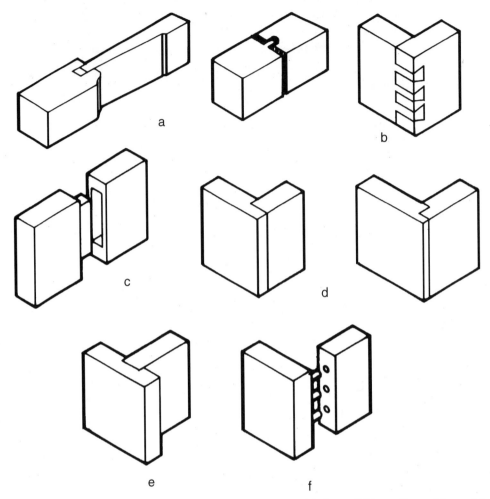

Figure 5.14   Wood joining methods include tongue and groove (a), dovetail (b), mortise-and-tenon (c), rabbet (d), dado (e), and dowel (f). *(Leon E. Korejwo)*

raised panels with beveled edges may be shaped to fit openings that are rectangular or curved.

A number of producers are now using doors and drawer fronts primarily made of plastics. A wood frame with a foam-plastic core and laminated plastic face is a common type. Complete assemblies consisting of vacuum-formed plastic faces assembled over a foam core are increasing in popularity because of the ease of molding grain and trim patterns. Such material can be cleaned easily.

Doors and drawers are fitted in one of three ways: flush with the frame (inset), partially overlaying the frame (offset), or completely overlaying the frame (full overlay). Doors are ¾ to 1¼ inches (1.9 to 3.175 cm) thick.

Drawers. Drawers frequently have hardwood lumber sides and backs and plywood bottoms. There is as wide a variation in drawer-front styling as there is in door design. Drawer construction also varies depending on cabinet quality. In some cabinets, the sides are connected to the front and back with multiple dovetail joints. In others, they are connected with lock-shouldered or square-shouldered joints. Drawer bottoms are dadoed into the sides, front, and back for more rigid construction. Some drawer units are one-piece molded polystyrene plastic (except for the front), with rounded inside corners for ease of cleaning. Exposing dovetail joinery on the drawer face is a styling trend that boasts fine solid wood cabinetry.

Cabinet hardware. Cabinet hardware includes pulls or handles, knobs, hinges, catches, and slides. Hardware is available in various materials, finishes, and designs to meet the needs of almost any style of cabinet. Many manufacturers permit customers to select the design and finishes of knobs and pulls or handles they desire.

Cabinet finishes. Most natural finishes on wood cabinets consist of a penetrating stain and seal followed by spray-applied clear finishes. Acrylic and polyester have largely replaced nitro-cellulose lacquer for this application. Other manufacturers use curtain coating or immersion finishing systems.

Finishing is a factor in cabinet quality and price. Some of the fine custom producers have 12 or more steps in the finishing process. The high-gloss finishes might be polyester or enamel. Achieving a high-gloss polyester finish is very time-consuming and expensive. High-gloss enamel is easier but is less durable. Glossy HP laminate is often considered quite acceptable at a much more acceptable price. In all cases, however, glossy finishes require more care than matte finishes.

Probably the most durable finish is produced by a plastic laminate. Such a finish is usually limited to slab doors or plain panels, although decorative moldings are sometimes applied over the laminate.

Steel cabinets are usually first treated with a rustproofing process called *bonderizing*. Primer is then applied and baked onto the steel parts. To complete the job, a smooth heavy coat of synthetic enamel is sprayed on, and the cabinet is baked in an electronically controlled oven. With quality steel and wood cabinet installations, all components of each individual kitchen are finished at the same time.

New materials and finishes are constantly being developed to improve cabinets and gain advantages in material availability. Therefore, the specifications for materials and finishes given here are subject to change at any time.

Reading manufacturer-supplied instructions

Understanding the system used to identify the cabinets and accessories you will install is important. The Kitchen Cabinet Manufacturers Association established a generic cabinet coding system more than 20 years ago to standardize the nomenclature used for identifying and specifying kitchen, bathroom, and other specialty cabinets. The National Kitchen and Bath Association has customized this coding system somewhat; this is the system explained here. The system has been adopted and used by most cabinet manufacturers, providing standardized definitions for sizes and types of cabinets. The system is based on an 11-character code that explains each cabinet category, type of cabinet, width of cabinet, and height.

The first character defines the general type of cabinet. There are six general cabinet categories, one accessory category, and one molding/trim category. The six general categories are W, all wall cabinets; T, all tall cabinets; B, all base cabinets; V, all vanity cabinets; D, all desk cabinets; P, all peninsula cabinets; and F, all furniture cabinets.

For some cabinets manufacturers, vanity and desk cabinets are interchangeable; therefore, the V designation is used in both applications. A D designation is applied only if sizing between the two systems differs.

Molding and trim pieces are identified by a separate code that describes each piece. There is no major category that sets them apart from the other groupings.

The second set of characters identifies the type of cabinet. For example, a BB is a blind base corner cabinet. A BC is a base corner cabinet. It may have fixed, adjustable, or rotary shelving, which is designated by a letter. A BD is a base cabinet that features a stack of drawers. A standard B is assumed to have a drawer above the door. A WO is a wall cabinet that has no doors; therefore, it is called an open cabinet. An M indicates a microwave cabinet.

The next two numeric symbols identify the width of the cabinet. This dimension is always listed because the widths are variable. Most manufacturers have 3-inch (8-cm) modules, from 9 inches (23 cm) to 48 inches (122 cm).

The next two numeric symbols identify the height of the cabinet. These two digits are used only if varying heights are available. For example, in wall cabinets, you can choose from heights of 12 inches (30 cm) to 30 inches (76 cm). Some manufacturers offer additional heights. Since base cabinets have one standard height that is used throughout the kitchen, no height dimension is a part of that code.

The last two characters identify any nonstandard configurations within that specified cabinet unit. For example, a D identifies a diagonal corner unit; GD identifies glass doors; D3 means three drawers; PC means pie cut; TO means tilt-out drawer head, and so forth.

Accessories to be added to the cabinet are designated following the cabinet code. Examples are BB for bread box; CB for cutting board; HU for hamper unit; MU for mixer unit, etc. Miscellaneous trim and finish pieces with no specific category heading have individual codes. VP is a valance panel. VP-C is a valance panel with contemporary styling; VP-T, traditional. A corbel bracket is indicated by CB, outside corner molding by OCM, and crown molding by CM.

Here are a few examples to demonstrate reading manufacturer-supplied instructions:

1. **W 30 12** indicates a wall cabinet, 30 inches (76 cm) wide and 12 inches (30 cm) high.

2. **W 30 15 24 D** indicates a wall cabinet, 30 inches (76 cm) wide, 15 inches (38 cm) high, and 24 inches (61 cm) deep (D).

3. **WP 30 24** is a wall peninsula, 30 inches (76 cm) wide and 24 inches high (61 cm).

4. **WB 24/27 30** is a wall blind, 24 inches (61 cm)/27 inches (69 cm) adjustable width, 30 inches (76 cm) high.

5. **WPC 24/27 30** is a wall peninsula, 24 inches (61 cm)/27 inches (69 cm) wide, 30 inches (76 cm) high.

6. **WC 24 30 PC** is a wall corner, 24 inches (61 cm) wide, 30 inches (76 cm) high, pie cut.

7. **TU 24/84/90/96** is a tall utility cabinet, 24 inches (61 cm) wide, 84 (213 cm), 90 (229 cm), 96 inches (244 cm) high.

8. **BD 12 D4** indicates a base drawer, 12 inches (30 cm) wide, with four drawers.

Evaluating cabinet construction

Regardless of the type and style of cabinets, quality construction is of utmost importance. If you are assisting the client in choosing new cabinetry, look for the following features as indicators of quality construction.

1. Front cabinet frames should be glued and fastened with steel reinforcements. They should be of the same wood species as the door to ensure a uniform finish match.

2. Wood doors should be manufactured from solid wood or a combination of solid wood and wood veneers.

3. All interior surfaces should be well finished, including adjustable shelves, which should be vinyl-clad to make cleaning easier.

4. Penetrating stains should be hand-wiped to enhance the depth and clarity of the wood grain.

5. Quality cabinets have doors that swing freely and latch securely and drawers that roll on metal tracks.

6. Shelves should be a minimum of $\frac{5}{8}$ inch (1.6 cm) thick to prevent warping and bowing. Shelves should also be adjustable to maximize storage flexibility. Hinges should be solidly mortised, concealed, adjustable, and covered by a lifetime warranty.

7. Drawer boxes should be full-depth, constructed of solid wood and feature either doweled or dovetailed joints that are backed by a warranty. The drawer suspension system should have a guarantee and feature captured, epoxy-coated slides with ball-bearing rollers.

Upgrading kitchen cabinets

It is very likely that the existing cabinets are structurally sound and perfectly serviceable. Cabinet refacing is a middle ground between redecorating and major remodeling. All types of cabinets can be refaced. Refacing existing kitchen cabinets can be an excellent, less-expensive choice, rather than replacing the entire cabinet system, with results that look essentially the same. The upgrading of existing kitchen cabinets can reduce the cost of a complete remodeling job; this money-saving option has gained widespread acceptance. However, if the homeowner doesn't like the present layout of the kitchen or if it seriously lacks storage space or countertop area, refacing is not a good idea.

Upgrading or restyling kitchen cabinets generally involves one or all of the following:

- Changing door and drawer fronts
- Applying new moldings
- Adding new hardware
- Sprucing up the cabinet surfaces

Replacing old doors and drawer fronts is one of the quickest ways to affect a kitchen's decor, often giving the area a new look. Stock doors and drawer fronts in a variety of styles and designs are available to fit standard openings. Flush-mount cabinetry doors and drawer fronts fit flush with the face frame, while surface-mount fronts fit over the frames. Surface-mounts are much easier to install. If an opening is out of square, the surface-mount simply covers it up. The flush-mount, however, may require that the front be cut out of square to match the opening. Surface-mounts also hide irregularities in the face frames, such as dings, holes, or poorly mortised hinges. They can cover most of the face frame and hide the bad spots without removing the entire cabinet.

Existing cabinet doors and drawer fronts are removed, the existing cabinets are refaced with solid wood or plastic laminates, and new doors and drawer fronts are installed. Many people choose to install new counters and update old appliances at the same time. Tear-out is minimal, and even with new appliances, counters, and flooring, a complete kitchen makeover can create a dramatic change.

Paint is the most inexpensive way to restyle cabinet doors and drawer fronts and offers a multitude of possibilities, ranging from solid colors to two tones and antiquing. Paint them with heavy-duty gloss enamel. Enamel paint is more durable than flat wall paint. Sand surfaces lightly between coats (usually two coats are required). Use only gloss or semigloss finishes, since other paints will not stand up to the wear requirements.

Staining cabinets should only be considered when the original wood has an attractive grain. Stripping wood of old paint can be a long and time-consuming task. It is often possible to obtain a "new" look by staining only the doors and the drawer fronts and finishing cabinet end panels, rails and stiles with paint. Varnished cabinets can be painted if the surface is properly prepared. Alkyd paints work best for painting varnished cabinets. Wood, metal, and previously painted cabinets may be painted any color. Do not paint plastic laminate. The best finish is accomplished by spraying. Such a finish is best applied in a shop which requires disassembling, transporting, and reassembly.

If a natural wood finish is preferred, resurface cabinets with $\frac{1}{8}$ inch (0.3 cm) hardwood veneered plywood instead of plastic laminate. Some contractors use $\frac{1}{4}$-inch (0.6-cm) -thick material, others use $\frac{1}{8}$-inch (0.3-cm) material, and still others use the paper-thin veneers that are commonly used in furniture construction. Choices of styles for the door and drawer fronts are just as wide open as they are with new cabinets.

Pulls, hinges, knobs, and other hardware designs can change the appearance of existing cabinets. The range of manufactured hardware styles is great and can create interesting design and period changes.

Molding and trim of either wood or plastic can add a great deal to the appearance. They are available in anything from screen moldings to ornate carvings. They can be used to frame a wall covering or plastic laminate insert or separate two painted colors. Adhesion with these sheet vinyls is caused by pressure. Once a strong bond has occurred, it is very difficult to break, and removal requires physically tearing the veneer from the surface and "rolling" the remaining adhesive off with the fingers.

A number of high-quality solid woods and laminates can be used, and the new doors and drawer fronts are the same as those used on new cabinets. Many refaced kitchens are indistinguishable from kitchens with all new cabinets. The downside to refacing cabinets is the lack of design flexibility; space cannot be reapportioned within the room. However, the refacing job can be combined with some minor cabinet modifications that can make a big difference in how efficient the kitchen is.

Plan for additional kitchen storage

Cabinets are, of course, the most common storage area, but there are many possibilities for imaginative and practical storage in the kitchen. Here are a few suggestions:

1. *Use the walls.* The space between wall studs is great for storage of many varied common household items—ironing board, shelves for canned goods, cleaning materials, etc. Leave open, close with bifold doors, curtains, or window shades. A full 24-inch (61-cm) depth is not necessary for pantry and broom closets.

2. *Plan around obstacles.* Do not let the chimney, clothes chute, heating ducts, or pipes interfere with storage plans. Locate a cabinet to house these necessary things, but let it also be just a bit larger for valuable additional storage—a broom closet, tray rack, or liquor cabinet.

3. *Add an in-between cabinet.* The space between wall and base cabinets is often wasted. Shallow cabinets with sliding doors are available from

several sources. Many of these fine units have lights built into them for even better value.

4. *Use roll-around carts.* Such carts can slide under a counter and provide extra storage space, as well as an additional work surface.

5. *Use sliding doors.* Cabinets with sliding doors may be installed between the underside of the wall cabinets and the countertops, to store the appliances close at hand. A heavy appliance, like a mixer, may be set on a rolling stand, making it easy to pull out and use.

6. *Use display shelves.* Open shelves can be used to display attractive dishware. Use walls to hang utensils, a spice rack, or a cutting board.

7. *Free-standing cabinets.* If the kitchen adjoins another room without a separating partition, such as a kitchen-dining room, additional storage can be obtained by installing sections of free-standing cabinets, which could include a peninsula or island base cabinet and a ceiling-hung wall cabinet.

8. *Add an island center.* An island activity center can be used as a work surface (cooking, mixing/baking, preparation/cleanup, or serving) and as an area for storage. Ideally, the island should contain everything needed in that activity—work surfaces, appliances, and storage for foods, utensils, and supplies. For instance, if the island is to be a real "cook's table," by all means design a ceiling rack above it with hooks to hold pans and large cooking tools. Add storage facilities for other basic utensils and supplies and perhaps a wooden, marble, or ceramic glass top section.

9. *Put in a drink or wet bar cabinet.* Entertaining is fun and much more convenient when the drink ingredients, glasses, and the rest of the necessary supplies are right at hand in a single storage unit. This area would also be a nice place to put a wine rack.

10. *Pull-out worktops.* If the kitchen is short on counter space, add pull-out work tops just above the drawers to solve the problem.

11. *Soffits.* The soffit area, the space between the top of the wall cabinet and the ceiling, can be utilized as extra space. It can be left open for plants. Add shelves or rails to display china, or close it completely, either with drywall or more cabinets. Many soffit areas are closed to hide duct work. Check to see if that is the case before tearing them out.

Countertops

The most-used "equipment" in the kitchen is the countertop. Residential kitchen countertop depth is normally 25 inches (64 cm) deep, front to back, to cover base cabinets and under-counter appliances. With typical 24-inch (61-cm) base cabinets, this counter depth provides a 1-inch (2.54-cm) overhang. Normally, they are $\frac{3}{4}$ inches (1.9 cm) thick. In most cases, they have an added $\frac{3}{4}$-inch (1.9-cm) build-down in front to conceal the joint where they are fastened to the base cabinets (for frameless cabinets, it is a build-up, because the doors come all the way to the top). The backsplash (wall surface between the countertop proper and the wall cabinet overhead) might fit against the back wall, approximately 4 or 5 inches (10 or 13 cm), or go all the way up to the wall cabinets. End splashes usually are fitted where tops fit against side walls.

Countertops are used in conjunction with every work center as a general work surface, as well as a storage area. The material used to surface the countertop must therefore fill a variety of requirements. Remember—space to put things on is just as important as space to put things in.

First, consider the decorative aspect. Since it is one of the most prominent features of the kitchen's design, the countertop should blend with the cabinets, yet add its own note of beauty, texture, and pattern to the kitchen (Fig. 5.15). At the same time, it must harmonize with the floor and wall colors. Adding front edge accents, such as wood edges, contrasting color, or rolled front edges, dramatically changes the look of the countertop.

As for the functional aspects of the countertop, remember the important fact that the countertop will take almost as much wear and abuse as the floor. Therefore, above all, it has to be durable, easy to clean, and resilient. When selecting the countertop material, the additional factors that should be considered are resistance to moisture, heat, sharp blows, knife cuts, scratching, staining, and fading.

The intended use of the counter area determines those factors most important to your client. The most effective kitchens have countertops that are not limited to just one surface and have several materials combined. In addition, discuss with your client which of the intended kitchen work centers requires specialized surfaces, such as heat-resistant materials near the cooking areas, easy cleanup surfaces near the sink, a wood chopping block next to the food preparation center, or a marble surface for candy-making, and so on.

Countertop materials

A number of countertop surface materials are available in a wide range of colors and prices. As the builder, you need to know the types of countertops available and all of their advantages and disadvantages. In addition, you need to know how to fabricate and install the various countertop types (Chap. 10), and how to provide proper substrate (base materials) for countertops.

The most common materials used for countertops include decorative plastic laminates, ceramic tile, solid surface composites, marble and granite, wood, stainless steel, and tempered glass tops. Any one of these surfaces can be installed throughout the kitchen; however, encourage a combination. In this way, different areas of the kitchen can utilize different types of surfaces.

Laminates. Laminated plastic is the most popular countertop material, since it is tough, easy to care for, long-lasting, and water-resistant and offers the widest possible range of colors, textures, and patterns (including wood grain, marbleized effects, simulated brick or slate, and geometric designs). It is also

Figure 5.15   Decorative aspects of a ceramic tile countertop. *(Merillat Industries, Inc.)*

cost-effective. These surfaces are not affected by soaps, detergents, or hot water. They resist stains, fading, and moisture and therefore are ideal for use in most areas of the kitchen. However, they should not be used for a cutting surface; a sharp knife blade will scratch them. Some other disadvantages are that although it withstands high temperatures, plastic laminate is not totally heat-resistant, so it can scorch. It should not be subject to direct contact with hot metal surfaces over 275 degrees F (135 degrees C). Most laminate has a dark backing that shows at seams.

There are two types of decorative laminates widely used for kitchen countertops: high-pressure decorative laminates and color-through laminates.

High-pressure decorative laminates are composed of a decorative paper that is bonded to several layers of phenolic resin-treated kraft paper, then covered with a translucent, melamine-resin paper that becomes clear during processing. The laminate material is then bonded to a substrate (plywood or particleboard) to form a countertop. Typically, laminate sheets are $\frac{1}{16}$-inch (1.6 mm) thick and range in width from 30 to 48 inches (76 to 122 cm). High-pressure decorative laminate countertops are relatively inexpensive and can be fabricated quickly. The majority of decorative laminate countertops are custom-produced. A countertop made with laminates can be tailored to fit any space. Custom laminate countertops are built by gluing sheet laminates to particleboard. Custom edges add a decorative molding to a standard laminate countertop, creating a customized design statement with thousands of design options.

There are two basic types of laminated plastic countertops: self-edged and post-formed. *Self-edged* tops offer a straight-line effect achieved by 90-degree corners at the backsplash and the front edge. They are flat with the edge of the same material and a separate 4-inch (10 cm) backsplash; separate strips of matching laminate are applied to the edges.

Post-formed countertops are one continuous sheet rolled at the backsplash and front edge, with a curved sweep from the bottom of the front edge to the top of the backsplash. Being all one piece, a post-formed top makes a smoother, cleaner surface. However, it cannot be curved because the material, already curved once in the forming, is not capable of compound curves. Self-edged tops have no such limitation. Post-formed countertops are made of sheet laminates glued to particleboard and come from the factory ready to install. They are manufactured in a variety of colors and styles, come in stock lengths, and are cut to fit your client's kitchen space. Premitered sections are available for two- or three-piece countertops that continue around corners.

Both self-edged and post-formed countertops can be had with a standard 4-inch (10 cm) backsplash or with a full-fixed backsplash. Both are also available as a flat top with no fixed backsplash. Where the backsplash is not an integral part of the countertop, many of the wall materials described in Chapter 11 can be used in its place. A good backsplash also has a practical side; if properly installed, it seals the area from moisture penetration, and therefore makes the wall a lot easier to keep clean. Post-formed, premolded, and prefabricated are the least expensive options; a custom top with a built-up lip and backsplash looks best but is more costly.

Color-through laminates are similar to standard high-pressure decorative laminates, except that the melamine color sheets are used throughout the material instead of the phenolic core kraft paper layer. The laminate contains color throughout, and joint lines are hard to detect after fabrication. Solid-color laminates do not show dark lines at the trimmed edges, but they chip more easily than traditional laminates and must be handled carefully. Color-through laminates can be used for the total countertop installation. They are also used for special edge treatments where the surface is engraved or routed to produce unique edge profiles or reveal other colors. This material is more expensive than ordinary high-pressure decorative laminate. The fabrication takes more care and attention.

The core stock of decorative laminate surfacing materials is either $\frac{3}{4}$-inch (1.9-cm) industrial-grade plywood or particleboard (also known as flakeboard). A $\frac{3}{4}$-inch (1.9-cm) -thick wood frame is attached to the bottom of the substrate material to give the countertop rigidity. The countertop is attached to the base cabinets by screwing into this wood frame. Laminated plastics usually are available in sheets a few inches larger than standard 4-by-8-foot panels to allow for the edge trimming that is necessary. When figuring the amount of material needed, remember to avoid joints as much as possible by using the longest lengths available. For example, if a countertop is to be built $9\frac{1}{2}$ feet (290 cm) long, a 10-foot length of base core and laminate is needed rather than joining an 8-foot piece with a 1½-foot piece. If the countertop has to turn a corner, it is probably best to join pieces. Provide for thermal growth expansion in long countertops, which may buckle if boxed in.

Ceramic tile. For years, ceramic tile was virtually the only countertop material found in the kitchen. Gradually, as cost became a factor, it was replaced by plastic laminates. Today, however, ceramic tile is still very popular. For elegance and durability, few countertop materials can touch tile. To produce a profitable job and a satisfied client, you need to have a solid understanding of the product, its installation methods (see Chapter 10), and care recommendations. Keep in mind that ceramic tile installation requires fewer tools and less working room than plastic laminate installation.

Ceramic tile is durable and heat-resistant. It is available in a wide variety of sizes, colors, textures, and patterns (along with decorative tiles that have colorful pictures and patterns on them for use as border or accent tiles), and therefore offer great design flexibility. Tiles can be magnificently decorative for counters, backsplashes, walls, or as display inserts in another countertop material. Because of their durability, large tiles often are set into laminates or solid surfaces next to the stove. A variety of edge treatments are also used.

The differences between raw materials, manufacturing methods, and surface finishes make some types of tile more durable in heavy-use areas than others. The firing method also affects the absorption rate of different types of tile, making some more appropriate than others for high-moisture areas.

Although you may not be specifying the tile selection, you should carefully review the type you will be installing on countertops, as well as the installation methods (Chapter 10). Setting ceramic tile is a skilled trade and may require the services of another contractor. The most commonly specified tile for kitchen installations are glazed tile, quarry tile, decorative tile, and mosaic tile.

Glazed tiles are available in high-gloss, abrasive, slip-resistant, and matte finishes. These tiles are an excellent choice for vertical surfaces; however, their susceptibility to scratching may make them a poor choice for kitchen countertops. Keep in mind that glazed tile should have a matte finish for better resistance to scratching.

Quarry tile may be very soft and irregular in shape, so breakage and installation time is increased. Quarry tile is extremely porous and requires a penetrating sealer to protect the surface. Unglazed tile is too porous for kitchen counter use. Nonporous glazed tiles do not soak up spills and stains.

Decorative tile (deco) may include a painted design or raised or recessed relief pattern. Deco tiles are recommended for use for vertical applications only, not as a countertop surface. With *mosaic tiles*, color goes all the way through the porcelain or clay material. Because they are baked at higher temperatures, they have a harder, denser body. Mosaic tiles are quite suitable for countertops. They are impervious, frostproof, dentproof, and stainproof.

For the installation project, be certain the type of tile specified is manufactured with all the specially designed edge and trim pieces required to put together a countertop. Installed correctly, ceramic tile is resistant to just about everything, including water, alcohol, and common chemicals, and, as stated earlier, it is not affected by heat. Ceramic tile is nearly indestructible in normal use; however, a sharp, hard blow can crack or chip the tile. It is noisy and has a hard surface. If a glass or ceramic dish is dropped on the tile, the object will most likely break. Tile should not be used as a cutting surface. Not only can the tile be scratched, the knives will also be damaged. Problems with the tiles themselves are rare—most tile countertop problems stem from improper installation of the underlying decking material. If an area is damaged, individual tiles can be replaced, rather than having to replace the entire countertop.

A major drawback of ceramic tile is that the grout that fills the spaces between tiles is porous and stains easily if not sealed. This problem can be minimized by sealing the grout lines.

While costs generally vary, ceramic tile countertops cost approximately $2\frac{1}{2}$ times more than standard decorative laminate countertops and are approximately equal in price to solid-surface countertops. Prices range from modest to extravagant, depending on style, accents, and accessory pieces.

Solid-surface materials. Solid surface materials are artificial composites made of acrylics or polyester resins blended with additives. These materials offer a number of advantages for the kitchen and have become increasingly popular in recent years. These synthetic surfaces have much of the visual elegant look and feel of natural stone for virtually seamless surfaces.

Solid-surface edge treatments are varied. A classical edge is achieved with an ogee cut with a router. Two layers may be different colors. The lower layer may be recessed to add depth.

The most desirable aspects of solid-surface materials are their workability and durability. These products are thick and strong enough to serve as the entire counter; no substrate is required. Many different manufacturers have a variety of solid-surface materials on the market and have produced their own varieties and blends of natural mineral fillers and resins to create an exceptionally strong and resilient solid-surfaced material. Each of these manufacturers varies in quality, durability, and composition; however, they share many of the same advantages.

All solid surfaces have no pores (making them resistant to most stains) or seams, so bacteria and mold have nowhere to collect and grow. They are very easy to care for (surfaces can be cleaned with mild soap and water), smooth, extremely durable, and water-resistant. Solid surfaces can withstand higher temperatures than most other countertop materials. They resist heat, and therefore will not burn. The surface is durable because the color is blended completely through the material, so minor scratches and stubborn stains (cigarette scorches) can be repaired easily by lightly sanding with 320- or 400-grit sandpaper to restore the original look. Deeper damage often can be filled with matching color fillers. Solid-surface countertops are more resilient than some other countertops. For example, delicate china and glassware are less likely to break if they fall on a solid-surface countertop than on other materials.

The design possibilities are endless with this adaptable, flexible product. Styling options include a wide variety of sink mountings, backsplashes, routed drain boards, different colored borders, and edge treatments, including integral units. The fabricator can make almost indistinguishable joints between sheets, create curved shapes, design intricate inlays with similar surfacing in contrasting colors, and sandblast patterns into the material.

More ornate selections imitate the random patterns found in natural materials; some even have deep, translucent, marble veining patterns. Solid surfacing is fabricated to the specifications of the individual design. Solid-surface sinks enhance the individualized look. They are made in many solid colors, a range of patterns or patterns that resemble natural stone (marble, granite, etc.). In comparison with laminates, colors are sometimes limited and costs go up for wood inlays and other fancy edge details.

For countertops, use $\frac{1}{2}$-inch (1-cm) material. Use $\frac{3}{4}$-inch (1.9-cm) material for built-up edge treatments. To form corners and long countertops, sheets are welded together with color-matched joint adhesive. Solid surfaces have characteristics of wood; they can be cut, shaped, sanded, and installed with woodworking tools. They allow for a variety of sink installations, including integral units.

All companies offer sheet goods in $\frac{1}{2}$-inch (1-cm) to 3-inch (2-cm) thicknesses. Some are also available in thinner form, approximately $\frac{1}{4}$ inch (.6 cm) for backsplashes or other wall applications. Other thicknesses are available and vary by company. All manufacturers recommend that unsupported overhangs should not exceed 12 inches (30 cm) with $\frac{3}{4}$-inch (1.9-cm) sheets or 6 inches (15 cm) with $\frac{1}{2}$-inch (1-cm) sheets. Solid-surface countertops are expensive (they can cost five times as much as a basic self-edged laminate countertop and at least three times more than a bevel-edged laminate countertop, according to industry figures) and require firm support from below. When properly fabricated, the seam between two pieces of the solid-surfacing materials is almost imperceptible. However, you should never promise an invisible seam. The quality of solid-surfacing installation is fabricator-sensitive. All manufacturers stress the importance of retaining only qualified or certified fabricators. You will work with fabricators, who take the solid-surface sheet stock and customize it for the home.

Handling solid-surface materials often requires the services of another contractor. It takes special training to shape, cut, and handle them. Most manufacturers of solid-surfacing materials offer seminars on working with their materials. In most instances, you need to take a course before the materials are purchased. Before establishing a relationship with a solid-surfacing fabricator, you should ask to see proof of training certification. Be aware that mixing solid surface manufactured sinks with that of another brand of solid surface countertop may void warranties.

Once the solid-surface countertop is delivered to the project site, you must handle it with great care. It can be easily broken in the field. Be sure it does not fall over, because it is often susceptible to breakage. Often, large solid-surface countertops are delivered to the site in sections that must be joined in the field. You may find it wise to use biscuits (three biscuits per joint) for most installations to allow for accurate alignment of the sections being joined.

The following materials can also be used as complete countertops but are probably more efficient and useful if used as small inserts in countertops of some other materials, such as laminated plastic, to provide specialized work space (mixing, cutting, chopping) or landing space on a counter for hot pots.

Stone. Marble and granite are beautiful natural materials for countertops. Clients must be aware that because it is a natural material, there will be slight variations from slab to slab. Generally, marble and granite are regarded as heatproof (marble counters never burn), water-resistant, easy to clean, and incredibly durable. Because of their fine, smooth, cool surfaces, they are considered the best surface for bread, pastry, and candy-making. Some of the disadvantages of these beautiful materials are that solid stone slabs are very expensive; stone tiles, including slate and limestone, are less-expensive alternatives. Oil, alcohol, and any acids (fruits and wine) can stain marble or damage its high-gloss finish. Granite, however, is usually resistant to these stains because of its extremely low absorption rate; it is also less prone to scratching than marble. Its coarse grain makes it more slip-resistant than marble. Marble counters need to be waxed and polished to prevent damage. Granite also can take a high polish. Repairs are difficult, if not impossible.

Marble, granite, slate, and other natural stones require specialized tools and skills for proper installation. The base cabinets have to be constructed with enough bracing and structural integrity to support the weight. Measuring granite countertops for installation is a precise process that can be completed only when the base cabinets have been installed. Accurate field dimensions are imperative, since granite countertops are prefabricated and delivered to the job site ready for installation. Installation costs are generally less for granite than for solid-surface countertops, because the fabrication is simpler and is completed at the factory. For most granite countertops, the optimum thickness is $1\frac{1}{4}$ inches (3.18 cm). The difference in cost over the more fragile $\frac{3}{4}$-inch (1.9-cm) slabs is minimal and the added thickness gives more strength for extensions, while reducing the risk of breakage during transport and installation. Keep in mind the weight of these countertops as you plan the installation.

Granite slabs for countertops can measure up to 4 to 6 feet (122 to 183 cm) wide and up to 9 feet (274.32 cm) long, allowing flexibility in countertop design. Should more than one piece be necessary, the slabs can be matched for color and grain consistency and then cut to butt squarely against each other. Plan seams at the most inconspicuous location. However, avoid seams in the vicinity of the sink cutout due to the possibility of moisture infiltration.

Marble is extremely brittle and must be handled like glass during installation. Marble is rated according to an A-B-C-D classification based on the fragility of the stones. A and B marbles are solid and sound. C and D marbles are the most fragile, but also the most colorful and decorative. The grade of marble affects its pricing: the more fragile and decorative it is, the more expensive. Clients should understand the durability aspects of marble before using it on a project. Many slabs of marble are available $1\frac{1}{4}$ inches (3.18 cm) thick. Other suppliers stock $\frac{3}{4}$-inch (1.91-cm) thick countertops. Some carry $1\frac{1}{2}$-inch (4-cm) slabs. In addition to slab countertops, marble tiles can be installed by a tile setter following specifications developed by the Ceramic Tile Institute. Marble is soft and porous, which means it will stain easily if it is not sealed with at least two coats of penetrating sealer. It must be frequently resealed.

Natural stones are not the counter choice for an active family. Suggest to your client a slab of one of these materials as a baking counter, rather than installing marble countertops throughout the kitchen, while using ceramic tile or other complementary countertop surfaces in the rest of the kitchen. A stone backsplash with a rough contoured top edge creates a dramatic effect.

Wood. Wood is handsome, natural, easily installed, and easy on glassware and china. Although not a widespread choice, wood countertops offer very pleasing visual and textural effects. Selected woods, either natural or laminated, are recommended for food-preparation areas and are excellent, durable cutting surfaces. Countertops made from laminated wood products are commonly referred to as butcher block. Butcher block consists of hardwood laminated under pressure and sealed with oil or polymer. Custom kitchens may feature the entire countertop in butcher block. When this type of surface is planned, you need to know what type of wood will be used, what type of finish is to be applied, and what water and heat protection the block will receive.

Maple butcher block, the most popular choice, is sold in 24-, 30-, and 36-inch (61-, 76-, and 91-cm) widths. The installed price is comparable to ceramic or top-of-the-line laminates. Smaller pieces are available for inserts. Constructed of oak, sugar pine, birch, alder, or other hardwood, they are almost completely indestructible, and even after years of use they can be renewed by sanding or planing. Butcher block is porous; if it is sealed, it is no longer real butcher block. Countertop installations should be of thick butcher block, which means the counters will weigh a great deal. It is probably more logical to insert a section of block adjacent to the sink or stove. Butcher block is the most desirable choice for a chopping or cutting surface because the surface is resilient enough to prevent stress on the wrist and does not dull knives. Because it is thicker than other materials, it raises the counter level $\frac{3}{4}$ inch (1.9 cm) above standard height.

Wood tops may be finished in several ways. The intended use of the block should determine the finish selection. Unfinished wood is most desirable if the entire counter surface is wood and local fabrication of seams or miters is required. Care and maintenance of unfinished wood includes protecting the wood from standing water, scraping with a steel scraper or spatula, and oiling the top weekly with mineral oil. For prefinished wood, the factory finish includes a penetrating sealer and a nontoxic lacquer finish. The combination of sealer and varnish ensures a waterproof surface and prevents scratches; however, it will not stand up to heat. No oiling is necessary, and a damp cloth may be used to wipe the board clean. Chopping on the surface does not damage the finish. This type of finish is appropriate for countertop sections, such as island tops or sandwich centers. Wood treated with urethane sealer is not appropriate for use as a chopping surface or in contact with food. The finish is, however, very good on countertops that will be exposed to moisture or liquids.

The disadvantages to wood countertops are that because they are porous, they can sometimes be difficult to keep clean, can soak up grease, are very susceptible to water damage, and can scorch from excessive heat. Hot pans should not be set directly on wood, and prolonged moisture should be avoided. However, if the wood is oiled occasionally, it will serve faithfully for years, and the surface can be easily repaired (sanded down) if it gets too chopped up. Additional drawbacks are the cost and upkeep. Doing an entire kitchen with real butcher block counters can be quite expensive. Unfinished blocks require weekly cleaning and finishing with mineral oil.

Laminated hardwood countertops are available in a wide range of standard sizes from 12 to 120 inches (30 to 305 cm). They are usually a full $1\frac{1}{2}$ inches (4 cm) thick and 25 inches (66 cm) deep and come with or without a 1-inch-thick (2.54 cm) backsplash. They come ready for installation on top of the base cabinets. In addition to full countertops, insert blocks are often installed in the kitchen work surface. Hardwood block inserts that come with a standard stainless-steel sink frame can be used in all countertops except mortar-bed-cement tile tops. They are also available in the drop-in type that can fit in a hole cut into the countertop.

Stainless steel. Stainless steel is heat- and moisture-resistant, but because of its cost, it is seldom employed for an entire countertop. As an insert material, stainless steel wears very well, is easy to clean, and will not crack, chip, or break. It may, however, show cuts, scratches, and water spots. Cutting on it damages the surface, as well as the knives. Ideal for areas around the sink, as a backsplash or a stove surround, stainless steel needs to be kept clean and dry to avoid a spotted look. It is important to use stainless steel over a solid wood base to insulate against sound. Stainless steel countertops are available with integrated sinks. If well cared for, stainless steel countertops can last a lifetime.

Tempered glass tops. Glass is primarily used as an insert for a section of counter. It is resistant to damage from heat changes or scratching, can be easily cleaned, and is suitable next to cooking surfaces or at a sandwich-making center. Ceramic glass comes in a range of sizes. It cannot be cut on the job.

As soon as you have signed the contract for an installation job, contact the countertop fabricator regarding the specifications for the surfacing materials and approximate date needed; therefore the fabricator can be prepared to schedule your job and produce the countertops. Generally, countertops are fabricated by countertop fabricators from measurements you supply. In most cases, they are not cut to fit in the field. For this reason, measurements that are supplied to the countertop manufacturer must be accurate for proper fit. Some manufacturers come to the site and take field measurements. It is generally recommended that measurements be taken only after all the base cabinets have been set. In situations in which the same kitchen configuration is being installed in a number of homes, countertops may be ordered in quantities without waiting for field measurements.

Depending on the manufacturer, the countertop material, and the level of difficulty required to construct the countertops, lead time for countertops can range from a few days to two weeks. It is important to establish a relationship with a countertop fabricator who is reliable and has a reputation for prompt delivery. You may need to use one fabricator for laminate and solid-surface countertops and a different fabricator for other countertop materials such as natural stone.

6

The work center appliances

Each of the work centers described in Chapter 3 is planned around one or more of the major appliances. The sink or preparation/cleanup area includes a disposer, a dishwasher, and a trash compactor; the storage area has a refrigerator/freezer; and the food preparation/cooking area contains the range, microwave, and ventilating hood. Since the appliances are usually the most important equipment in the kitchen, they should be selected with great care and consideration. Most new appliances offer a wide choice of sizes and colors and come in either built-in or freestanding designs.

While the homeowners or customers make the final decision on the major appliances that make up the various work centers, the remodeling kitchen contractor may be able to influence their selection. The remodeler or installer should be knowledgeable in appliance sales, customer education, and maintenance of appliances.

Quality standards for gas appliances are set by the American Gas Association (AGA), while the National Electrical Manufacturers Association (NEMA) and Association of Home Appliance Manufacturers (AHAM) set the quality standards for electrical appliances. The Underwriters Laboratories, Inc., (UL) certifies that electrical appliances have been submitted by the manufacturer for testing and have met standards regarding life, fire, and casualty. The Housing and Urban Development agency and Federal Housing Authority (HUD/FHA) require that all gas appliances carry the AGA seal of approval and all electrical appliances carry the NEMA or AHAM seal.

Let's take a closer look at the three major work centers and how the remodeling contractor can help the customers select the appliances that can make their "dream" kitchen come true.

Cleanup centers

As every homeowner knows, the cleanup center is used before, during, and after meals, so it is a good idea to locate it as centrally as possible. The center consists of the sink, garbage disposer, trash compactor, and automatic dishwasher, as well as adequate counter space on both sides (Fig. 6.1).

Sinks

No kitchen component does more jobs more often than the sink. Actually, it is probably the single most important piece of equipment in the kitchen. Studies show that up to 50 percent of all kitchen time is spent in the cleanup area. Locating the sink in the best possible spot is vital to the whole kitchen plan. For instance, if there is a window at least 40 inches (102 cm) above the floor (enough to allow for a backsplash), your client may want to place the sink under it. If a peninsula divides the work area of the kitchen from the dining area, it makes an excellent site for a sink. Gourmet cooks often favor putting the sink in an island work counter in the middle of a kitchen (Fig. 6.2).

Sinks are available in five basic types:

1. *Porcelain enameled cast-iron sinks* are manufactured with a heavy wall thickness of iron (Fig. 6.3a). A baked-on enamel finish is applied to all exposed surfaces. The enamel on a cast-iron sink is four times thicker than

Figure 6.1 A pull-out surface provides even more counter space for the cleanup center. (*GE Appliances*)

Figure 6.2   A sink added to an island work counter is a nice addition for any cook. (*Merillat Industries, Inc.*)

other types of sinks. It provides much greater resistance to cracking, chipping, and marring. Cast-iron sinks are available in colors or in white. The solid, heavy construction makes them less subject to vibration and, therefore, extremely quiet when installed with a disposal.

2. *Porcelain enameled-on-steel sinks* are formed of sheet steel in one piece and are sprayed and fired to produce a glasslike finish, much like the surface found on cast-iron sinks. But because of the unique physical characteristics of the material used, the finish on porcelain enameled-on-steel sinks is only one-fourth as heavy as that on cast-iron sinks. These porcelain sinks are available in colors and in white.

3. *Stainless-steel sinks* are the lightest type and provide a lifetime of service; they have been the most popular in remodeling (Fig. 6.3b). The surfaces are easy to keep clean, are stain-resistant and do not fade, chip, crack, or rust. Many stainless steel sinks have been manufactured with a little extra "give" to reduce dish chipping and breakage. Many manufacturers feature undercoatings that are used on the underside of the basins to reduce disposal noise and vibration while minimizing condensation and maintaining consistent water temperature. Since stainless steel is the finish, this type of sink is not available in colors, but the natural finish blends well with most color schemes. Stainless-steel sinks are graded by gauge

(thickness) and nickel and chrome content of the steel. They might be 18- or 20-gauge steel. The 18-gauge is heavier and more satisfactory. They sometimes are designated with two other numbers, which refer to the mixture of alloys. For example, 18-8 stainless steel means that there is an 18-percent chrome content and an 8-percent nickel content. Nickel gives the steel the ability to withstand corrosion, while chrome enhances the sink's ability to stand up and keep its finish over the years. They are drawn from 18- or 20-gauge stainless steel, the highest quality available for residential sinks.

4. *Vitreous china sinks*, typically a common bathroom component, are being used in the kitchen. These are highly ornamental, sculpted sinks, and are very expensive. They are fired at intense heat to produce a durable, high-gloss finish that is resistant to scratches and abrasion. The smooth, glazed surface does not rust, fade, or discolor. They are available in many original styles with colorful patterns and textured surfaces (Fig. 6.3c). This type of sink is very nice for decorative purposes, but it is subject to chipping and cracking.

5. *Artificial composites* have been combined to form very popular, durable kitchen sinks. Many different manufacturers have produced their own varieties and blends of natural mineral fillers (such as quartz) and resins to create an exceptionally strong and resilient solid-surfaced material (Fig. 6.3d).

Figure 6.3a   Self-rimming cast-iron sink with porcelain enamel finish (*Kohler Company*)

Figure 6.3b This stainless-steel sink features a smaller functional bowl on the left and a larger bowl on the right, providing ample work space. (*Elkay Manufacturing Company*)

The actual contents vary among the different products on today's market, as well as among different manufacturers. The very strong, nonporous materials are easy to clean and are highly resistant to tough stains (including fruits, ink, coffee, and other household products), scratches, chips, and burns. They come in various colors and patterns and some are formed like a fine stone, providing the elegant look and feel of natural granites and marbles. They are heat- and fade-resistant. If minor cuts and scratches were to occur, they can be easily removed. This type of kitchen sink can be finished to a matte or polished surface, with unlimited design possibilities. They are solid with the color and pattern all the way through. Today's solid-surface countertops can be coupled with a molded integral sink for a sleek, sculpted look. The sink color can either match the countertop exactly or complement it.

Sinks are no longer limited to one shape and one size. They come in an almost limitless number of sizes, shapes, and bowl combinations. Bowl designs can be used to create a variety of custom-designed sink configurations. There are single-bowl, double-bowl, and triple-bowl sinks, all of which are available in a variety of depths and widths. Let's take a closer look at some of the shapes (Fig. 6.4):

1. *Single-bowl sinks* are usually used where space is limited, for a family that does an average amount of cooking and has a dishwasher. The bowl should

Figure 6.3c A tile-in fireclay sink, similar to vitreous china products, offers original style with colorful patterns and textured surfaces. (*Kohler Company*)

be large enough (usually from 24 to 30 inches—61 to 76 cm—wide) with 24 inches (61 cm) being standard, to handle big pots and pans.

2. *Double bowls* permit one side to be used for soaking pots and pans while the other is employed in the preparation of food. Both bowls might be the same size, or one could be shallow and the other deep. While most people prefer the double-bowl sink, it is wise to remember that one large sink, 24 inches (61 cm), takes less space than an average double one, 33 to 42 inches (84 to 107 cm), for example, with 33 inches (84 cm) being standard. One large sink is quite satisfactory if there is a dishwasher. A variation of the double bowl is the *side disposer* sink. It occupies the same space as a double-bowl sink, but uses it for two different-size basins—a big one for dishwashing and a smaller one for food preparation. The small basin's shal-

low depth lets the remodeler install a disposer without changing the drain-pipe and frees more space inside the base cabinet.

3. Triple-bowl sinks are excellent for kitchens where more than one person works at the sink area at the same time, or if the homeowner does a lot of freezing, canning, and baking. A typical triple-bowl sink may have a small vegetable sink (about $3\frac{1}{2}$ inches—9 cm—deep) between two standard bowls. The standard bowl depth is about $7\frac{1}{2}$ inches (19 cm), but it can vary somewhat among manufacturers. The average triple-bowl sink ranges from 40 to 48 inches (102 to 122 cm) long.

4. The extra-deep sink, approximately $13\frac{1}{2}$ inches (34 cm) square and $15\frac{1}{2}$ inches (39 cm) deep, is a good choice in small kitchens and family rooms and in homes without a separate laundry room. The depth of the basin makes it equally suitable for repotting plants.

5. *Vegetable sinks* have one bowl of standard depth while the other is rather shallow. It may come with or without a cutting board.

6. *Corner sinks* contain two standard bowls in a pie-cut angle configuration.

7. *Round sinks*. Small, round sinks are very popular, but they are too small for normal kitchen use. Typically, they are best used as a second kitchen sink, hospitality sink, small bar sink, etc.

Figure 6.3d   A typical top-mount, double bowl, solid-surfaced sink. (*Elkay Manufacturing Company*)

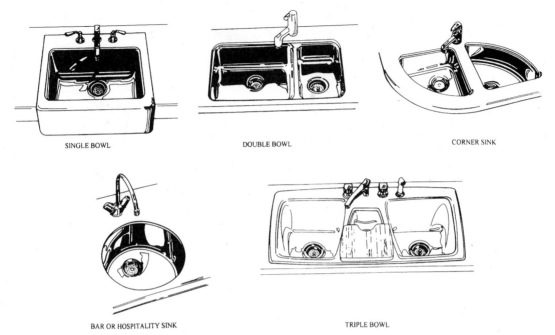

SINGLE BOWL   DOUBLE BOWL   CORNER SINK

BAR OR HOSPITALITY SINK   TRIPLE BOWL

Figure 6.4   Sink design varieties. (*Leon E. Korejwo*)

8. *Bar sinks* offer more conveniences than just for entertaining. For example, they are useful when there is more than one cook in the kitchen. A family with children can easily use an auxiliary sink as a beverage center, potting sink, or wash-up station.

There are several different installation styles to choose from when deciding on a kitchen sink:

1. An *undermount (or rimless)* sinks fit tightly, or flush, beneath any solid-surface countertop material for an integrated look (Fig. 6.5a). There are all-in-one counter/sink combinations with the sink formed anywhere the homemaker desires. Without any seams or ridges between the countertop and sink, its design is sleek and clean. These sinks offer unobstructed countertop surfaces, giving more options for working with color. They can be seam-mounted into the countertop surface, so there is no rim to catch and trap dirt and water. This installation creates a dramatic design and makes cleaning easy.

2. *Top-mount (or self-rimming)* sinks are the easiest to install (Fig. 6.5b). They mount directly over the countertop, with no special tools or techniques required.

3. *Tile-in* sinks feature flat edges and square corners to fit flush with a tiled surface (Fig. 6.5c).

4. *Seamless* sinks are artificial composite sinks that can be installed under a countertop made of a matching product to produce a new seamless effect (Fig. 6.5d).

5. Flush mounts are usually enameled steel sinks that are held in place through the use of a separate, stainless-steel sink rim and a set of special screw clamps (Fig. 6.5e).

Frequently, if space permits, it is a good idea to have two separate sinks in the kitchen for two work centers or an extra small sink for bartending, plant watering, or fingerpainting. Also consider installing a sit-down sink. This sink puts the users close to the work and takes them off their feet at the same time. The sink can be installed at the most convenient height. Where space is at a premium and a double-bowl type is desired, there are sinks available in which one of the bowls is shallow enough (5 inches — 13 cm) to fit over a built-in dishwasher. Another good space-saver is a single-bowl sink that has a small, raised

Figure 6.5a Undermount, double-bowl, stainless steel sink. (*Elkay Manufacturing Company*)

Figure 6.5b   Typical top-mount stainless steel sink. (*Elkay Manufacturing Company*)

Figure 6.5c   Typical tile-in sink. (*Kohler Company*)

Figure 6.5d A solid-surface countertop can be coupled with a molded, integral sink for a sleek, sculpted look. (*Formica Corporation*)

bowl in one corner to house the food waste disposer. Other sinks are made to fit special needs or problems (see Chap. 4).

Automatic dishwashers

Dishwashers are available as undercounter built-ins or as convertible units. The built-in or permanently installed models are standardized in size to fit under the standard kitchen countertop with base cabinets (Fig. 6.6a). This model is usually located next to the sink and has permanent water and electrical connections. A convertible dishwasher comes on casters, but can be installed under the counter later if the owner decides to make a permanent arrangement (Fig. 6.6b). Both the freestanding and built-in models are typically 24 inches (61 cm) wide. In general, the principles of operation and the resultant servicing

Figure 6.5e A typical flush-mount sink. (*Kohler Company*)

problems are similar in these types, except that the portable and convertible models require electrical and water connections to be remade each time they are used. The presence of a flexible attachment service cord and hoses with these types introduces the possibility of openings in the cord and a poor connection at the outlet, not usually problems with the undercounter models.

The basic principles of operation are fairly simple. In almost all models, the washing action is accomplished by means of a strong spray of hot water coming up from the bottom. It hits the dishes from the bottom directly and from the top and sides indirectly by bouncing off the inside of the tub. There are two methods of obtaining the necessary strong spray of water. In one, a revolving spray or wash arm in the bottom, with jets (perforations) from which water emerges, provides the washing action. The other method, often called the splash type, involves the use of an impeller blade (vane) in the bottom, which hurls the water at the dishes and cookware. In recent years, the spray-arm action type of dishwasher has become much more popular and is the type we consider here.

In a typical spray-type dishwasher with a wash arm, the pump and motor assembly is located in the sump at the bottom of the dishwasher. When the pump motor rotates clockwise, the lower or drain pump impeller, rotating backwards, does not pump. The upper impeller, rotating forward, picks up water from the sump and propels it upward into the spray arm and spray column.

When the motor turns counterclockwise, the upper impeller windmills and the lower impeller pumps the water and food particles in the sump out of the drain hose. The drain line loop is quite important because it prevents the water from draining out by gravity during the wash or rinse cycles and prevents the drain impeller from forcing any water out while it is windmilling during the wash or rinse cycles. An equivalent to the loop height, which is about 30 to 32 inches (76 to 81 cm), is provided on mobile models by connecting the drain hose to the faucet coupler.

Water is admitted to the dishwasher by a timer-controlled solenoid valve that controls the cycles of the dishwasher. Detergent is usually provided automatically from a detergent dispenser. In some models, a rinse-additive dispenser releases a small amount of additive into the final rinse to help protect glassware from the spotting caused by minerals in the water. The rinse additive acts as a wetting agent, causing the water to run off the dishes without spotting.

Air vents located at the bottom rear of the tub and in the door provide for the circulation during the drying cycle. This drying is accomplished by means of condensation. In this process, the room air cools the outside of the tub in the same way that a cold winter day cools the windows of a house. The windows sweat on such a day; the windows are wet but the warm air in the room is dry because most of the moisture in the room has condensed on the cool windows. The same thing happens in the dishwasher. Room air is cooler than the air in the dishwasher, so it cools the tub walls, and the moisture inside the dishwasher condenses on the cooler walls. Drops of water may be on the inside of the door when it is opened, but the dishes are virtually dry.

To accomplish its task, the dishwasher performs three basic functions. The first function is the wash/rinse, in which the dishwasher fills with water and recirculates it. A typical fill admits about 6 to 9 quarts of water, flowing at a rate of about $1\frac{1}{2}$ gallons per minute. Depending on the cycle, detergent or

a   b

Figure 6.6 A typical built-in dishwasher (a); undersink offset dishwasher (b). (*Leon E. Korejwo*)

a rinse additive may be put in at a specified time. The second function is to drain. While there are two types of drainage systems in use with dishwashers, the pump drain is used a great deal more than the gravity drain system. In this function, the pump motor stops, changes direction of rotation, and pumps the water out of the dishwasher. The third function is to dry. The heater may be on all the time or may cycle on and off to provide a lower temperature, depending on the cycle selected.

All dishwashers have one or more wash cycles, one or more rinse cycles, and a drying cycle. Today, dishwashers come with a variety of cycles and options that give the flexibility to handle almost every dishwashing need. Today's dishwashers come with computer monitors with digital panels displaying information about how the dishwasher is functioning, while showing what is going on in the dishwashing cycle. Some feature sensors that sense water temperature and a wash load's soil level and adjusts the wash to suit the dish load, while the computer monitor shows cycle time in minutes and counts down during operation to show time remaining in the cycle. They even alert the user to possible problems. Push button, rotary, or touch controls are easy to read and easy to operate.

More elaborate controls on some units make it possible to obtain the following functional cycles in addition to the regular ones:

- A *dish-warmer cycle* lets the householder take advantage of the drying stage of the regular cycle to preheat dishes for serving hot foods at mealtime.

- A *sanitizing cycle* heats water to a factory-set temperature to sanitize dishes.

- A *heavy-duty cycle* extends the normal cycle or adds an extra wash cycle for tough jobs.

- A *soak / scrub* (for pots and pans) *cycle* briefly wets the dishes, then a soaking solution softens the food before regular washing.

- A *quick cycle* provides a short wash for lightly soiled dishes.

- A *rinse-and-hold cycle* rinses dishes in a short, no-detergent cycle, preventing food from souring and drying on dishes while a full dishwasher load accumulates.

- A *delay-wash cycle*, allows the operator to program the dishwasher to begin the wash cycle up to 14 hours after the controls have been set.

- A *power-saver cycle* conserves energy by letting dishes air-dry rather than using the heating element. Sometimes the power-saver cycle also bypasses the heat booster that maintains a minimum water temperature during washing.

Food waste disposers

A food waste (garbage) disposer installed under one sink eliminates messy food waste and simplifies cleanup after meals, since the homeowner can peel fruits and vegetables, scrape plates, dump coffee grounds, and empty cereal bowls directly into the sink. The good-quality disposer can handle bones, fruit pits, egg shells, shrimp shells, in fact, almost any food waste. Some of the more efficient models have cutter blades that can handle fibrous waste such as corn husks, celery stalks, and artichoke leaves. Disposers are so efficient in handling food waste that today many cities require their installation in all new housing starts or major improvements. But just as some cities require them in all new kitchen installations, a few areas prohibit or restrict the use of disposers because local sewage systems are not able to handle the extra waste. Therefore, check the local plumbing code before installing a disposer.

Speaking of waste, research by the United States Public Health Service proves that septic tank soil absorption systems that meet the HUD/FHA minimum property standards can handle the additional loads from food waste disposers. The addition of ground food waste may reduce the time between tank cleanings by approximately one-third.

There are two basic types of disposers—batch-feed and continuous-feed. The batch-feed is controlled by a built-in switch. Waste is placed in the chamber, the cold water is turned on, and the lid is put in place. In some cases, the lid must be turned or positioned a certain way to start the motor. In others, a sealed-in magnet activates a hermetically sealed switch so that the disposer starts as soon as the lid is dropped into place. A continuous-feed model means food waste can be put into the disposer as it is operating. These types do not have a locking cap and are activated by a switch that might be in the back-splash above the counter. Because the batch-feed type operates only with the cover on, it may be safer when children are around, but it may prove to be slower. Some models have interlocking doors with automatic reverse that prevents operation when the door is open.

All modern food waste disposers consist of four main components: an electric motor; two grinding elements, including the rotating shredder or flywheel and impeller plus the stationary grind ring; grinding chamber; and sink-fitting mounting arrangement.

Motor sizes normally range from $\frac{1}{3}$ to $\frac{3}{4}$ horsepower. The larger the horse-power, the more torque is created, reducing the incidence of jamming. Motors are generally either capacitor-start, which brings the rotating shredder up to full speed almost instantly, or split-phase, which takes longer to come up to speed. The split-phase motors also draw excessive amperage on house current when turned on. Most split-phase motors have a centrifugal starting switch to turn on the rotor, while the capacitor-start type usually employs a relay.

Generally, a primary seal is installed to prevent water and waste from entering the bearing and motor areas. A few manufacturers also use a plastic protective shield against the corrosive action of detergents and electrolysis. This detergent shield is generally made of polypropylene and is resistant to electrolysis and corrosion. In some other makes, the grind chamber itself is given an epoxy coating.

A look at most disposer motor housings reveals a rotor shaft protruding with the rotating shredder or flywheel attached to it with a positive nonslip drive. A solid grinding or cutting element can also be attached and locked to the shaft, providing direct drive. A stationary grind ring is anchored into position by press-fitting. It has a predetermined internal clearance, allowing the rotating shredder or flywheel to revolve within its circumference. The internal clearance is equal to the thickness of a cigarette paper, providing a steady, consistent grind size. While reverse switches operate on different principles in different makes, all reverse the disposer every time the wall switch is turned on. This feature doubles the life of the cutting elements.

Regardless of the type of motor, the basic principle of operation for all garbage disposers is centrifugal force. As we already know, food waste drips through the opening into the grind area. With the disposer operating at approximately 2700 rotations per minute (RPM), it throws food by centrifugal force against the walls of the grind chamber, where the stationary grind ring is located. The lugs or impellers on the rotor shredder or flywheel channel the food waste into the grind ring. In essence, the waste is grated until the particles are reduced in size. When they become small enough, they slip between the teeth of the grind ring and through the holes of the rotating shredder. While this principle of operation is basically the same for all disposer makes, some models have movable cutting impellers or lugs that lie flat, then rise into position when the disposer is turned on. The manufacturers claim various advantages for both types, but essentially they do the same job. When the particles are small enough to fall beneath the rotating shredder or flywheel, the water carries them out through the drain line.

One of the most important operating rules with any garbage disposer is that it must be used with a full flow of cold water only, never hot water. Cold water congeals grease into globules that flow easily through drain lines, while hot water makes grease more fluid and prone to coat the drain line. The cold water must be turned on before the disposer is made operative and left on for a full 30 seconds after the disposer is turned off. Also, disposers should not be used with a grease trap in the drain line. If present, the trap must be either removed or bypassed. If the grease trap has a removable baffle, its removal solves the problem. Some disposers are self-cleaning, others are not, depending on design and material. It is a good idea to flush out all units occasionally, thereby ensuring that they remain odor-free. Flushing can be done by filling the sink with water, turning the disposer on, and removing the stopper. The action of the swirling water removes residue food wastes.

To free jams caused by some refuse, antijamming devices are built into the disposer. Units with rigid impeller hammers usually have an automatic or self-reversing feature. Models with swinging impeller hammers often have a manual reverse switch to prevent jamming. All disposals have a reset overload button incorporated in the body of the disposer that protects the motor from overheating and allows the householder to get the disposer back into operation when stalled.

Some manufacturers have created systems to dispose of waste right where it is produced—through the built-in waste disposal chute right into the waste bin below the sink. No cabinet doors need to be opened. A waste disposer can easily be accommodated in addition to the waste bin for maximum convenience. Some have elaborate systems with plastic pullout bins that separate them into different classifications (Fig. 6.7), allowing plenty of space for tins and aluminum foil, glass, paper, and plastic materials, to be sorted for recycling, as well as compostable waste, room to store detergents, etc.

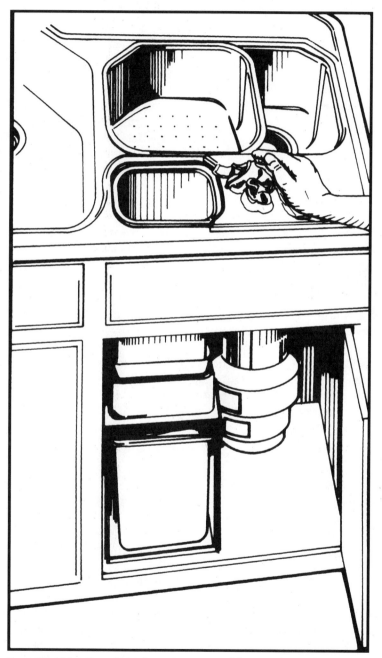

Figure 6.7   A handy undersink waste storage with waste disposal unit. (*Leon E. Korejwo*)

Trash compactors

Waste compactors are a favorite in many households. They reduce the volume of trash leaving the home and simplify waste disposal. These units are designed for compacting dry garbage into small disposable packages, usually 20 to 25 pounds in weight. Some models can compress 14 average-sized garbage bags of refuse into one compactor bag. Dry garbage is compacted under pressure of 2000 to 5000 pounds. Though some manufacturers claim a compactor can handle a normal amount of wet garbage, it is best to put such waste into the disposer. Also, smelly items such as fish and poultry trimmings and combustibles such as chemical containers, oil, and rags should be avoided.

Compactors are available in freestanding and built-in models. The former are usually top-loaded while the built-in are drawer-loaded. For built-in installation, the compactors come in the form of a base cabinet, usually 15 to 18 inches (38 to 46 cm) wide. Space-saving models are available 12 inches (30 cm) wide, while 15 inches (38 cm) is standard. All models offer an invaluable safety feature that locks the unit when not in use. The compactor will not operate unless the keylock is in the "on" position, the unit is completely closed, and the operating switch is on. Additional features of today's compactors include hands-free toe-opening drawers, automatic antijam features, bag storage, and rear wheels for easy movement. Most compactors are available in heavy-gauge steel construction, black or white, and some manufacturers have constructed models to accept panels that match cabinet facings.

Several models use a charcoal filter arrangement to purify the air in the compactor, while others automatically release a measured amount of deodorant spray into the trash container every time the drawer is closed. Some trash compactors feature a reusable vinyl caddy with handles that facilitate trash removal. In others, the trash bag rests in a heavy reusable plastic container, and the whole unit can be taken to the trash can. The waste compactor requires no plumbing or special wiring and can be used wherever it can be plugged into an ordinary 120-volt outlet.

Hot water heaters and water softeners

Hot water heaters and water softeners could also be considered cleanup center appliances. As mentioned earlier, a continuous supply of hot water is a necessity if the dishwasher is to operate properly. Soft water can help in the cleaning process of a dishwasher, too.

Refrigerator/storage center

The main appliance of the refrigerator/storage center is the refrigerator or refrigerator/freezer. In recent years, advances in design and insulation have made it possible to put a refrigerator of surprisingly large capacity into the floor space formerly required for a unit with much smaller capacity. This is primarily due to the discovery of new urethane insulation materials, which permit the walls of the unit to be thinner, thus increasing the interior capacity.

In addition to the refrigerator, this work center may also house a separate freezer, ice maker, and possibly a water center.

Refrigerator

Besides the general cold-storage space for fresh foods, the most elementary refrigerator usually contains a small freezer compartment in which frozen foods may be stored and ice cubes may be made. Since storage space is so important, most modern refrigerator manufacturers pay special attention to shelf and compartment design, featuring gallon-plus door bins and spill-safe sliding glass shelves that slide out for easy access to food, as well as for fast cleaning. Well-lighted, extra-large crispers and meat keepers with clear fronts should have tight covers to keep fresh vegetables, fruits, and meats from drying out. Many models feature dairy shelves with magnetically sealed compartments that are fully adjustable for small items that need a fresh environment. Another popular feature is the refrigerator with temperature zones. Microprocessors give the user the capability of setting the temperature to ensure that food is stored at its optimum level with several different storage compartments, including adjustable-temperature meat keepers. Another elaborate feature is an audible alarm that sounds if the door or drawer is left open for more than 15 seconds. This alarm can be disarmed for cleaning the unit.

It is important for the family to consider what they generally like to put into the refrigerator when they shop—watermelons, turkeys, tall bottles, and other large items can influence their selection. Additionally, some manufacturers have created designs that are so functional that some units can be placed anywhere in the home. They can be placed in the den, family room, master bedroom, or in an entertainment area.

As to the amount of refrigerator space needed, experts agree that a minimum of 6 to 8 cubic feet should be allowed for a family of two. Add one cubic foot for each additional person. If the family entertains regularly, add another 2 cubic feet to the required amount. Incidentally, this footage is for fresh-food refrigerator space only; it does not include the freezer compartment.

Actually, a refrigerator is designed to cool the food placed in it and the air that surrounds this food inside the compartment. Whether or not the refrigerator also provides long-term storage of frozen food depends on the model the householder selects. To function properly, regardless of the model, the refrigerator must have some means of removing the warm air that is brought in by the food or that enters through the door and walls. In an electric refrigerator, the heat is absorbed by a liquid refrigerant. Gas refrigerators are cooled with ammonia. The refrigerant is held in evaporator coils. The location of this freezer section is a key factor in determining the model of the refrigerator.

Basically, all refrigerators fall into three general types as outlined here:

1. *Compact* units can be as small as 2 cubic feet for use in an office, family room, nursery, or den. Some have a cart so that they can be rolled out to the

pool or patio. There are also undercounter and wall-cabinet compact refrigerators, usually about 7 cubic feet in size, that fit in with standard kitchen cabinets. These units are popular for apartments and vacation homes. Compacts usually have a manual defrost.

2. In the *conventional upright* model, the freezer compartment is enclosed by an inner door inside the fresh foods section (Fig. 6.8a). Temperature in the freezer stays at 10 to 20 degrees, adequate for short-term storage. Generally this model is available in 9- to 13-cubic-foot sizes and is the lowest in price of standard types.

3. *Two-door standard upright* refrigerator/freezer combinations have separate doors for each compartment. In many two-door units, the freezer door is at the top of the appliance, sometimes referred to as top-mount. Other designs provide a freezer space alongside the refrigerator space, while still others have the freezer compartment in the lower section, sometimes referred to as bottom-mount. The two-door, side-by-side model seems to be gaining favor (Fig. 6.8b). This type offers storage at a convenient height for both refrigerator and freezer. Doors do not require as much space to open because they are narrower than other types. A few models even have three doors, with the third compartment designed for ice cube production, water cooling, etc.

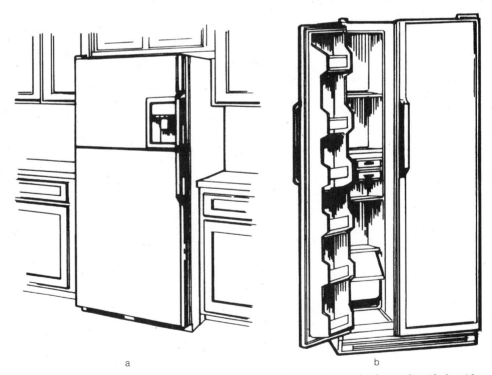

a   b

Figure 6.8 Conventional upright two-door refrigerator/freezer (a); standard upright side-by-side refrigerator/freezer (b). (*Leon E. Korejwo*)

Both the compact and the various standard types can be had in freestanding or built-in models. The latter are frequently preferred because of their ability to fit into the overall design of the kitchen. However, the built-in models are quite a bit more expensive than the freestanding types.

Refrigerator/freezers vary in size due to several design types. They vary in width from 31 inches (79 cm) to 48 inches (122 cm) and may be as large as 60 inches (152 cm) wide for some side-by-side styles. Freestanding, uprights, or top-mounts may be from 33 to 36 inches (84 to 91 cm) wide, 69 to 71 inches (175 to 180 cm) high, and 33 to 34 inches (84 to 86 cm) deep. A bottom-mount can be 33 inches (84 cm) wide, 34 inches (86 cm) deep, 68 inches (173 cm) high. Side-by-side designs vary from 33 to 37 inches (84 to 94 cm) wide, 69 to 71 inches (175 to 180 cm) high, and 33 to 35 inches (84 to 89 cm) deep. Freestanding units are deeper, their doors extend beyond the 24-inch (61 cm) cabinet depth, and their handles add a few inches that must be considered in the floor plan. When the doors are opened, the hinge location and door thickness increase width and interfere when close to corners. Take into account air clearances for cooling/breathing, usually 1 inch (2.54 cm) at the top and $\frac{3}{4}$ inch (1.9 cm) on each side. Freestanding upright freezers as large as 32 inches (81 cm) wide, 29 inches (74 cm) deep, and 70 inches (178 cm) high are available. Built-in refrigerator/freezers of the side-by-side variety range in size from 33 to 36 inches (84 to 91 cm) wide, 68 to 70 inches (173 to 178 cm) high, and are 26 inches (66 cm) deep without the door handle, or 28 inches (71 cm) deep with door handle. Air clearance is usually 1 inch (2.54 cm) at top, $\frac{1}{8}$ inch (.317 cm) each side, and $\frac{1}{2}$ inch (1.27 cm) in back. Built-in units of flush mount design are made to fit more tightly. An installation, for example, described as 24 inches (61 cm) deep, 84 inches (213 cm) high, and 48 inches (122 cm) wide actually requires a rough opening of $47\frac{1}{2}$ inches (121 cm) wide and $83\frac{3}{4}$ inches (213 cm) high. These designs use specific rough opening dimensions and the manufacturer's specifications must be consulted for the final plan. If preliminary plans do not allow for a specific model and make, a good average is 36 inches (91 cm) wide, until a final selection is made.

Some refrigerator/freezers are somewhat deeper than the usual 24-inch (61 cm) depth of the base cabinet assembly. This extra depth may create some problems of access to base cabinets, particularly drawer cabinets, where a refrigerator is located close to an inside corner of the base assembly. Of course, flush-mount refrigerators are built with the condenser coils (usually on the refrigerator back) under the refrigerator cold-food compartment, which allows the refrigerator to be pushed back against the wall to save room space. A fan cools the condenser coils when the refrigerator is running.

As mentioned in Chapter 3, careful attention must be paid to the manner in which the doors are hinged, particularly when the refrigerator door is to be swung open against a wall (Fig. 6.9). Some doors are hinged so that when the door is opened 90 degrees, the space occupied is 4 or 5 inches (10 or 13 cm) wider than the refrigerator cabinet width, the added space due to the thickness of the door. Other designs keep the 90-degree door swing within the width

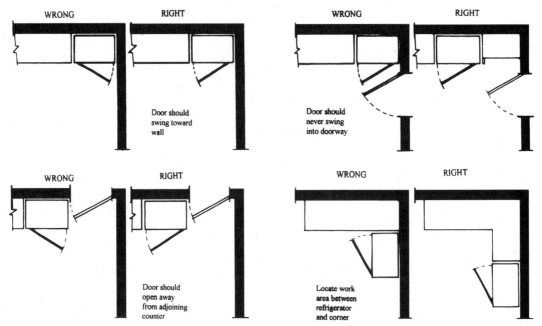

WRONG   RIGHT   WRONG   RIGHT

Door should
swing toward
wall

Door should
never swing
into doorway

WRONG   RIGHT   WRONG   RIGHT

Door should
open away
from adjoining
counter

Locate work
area between
refrigerator
and corner

Figure 6.9   Refrigerator-door swing considerations. (*Leon E. Korejwo*)

of the refrigerator, but this arrangement may cause some difficulty in removing shelves, vegetable crispers, meat keepers, etc. Some manufacturers allow the user to choose a left or right door swing. This option must be specially ordered when requesting the unit.

Some refrigerator/freezers have features such as automatic ice cube makers and door water cooling facilities. Both these features necessitate the installation of a water supply for the refrigerator—usually a small-diameter copper tube is all that is required. While these devices are usually in a refrigerator compartment, some models are constructed so that cooled drinking water and ice cubes may be obtained without opening the door of either the freezer or fresh-food compartment. Today's ice makers allow the user to select the type of ice—from finely crushed to cubes—and chilled water. By the way, many of the automatic ice makers, which eliminate the tray-filling job, are designed so that they can be installed either at the time of purchase or later on. In the latter case, be sure to consider the cold-water line in planning the kitchen, with an easy-to-reach shut-off valve. Nearly all new refrigerators are electric and operate off standard 120-volt wall outlets. When designing a kitchen, account for the space needed to install and remove the refrigerator for service and cleaning. A separate fuse circuit allows the unit to be shut off for defrosting or service without moving the unit to unplug it.

Freezers

A separate freezer, when used properly, can help a large family save money by buying food in quantity. The freezer seems to fit nicely with the trend toward busy family lives. In addition, many families enjoy freezing fresh foods and preparing foods to freeze for later use as well.

There are two types of separate freezer units—upright and chest. The former type, which is more popular, has shelf space that is both easy to see and easy to use. Some uprights now also offer a "no-frost" feature, as well as door locks, signal lights to warn of current interruptions, adjustable shelves, and rollers or wheels for easy cleaning. A chest-type freezer costs less to operate than an upright, but it takes more floor space and makes reaching foods at the bottom more difficult.

Both types of freezers are available in sizes from 3.2 to 30 cubic feet. The most popular size is 15 cubic feet. Generally speaking, allow 2 cubic feet for each member of the family. Sometimes freezer capacities are quoted in pounds. To convert to cubic feet, remember that 35 pounds of produce can be stored in 1 cubic foot of space.

As previously mentioned, separate freezers are not always installed in the kitchen. Frequently they are found in a basement, attached garage, or in some other secondary space in the house. But there are some facts to remember about the installation of separate freezers. For example, the depth of upright units with the door open may be nearly 60 inches (152 cm), and the height of chest-type units with lid open may be nearly 63 inches (160 cm). Condenser units on some models require top ventilation. Some models require $2\frac{1}{4}$ inches (6 cm) side clearance for door lock to allow key insertion. Therefore, be sure to check the manufacturer's specifications very carefully before installing a unit.

All home freezers and combination refrigerator/freezers (except for those of the conventional refrigerator) pull down to 0 degrees F. This can be accomplished by setting the control to colder. However, it is not mandatory to have 0 degrees F if food is kept for less than a year. Foods can be kept for three to four months at 10 degrees F without harm and from six to nine months at 3 to 8 degrees F. Even at 0 degrees F or less, food kept more than a year deteriorates in appearance, taste, and nutritive value.

Most refrigerators, refrigerator/freezers, and freezers are finished with a porcelain-enamel finish, and a variety of colors are available. Several manufacturers make provisions for the installation of a decorative panel on the door surface. This panel may be replaced from time to time so that the refrigerator appearance may be modified to comply with changing decor requirements. These door panels can match the cabinetry and are compatible with virtually any style and look of cabinetry (Fig. 6.10). These units can be used with framed or frameless cabinets.

Good air circulation is also vital to the efficient refrigeration and preservation of food. If the air is restricted from circulating to all parts of the cabinet, the food in the lower area won't be refrigerated sufficiently. Efficiency may be affected by overloading, blocking air circulation, or by storing hot items.

A loose door seal allows warm air to leak into the cabinet and causes excessive sweating during warm weather. If the door seal is good, the customer must be informed that a certain amount of sweating is normal and that the condition can be kept to a minimum by following these recommended procedures:

Figure 6.10 As seen here, the ideal built-in refrigerators also complement the cabinetry. (*SubZero*)

1. Reduce moisture by keeping all liquids and moist foods covered.

2. Keep the number of door openings to a minimum.

3. Defrost per manufacturer's recommendations.

During very hot, humid weather, the refrigerator runs longer. If the temperature control knob is set too high under these prevailing conditions, the on cycle becomes too long. On models that have a cooling coil in the fresh-food section, the entire cooling coil becomes heavily frosted. Heat cannot penetrate the thick frost surface, so the result of this condition would be high temperatures in the fresh-food compartment. The solution to this problem is to turn the thermostat to a lower setting until the cooling coil has defrosted completely. Then the control knob should be kept at a low setting to allow an off cycle. Each on cycle begins with a cooling coil free of frost, and better fresh-food temperatures result.

Ice makers

Ice makers are now available as separate appliances. They are usually available as a base cabinet, ordinarily 15 or 18 inches (38 or 46 cm) wide. They are intended for use in bar areas or where an abnormal supply of ice cubes is necessary. Ice makers must be connected to a 120-volt ac electric power source and to the cold-water supply. Compact or undercounter ice makers might be 12 to 24 inches (30 to 61 cm) wide.

Water centers

Water centers are a wonderful up-and-coming feature for many homeowners. These machines can produce near-boiling water (190 degrees F) instantly, for making coffee, tea, soups, cereals, gelatins, and sauces with no waiting for microwave ovens or boiling kettles. Steaming water is available on demand for cooking pasta, baby food, and soaking hard-to-clean pans. Some feature lever locks for a continuous flow in filling large pots.

The cold dispenser produces instantly refreshing 40-degree water for cool drinks, also eliminating the need for ice cubes when chilling drinks. Some manufacturers have actual centers installed into the cabinetry that perform a type of reverse osmosis purification, removing major pollutants and contaminants to provide for pure water, whether it is hot, cold, or at room temperature. Many systems can be combined with water filtration systems, for maximum protection against lead, asbestos, chlorine, and other contaminants.

Food preparation/cook center

There are a variety of processes used in cooking. The more common of these processes are boiling, poaching, frying (including sautéing, pan-frying, and deep-frying), braising, broiling, baking, and roasting. Most of these cooking processes can be performed using two basic cooking devices—the oven and the surface cooking unit—but many other cooking devices described in this chapter are available. Some of these additional devices are frequently included in the range.

Ranges, cooktops, and ovens

A wide variety of designs and types of ranges are on the market today (Fig. 6.11). Ranges combine a cooktop and one or more ovens. They are available in a variety of styles and configurations. The heat sources vary, as does energy efficiency. If helping your client decide on these appliances, evaluate the available energy source, aesthetic desires, price ranges, and cooking style.

Among the more popular ranges are these styles:

1. *Freestanding ranges* rest on the floor and are completely independent of the adjoining walls, cabinets, etc. Such a range usually has side panels with a finish that is the same as the ones to the front. However, it is usually installed as part of a continuous kitchen assembly of appliances, cabinets, and counters. Many freestanding ranges now are available with squared sides and corners that give a built-in appearance. The freestanding counter-height ranges are available with one or two ovens and/or broiler below the surface cooking unit. The high-level type have the oven above the surface unit, and the high-low (bilevel models) style has ovens above and below the cooktop. Eye-level ovens and high broilers are real back-savers, but if the homemakers are shorter-than-average height, lifting foods in and out of an eye-level oven may be rather difficult. The controls may be in the front, along the side, on

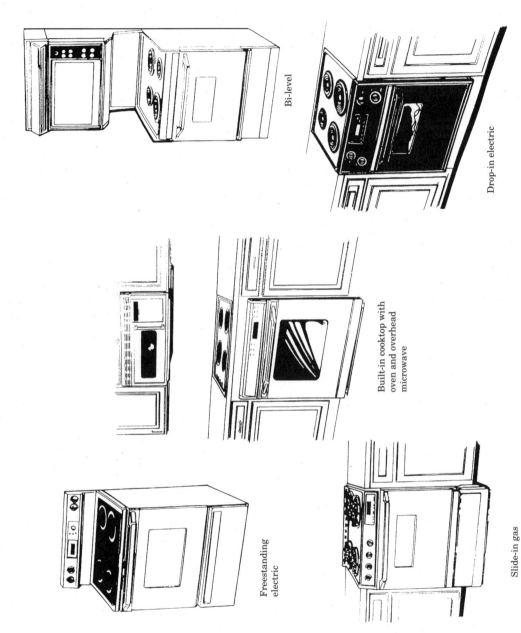

Bi-level

Drop-in electric

Built-in cooktop with oven and overhead microwave

Freestanding electric

Slide-in gas

Figure 6.11 Types of ranges. (*Leon E. Korejwo*)

top, or on a backsplash. As for widths, freestanding ranges vary from a minimal 20 inches (51 cm) to as large as 60 inches (152 cm), with 30 and 36 inches (76 and 91 cm) being considered "standard."

2. *Built-in ranges* are designed to be fully integrated into the kitchen and counter area. The various forms of built-in ranges are constructed so that connections to adjoining cabinets and counters are relatively tight, thereby eliminating many of the cleaning problems that exist with freestanding ranges. For example, the *slide-in* or *slip-in range* may be considered a transition design between a freestanding range and a built-in range. In fact, some slide-in ranges can be used in either manner. Slide-in ranges rest on the floor, have raised edges, are usually supplied without side panels, and square up quite tightly with countertop and cabinet fronts. (When side panels are available, the range may be transformed into a freestanding range.) The configurations of these ranges and their sizes are very similar to those of freestanding ranges.

   The *drop-in range* differs from the slide-in range in that it does not rest on the floor but on the adjoining counter or on a special base cabinet. In some instances the range may extend down to the toe space and rest on a framework behind the space. Some units that extend back to the wall have a back panel above the range top. Others are fitted into a cut-out in the cabinet countertop.

   A *stack-on range* is essentially a drop-in range in which the oven has been placed at eye level rather than below the surface cooking units. The range is supported by flanges that rest on the countertop.

3. *Built-in ovens and cooktops.* In recent years, separate built-in ovens and cooktops (surface cooking units) have become increasingly popular (Fig. 6.12). The popularity of the built-in oven stems from two factors: its eye-level placement is easier to use, and built-in installations eliminate a considerable amount of exterior surface cleaning that is necessary with the freestanding range. A disadvantage of separate ovens and cooktop units is that more kitchen space is needed. Some separate oven appliances actually have more than one oven in the appliance. A microwave cooker may be added to this unit, or, in some instances, two standard ovens are included. The standard built-in oven is generally designed to fit into 24-, 27-, 30-, and 36-inch (61-, 69-, 76-, and 91-cm) wall oven cabinets. Built-in ovens have both a cutout size and an overall size, which is the width of the front frame, and almost every model is different in both dimensions. Remember not to install the built-in oven too high. High mounting makes it difficult to remove pans and can result in burns.

   Built-in cooktops are available in several forms. Some cooktops are available as drop-in units with backsplash panels, but the majority are designed for installation in cutouts in counters. In this case, the unit may rest directly on top of the counter or it may be fastened into the counter with a Huddee or steel-rim or similar device. Cooktops are available with various numbers of burners, two to six burners on the average, with four being the most common.

Figure 6.12   Built-in convection double ovens. (*GE Appliances*)

Separate units of two burners each range in standard diameters of 9 to 15 inches (23 to 38 cm). Placement of the burners and the controls also varies.

Today's cooking devices have such features as electronic touch pads which allow the user to select cooking functions, oven temperature, and cooking time; automatic clean and bake indicator lights; interior oven lights; removable full-width storage drawers; LED electric clock with timer; metal/glass backguards; adjustable oven racks; 3.7 cubic-foot ovens; broiling systems; grills; griddles; filter vent covers; and drip pans that can easily be cleaned in the dishwasher.

Stainless-steel framing, spill-control edges, rounded corners, multiple oven racks, built-in spice racks, warming drawers, and touch pads, or childproof controls are just some of the features available today. Other top-of-the-line amenities include flavor-generator plates that circulate smoke around food to lend an outdoor-grilled taste, and electric-spark reignitions, which automatically relight a burner if it is extinguished by any means other than the control.

If the homeowner is a gourmet or real serious cook, it may be wise to suggest a *commercial restaurant range* (Fig. 6.13). This range offers large capacity and durability. It sits flush against the rear wall and standard-depth kitchen cabinets. Some models are equipped with as many as 12 burners and two extra-wide ovens. The burners are usually 11 inches (28 cm) in diameter to handle large frying pans and stock pots. Widths vary from 23 to 68 inches (58 to 173 cm), and depth is usually at least 6 inches (15 cm) greater than that of conventional freestanding ranges. The standard commercial ranges are available in 30 inches (76 cm), which are perfect for any kitchen.

Figure 6.13 A 36-inch, professional-style range with a single oven. (*Five Star*)

These ranges are designed to make cleaning easier and operation safer. They deliver incredible heat for cooking food quickly without burning. They feature separate broilers, deep ovens, multiple burners, professional-style griddles, warming drawers, lift-off oven doors, warming lamps and racks, and flip switches to go from standard to convection baking or roasting. Some manufacturers carry full lines of professional equipment, including ranges with standard ovens, convection, and self-cleaning ovens; cooktops; and range hoods. They are designed to deliver heat up to 14,000 Btu. There is little choice of finish—they usually come in black or stainless steel only.

Most types of cooking appliances just mentioned are available in either electric or gas (drop-in is only available in electric). Utility rates in the area dictate which is the most readily available and economical to use. In most areas, gas is considered less expensive, but it depends partly on the methods used by the homemaker. Electric cooking is extremely efficient if used properly. If the homemaker has never cooked on a standard electric unit, the smooth top may require a major adjustment. It can have five or more heat settings or be entirely thermostatically controlled.

Most cook-center appliances are finished in either stainless steel, brushed chrome, porcelain enamel, or a combination of these finishes. The porcelain enamel finish offers the possibility of incorporating color into the kitchen appliance; however, porcelain enamel can chip. Stainless is less colorful, but it cannot chip, although it can dent. Abrasive cleansers should not be used on either finish. Interiors of ovens are generally finished in porcelain enamel. A special finish is required for catalytic-process continuous-cleaning ovens.

Let us take a closer look at the two major parts of any cook work center—the surface cooking units and the oven—and how they are heated. Cooktops feature a variety of heat sources and are available in numerous styles (Fig. 6.14). Many models afford interchangeable cooking modules that include burners, deep fryers, grills, griddles, and quartz halogen/radiant tops. All cooktops require a ventilating unit—either a downdraft vent system that is built into the cooking surface or a less-expensive overhead hood.

Electric cooktops

Electric surface units are usually equipped with heating elements of the metallic-sheathed resistance wire and consist of two elements built in a flat spiral. The two elements may be placed side by side in a continuous spiral or one element may be wound in the central portion of the burner, while the second element forms the outer periphery. This latter form is especially useful when it is necessary to place a small pan on a larger burner. The intensity of heat of the burner is usually controlled by an infinite heat-type switch.

In some instances, one or more units may be equipped with a thermostatic control. At the lower temperatures, the resistance heating elements do not glow; they appear much the same as burners that are not in use. For this reason, most surface cook units are equipped with indicator lights that show that

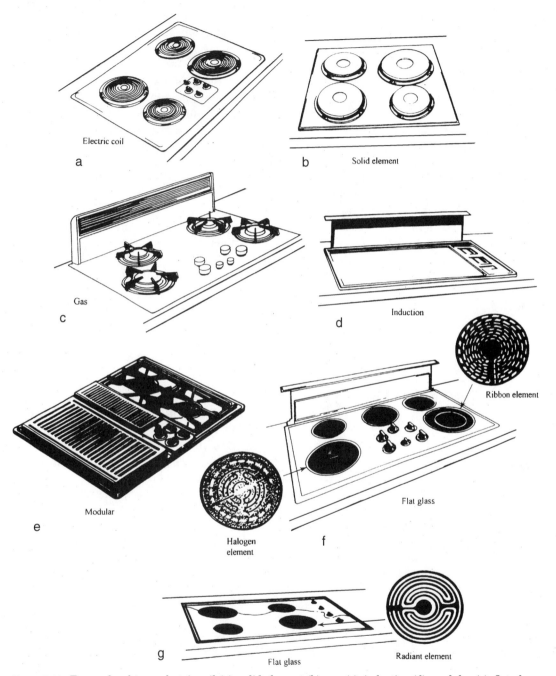

Figure 6.14 Types of cooktops: electric coil (a); solid element (b); gas (c); induction (d); modular (e); flat glass with ribbon element and halogen element (f); flat glass with radiant element (g). (*Leon E. Korejwo*)

a burner unit is operating. Indicator lights may be attached to each burner or to pushbutton controls, or one light may be used to indicate that one or more burners are on.

In the typical electric surface unit, the heating elements are mounted in an opening in the top surface panel of the appliance. Under the top surface panel is a space used for certain necessary wiring connections to the burner and also for the positioning of reflector pans, drip cups, and drip trays. The primary purpose of this arrangement is to facilitate cleaning of the appliance. When spills or overflows occur, they pass down through the opening in which the heating element is located and are collected in the drip cups and drip trays. Access to the space is through the heating element opening or by means of a top surface panel that can be tilted upward.

Electric coil burners afford low, even heat—they heat and cool slowly. Heat adjustment does not occur immediately, so reducing a boil to a simmer, for example, takes longer than with other types of elements.

Surface cooking units should not be located close to an internal corner of a kitchen counter assembly. A person standing at a counter that is adjacent and perpendicular to a range or counter containing a cooktop can inadvertently extend an arm or elbow over a burner and be injured. This hazard potential can be reduced by keeping surface units 12 inches (30 cm) away from the internal corners of counter assemblies.

Gas cooktops

Gas burners heat and cool quickly and are designed to give a circular flame pattern, the size of which can be controlled to some degree by the amount of gas delivered to the burner. The burners are usually controlled by a throttling valve that resembles a rotary switch in appearance. The intensity of the flame can be controlled, on a graduated basis, from low to high. Some burners are equipped with special simmer units that are, in effect, a very small burner located in the center of the main burner. Burners are lighted by a centrally located pilot light or by electric ignition (although older models had standing pilot lights, modern designs use energy-saving electric ignition). When the gas supply of a burner is turned on, a portion of the gas is guided through a small open-ended pipe toward the electric ignition. Once the gas reaches the electric ignition, the burner is ignited. These ignitions lower energy consumption by as much as 40 percent. A standard household gas cooktop generates between 8000 and 10,000 Btu per hour per burner. The heavy-duty burners of commercial-style models generate up to 15,000 Btu.

Gas burners are mounted in the top surface panel of the appliance in the same manner as are electric elements. Most gas burners are designed so that the burner grill, the burner, and the piping can be removed easily for cleaning.

There are two designs of gas burners, standard and sealed. The standard burner uses a grate attached to a cooktop that lifts away for access to the burners, which are mounted on a subtop. The sealed burner is designed to prevent spilled liquids from running under the cooktop.

Flat-surface cooking

A smooth, ceramic-glass cooking surface unit is basically a flat, glasslike plate approximately $\frac{1}{4}$ inch (.64 cm) thick. On some units, a separate place is provided for each unit. Other manufacturers use a one-piece plate to accommodate several units. The heating elements are placed immediately beneath the glass plate. Some manufacturers recommend that the unit be used with glass cooking utensils that are specially ground on the bottom so as to fit closely to the surface of the cooking unit. To prevent scratching and discoloration, other manufacturers suggest only flat-bottom pans of any material. In any case, the unit performs most effectively when the pans have flat bottoms that fit closely to the surface of the unit.

Smooth-top cooktops possess many of the same characteristics as electric coil units; they are, in fact, composed of electric-coil burners covered by a glass ceramic surface. Slow to heat and cool, some units retain heat for up to an hour after they are turned off.

The major advantage of this type of surface cooking unit lies in the ease with which it can be kept clean. There are no drip cups or drip pans. Spills can be wiped clean from the surface. One interesting feature of the ceramic-glass material is that it readily transmits heat upward directly to the cooking utensil, while at the same time little heat is lost laterally through the panel.

Controls on some ceramic cooktops are more sensitive than those of the usual burner. As in the ordinary range, the controls can be set for a typical range of temperatures, but the burner contains a sensor that shuts down the unit when temperatures exceed the selected level. The heat-up and cool-down times on ceramic-glass cooktops are longer than on the conventional electric types.

Although gas, electric-coil, and smooth-top models are the most common designs, there are several other cooktops available, but the equipment and effects are different.

1. *Radiant heating units* feature a wire that radiates heat through a glass-ceramic surface. The heat comes from brightly glowing elements positioned beneath the smooth surface. Even heat is directed straight up to the pan for quick cooking response. Radiant heating units are economical to operate.

2. *Magnetic induction units* also offer glass-ceramic surfaces, which have the added advantage of remaining cool throughout the cooking process. Heat is generated by an electromagnetic field that occurs when the pan comes into contact with the cooking surface. These cooktops are extremely energy-efficient, but they work only with iron, steel, or steel-alloyed cookware.

3. *Halogen units* have a halogen bulb that radiates heat through a glass-ceramic cooktop. They afford fast heating and precise temperature control, providing an instant "on." Cleanup is easy because elements are positioned beneath the smooth surface. Although halogen light is visible immediately, these units do not heat any faster than radiant units.

    Halogen burners work somewhat like incandescent light bulbs, heating from electricity passing through a tungsten filament in a quartz tube.

Halogen gas in the tube combines with evaporated tungsten, but heat from the filament causes the tungsten to redeposit on the filament so it doesn't darken as a light bulb does, and it lasts longer. This heat is mostly infrared light in wavelengths that pass easily through a glass ceramic cooktop. The burner lights up immediately with a bright, red glow. The advantage is that the user can see the heat, and it heats up much faster than other glass tops. In addition, with a halogen burner, the user does not need perfect contact between the burner surface and the bottom of the cooking vessel.

4. *Solid-element cooktops* use a metal disk. Solid cooktops have declined in popularity and have been replaced by more efficient units that heat up and cool down faster than previous models, although, because they are fused to the cooktop, they tend to stay cleaner. Flat bottoms are required for the pots and pans.

5. *Modular cooktops* let the users customize the cooktop with flexible modules to suit their taste. Simply insert the modules into either side of the cooktop and remove for easy cleaning and storage. They are available in grill modules, griddle accessories, ribbon modules, and coil modules. Gas module units are equally versatile.

Ovens

Cooking time and energy efficiency are the most obvious differentiations among microwave, convection, and conventional ovens. Conventional ovens are powered by gas or electricity and employ a radiant-heat system. Heating elements are placed inside the oven walls and heat radiates into the cavity. Because heat rises, stratification (the difference between the temperature in the upper and lower sections of the oven) can be a problem.

Convection ovens, on the other hand, are equipped with a fan that circulates hot air around the oven cavity to maintain a uniform temperature. These ovens are more energy-efficient than conventional radiant-heat ovens (they can reduce electrical energy usage by almost half) and can reduce cooking time by up to 30 percent. Many models are equipped with computerized controls that convert traditional recipe temperatures to the lower convection temperatures. Ideal for roasting and baking, convection ovens are less effective for deep-dish entrees. Some models also have settings for convection dehydrating and defrosting.

Foods cook quickly with high-frequency microwaves but do not brown unless the microwave incorporates a separate browning element. Microwaves are energy-efficient; they employ from 15 to 70 percent less cooking energy than conventional ovens.

Conventional ovens   Baking and roasting are the basic cooking operations performed in the conventional oven, but many ovens are also equipped with broiler units. Some may incorporate microwave cooking units. Many models are self-cleaning; others have such features as electronic timing control; removable, counter-balanced glass doors with windows; built-in sensors that automatically calculate and adjust heating/cooking times and power levels for favorite foods; and convenience cooking controls that allow the user to defrost, cook, or reheat at the touch of a button (or remote control). Some ovens feature a unique airflow system where air circulates around the oven to prevent heat buildup on the outside and allow more balanced heat distribution on the inside.

In electric ovens, the heating elements are usually metallic-sheathed resistance coils. Metallic-sheathed coils are self-cleaning and generally withstand more severe use than the unsheathed resistance coils. The resistance heating units are frequently installed as plug-in units so that they can be removed and replaced relatively easily.

Controls in the typical conventional oven, whether gas-fired or electric, are designed to provide thermostatic control of the temperature in the baking and roasting range, that is, from 140 to 550 degrees F. Many oven units are equipped with timers that operate the oven for a given period of time. Others have programmed timers that start oven operation at a predetermined time, operate the oven at a selected temperature for a period, and, upon completion of the cooking operation, hold the temperature of the oven at a uniform keep-warm temperature until the operation is terminated by the user. Programmed cooking lets the cook leave the house while dinner is being cooked.

Some ovens are equipped with a probe thermometer connected to a thermostatic control unit. This device is particularly useful in cooking meat. Once the temperature of the meat at the point of the probe reaches a predetermined level, the cooking process is stopped and the oven is maintained at a warm level.

Broilers, as already mentioned, usually are incorporated in ovens. The broiling process itself is a rapid, dry, radiant-heat cooking process used primarily for cooking meat, poultry, and fish. The item to be cooked is placed $1\frac{1}{2}$ to 2 inches (4 to 5 cm) away from (usually beneath) the broiler heating unit, and the heat controls are set at the highest temperature. Most gas ranges have separate broiling compartments, so one oven is sufficient unless the users have a large family or do a lot of entertaining. Two electric ovens are desirable so that the homeowner always has one available for broiling. Two small ovens are usually more useful than one large one.

The broiler's heating burners or elements are located at the top of the oven, and the controls are incorporated in the oven controls. Special separate broiler units are also available and are sometimes incorporated in ranges or ovens. In this case, broiling can be done at the same time the oven is being used for baking. Rotisseries are also found in many conventional ovens.

Most modern gas or electric conventional ovens employ one of two systems to clean themselves: *pyrolytic* (self-cleaning) and *catalytic* (continuous cleaning). The pyrolytic method of oven cleaning is a chemical decomposition of cooking soils by the application of high heat. When the oven needs cleaning, controls are set that lock the door shut for safety, and the oven is heated to a point between 850 and 1050 degrees F (490 and 921 degrees C). At the end of the complete cycle, which takes 2 to $3\frac{1}{2}$ hours, after a cool-down period, all that remains of the food soil is a powdery ash that is easily removed with a damp cloth.

Advocates of the pyrolytic, or self-cleaning, system claim the following advantages: All six sides of the oven interior are cleaned completely; no special care is required to prevent scratching or damaging oven surfaces, as with the catalytic finish. In addition, the extra insulation provided for the cleaning cycle results in a cooler range during normal cooking operations.

In a catalytic, or continuous-cleaning, oven, there is a catalyst in the porcelain enamel of the interior oven panels that helps to oxidize spatter as it hits the surface. The oven never gets terribly soiled, unless there is a spillover, but it may look entirely clean. Grease that is deposited on oven surfaces from meat tends to disappear as the oven is used for baking. The catalytic system does not require extraordinary temperatures, so no door latch is needed, and the oven is not taken out of use for cleaning. But manufacturers using the continuous-cleaning method emphasize three points:

1. Heavy spillovers will not be cleaned without first wiping up the excess with a damp cloth.

2. Harsh abrasives and scouring pads should not be used on excessive soil or stubborn stains.

3. Certain types of food stains may not disappear in one operation but fade away during subsequent use of the oven.

Proponents of the catalytic approach point out that the oven is always available for cooking since cleaning is continuous under normal usage. This continuous cleaning prevents any buildup of dirt and very little basic redesign is required since high temperatures are not employed. The consumer gains an economic advantage because the initial cost is lower for the catalytic oven than for the pyrolytic.

Convection ovens   The convection oven is a modern cooking device for home kitchens. It is really like the conventional ovens just described except that it uses increased air circulation within the oven to speed up the cooking or baking process. The air inside a conventional oven is almost static, and cooking depends on the gradual conduction of the heat from the outside of the food to the center. In the convection oven, both upper and lower elements are on during baking, and a fan keeps the air circulating evenly around the food, making the whole process more efficient. Cooking temperatures can be reduced by 50 degrees F and cooking times shortened up to about a third, so the convection oven is both a time- and an energy-saver. The convection system continually circulates heated air around food to keep temperatures uniform for fast, even cooking (Fig. 6.15).

Convection ovens have been standard equipment in restaurants and commercial bakeries for many years because more food could be crowded into them, side to side and top to bottom, while still achieving even baking and browning results. In the case of meat cooking, less shrinkage of the meat meant thousands of dollars of savings on a commercial or institutional scale.

To take advantage of the time and energy savings of the convection oven, the cook has to adjust familiar cooking habits only slightly. For favorite recipes,

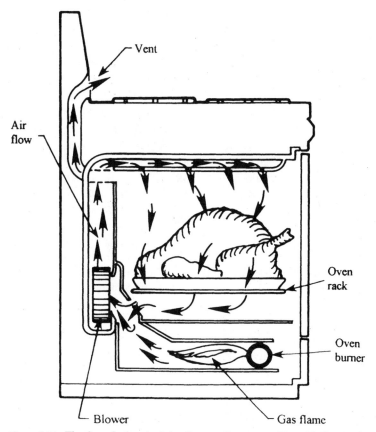

Figure 6.15 The forced-air principle of convection ovens. (*Leon E. Korejwo*)

simply lower the usual temperature by 50 degrees F and then check for done-ness after about two-thirds of the customary cooking time. Food may be cooked in either glass or metal containers. An especially desirable feature of the con-vection oven is that it lends itself to fast cooking of foods from the frozen state—a real plus for a working person who forgot to take the meat out of the freezer. Frozen convenience foods, such as TV dinners, cook in half the recom-mended time with a 25-degree F reduction in temperature.

Microwave units Microwave units are generally called "ovens" because they look like ovens. Their use is, however, not confined to oven cooking. With a conventional range, roasting and baking are done in the oven, and frying and boiling are done on top of the range. Many foods, such as bacon, eggs, cereals, and vegetables, that are usually done on top of the conventional range can be done in a microwave unit.

While microwave, or electronic cooking, offers many advantages, the most out-standing is its time- and energy-saving features. Food can be cooked much more rapidly by microwaves than by the heat generated in a conventional oven or even in a convection oven. Microwaves are very short waves of radio frequency.

When these waves penetrate the material to be cooked, movement is created between molecules, and this action cooks the food. The cooking action takes place almost simultaneously throughout the object being cooked, rather than progressively as in cooking by heat penetration. In addition to rapid cooking, the microwave unit is very useful for quick thawing of frozen foods. Still another important advantage of microwave cooking is that it is clean, both inside and out. Because the dishes stay cool, there are no grease vapors caused by sizzling fat, nor are any grease splatters baked on. When a roast is cooked, splatters do occur as in conventional cooking, but because oven walls are cool, the splatters stay in a semiliquid state and are very easily cleaned off.

The difference in the electronic cooking action and its rapidity usually results in a piece of meat becoming cooked without the surface being browned. Most people are accustomed to seeing a browned crust on roasted meat. To overcome this aspect of microwave cooking, many manufacturers have included a browning unit in the microwave oven. By the way, contrary to popular belief, food prepared in a microwave oven is not cooked from the inside out, but is cooked all the way through at the same time, with more cooking being performed on the exterior of the food. It is, therefore, possible to prepare a rare, medium, or well-done roast in the electronic oven.

Generated by a magnetron, microwaves travel in straight lines much the same way as light. The microwaves follow the waveguide channel into the oven. At the end of the waveguide, the microwaves strike the stirrer, which is a slowly rotating fan. The fan blades reflect the microwaves, causing them to bounce off the walls, ceiling, back, and bottom of the oven. (Some manufacturers use a slowly revolving shelf or turntable rather than a stirrer.) The microwaves enter the food from all sides, accomplishing cooking inside and outside from all directions. The food, in an appropriate utensil of glass, ceramic glass, china, pottery, paper, and some plastics, is positioned about 1 inch above the bottom of the oven. The microwaves pass through a special glass shelf and are reflected off the bottom of the unit to enter the bottom of the food. Since the cooking utensil is made of materials that transmit the electronic energy, it does not heat up. Ordinary metal cooking ware should never be used in the electronic oven as they would reflect waves away from the food to be cooked.

Some concern has been voiced about leaks of microwave energy creating hazards. It is important to point out, however, that microwave rays are longer than visible light, and their effect is to create heat. They should not be confused with cosmic rays, gamma rays, X-rays, and ultraviolet rays, which are all shorter than visible light and, when strong enough, can alter living cells. Also, microwave ovens are required to comply with rigid federal regulations on leakage. Some heart pacemakers are affected by microwave energy, so if there is a pacemaker wearer in the family, consult with the individual's physician before installing a microwave oven in the kitchen. The Federal Communications Commission has assigned three frequencies for the operation of microwave units and other related types of equipment. They are 915 megahertz (MHz), 2450 MHz, and 5800 MHz. While little difference in overcooking results can be found at the various frequencies, most microwave ovens produced today operate at 2450 MHz.

While microwave units are most frequently sold as portable appliances for use on the countertop or table top (Fig. 6.16), electronic cooking may be combined in a single oven with a conventional electric element or as the second oven in a freestanding or drop-in range. When combined with a conventional element, a 240-volt hookup is required to microwave and brown simultaneously. Where the two types of cooking are combined in a portable model operating on 120 volts, the two operations, as mentioned previously, can be done simultaneously. The smaller portables may not have the conventional browning element. In some ovens, browning of meats is done on a plate made of a special material that can attain an unusually high temperature. Portable units are available in sizes that fit into 24- and 30-inch (61- and 76-cm) cabinet spaces. The standard models are usually 30 inches (76 cm) wide, but sizes vary by the brand. All models can be used as countertop microwave ovens, installed under kitchen cabinets, or built into the wall, leaving counter space free for other uses. There are several convenient locations for a microwave oven in a kitchen:

1. Stacking a microwave unit and a conventional oven as a built-in double-oven unit makes a useful working arrangement (Fig.6.17). Where the kitchen plan requires this space-saving setup, the microwave oven should be on top. This is handier because food is taken in and out of the oven repeatedly and in short intervals, whereas food is generally placed in a conventional oven for longer periods, requiring few return trips.

2. Microwave units are available that replace range vent hoods. Exhaust fans are incorporated in their design, allowing the area above a cooktop to be used more efficiently.

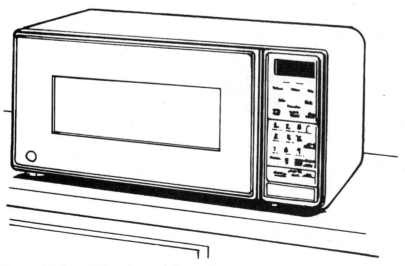

Figure 6.16 A portable microwave is usually placed on a countertop. (*Leon E. Korejwo*)

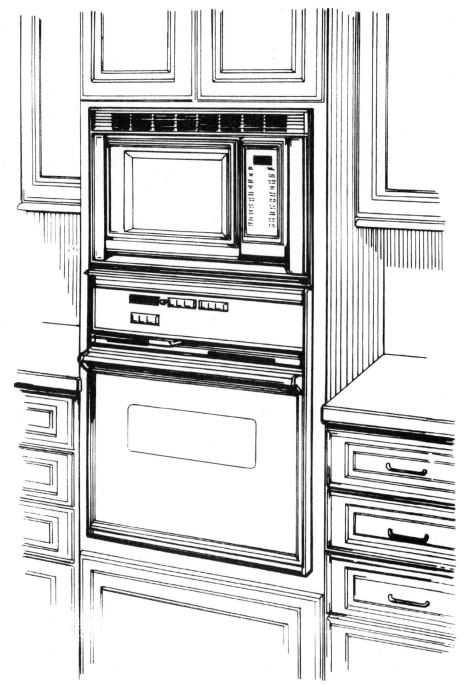

Figure 6.17 Stacking ovens: microwave oven (top) and convection oven (bottom). *(Leon E. Korejwo)*

3. The microwave oven can be enclosed in a space between the bottom of the existing cabinet and the countertop. This level makes opening the door comfortable, especially for a cook of average height or less. Frequently, the location of the oven can be away from the regular cook/work center. But, with any built-in installation, be sure there is clearance at top and back for ventilation. Note: Steam accumulates quickly in a microwave cavity.

4. A special shelf can be constructed under a high wall cabinet to house the microwave oven. This height works well for anyone 5 feet 7 inches or taller. If the oven is used frequently for entertaining, it should be located near a dining serving counter and adjacent family room.

5. A portable microwave unit can be used on a cart or small movable island and wheeled to any area of the house or outside. The portable oven should always be plugged into a grounded 120-volt ac outlet.

Today's microwave ovens come with many features, including 1.1-cubic-foot interiors, 850 watts of microwave power, two-speed vent systems, timed defrost, auto defrost, two program memory levels, 12 hours auto-start, popcorn pad, removable glass turntables, 10 variable power levels, electronic touch control, electronic digital display and clock/timer, cooktop light, and two-speed, high-capacity exhaust fan to replace a vent hood. Rotating turntables provide the most even, consistent, microwave cooking results. Some ovens feature one-touch preprogrammed food pads that let the users program their own customized times, which are stored in permanent memory.

Other cooking devices

While cooktops and ovens are basic tools of the cook center, some ranges contain such features as a grill and a warming drawer. These items are also available as separate appliances for built-in installation.

Grills Cooking by grilling is similar to broiling because it involves cooking by radiant heat; however, the item to be cooked is above the heat source rather than below it. The grill is located in a horizontal position above the source of heat, and the item to be cooked is placed on top of the grill. A considerable amount of smoke can be generated in this process as the drippings from the cooking process may fall on the heating unit and be burned. Ventilating hoods or similar smoke-gathering devices are essential.

As already mentioned, griddles are sometimes incorporated in ranges or cooktops along with standard surface units that can be used for cooking meats, pancakes, and the like. In essence, the griddle serves as the pan as well as the heating element for certain types of frying operations. Some manufacturers offer griddles as optional cooking devices for range-top cooking. Frequently, a grill cooktop can be interchanged with ceramic-type or standard electrical elements. Other cooktop options include deep-fat fryers and rotisseries.

Barbecue grills Modern barbecue units, which may be installed in standard wood or metal kitchen cabinets, bring barbecue cooking conveniently indoors for year-round enjoyment. Prefabricated barbecue units are available with regular charcoal firing, as well as with gas burners or electric cooking elements. In the gas models, radiant ceramic "coals" above the burner hold and evenly distribute the cooking heat. The electric models usually have a layer of ceramic below the element to distribute the heat. The drop-in units, which are most commonly used, can be had in the following models, in units of different widths from 24 to 48 inches (61 to 122 cm):

1. Top controls, countertop cabinet model

2. Top controls, masonry model

3. Front controls, countertop cabinet model

4. Front controls, masonry model

Barbecues can also be part of surface units. Some cooktop grills have a charcoal plate that gives food a charcoal-like flavor. This special charcoal plate keeps its flavor in spite of occasional washing in a dishwasher.

Most of the standard drop-in units are encased in insulated shells and can be safely installed in wood or metal kitchen cabinetry or built into brick, stone, or concrete block, indoors or out. Some units are designed for masonry only, with open sides, back, and bottom, so make sure to buy the right type for the desired purposes. Most units also offer a motorized rotisserie as an extra. To exhaust cooking aromas and charcoal smoke adequately from indoor areas, a hood and fan is a must.

Food warmers While there are several different types of food-warming appliances, the most popular are still in the form of a drawer approximately 12 inches (30 cm) high and 24 inches (61 cm) wide that can be installed in the base cabinet assembly. An opening must be provided in the base of the oven cabinet for the appliance. A 240-volt power supply is normally required. Wood trim is usually available for these drawer food warmers, so they accept paneling to match the cabinets.

Another popular food warmer, especially in a food serving center, is a wired, rectangular, glass-ceramic plate that is recessed into the countertop. This unit has the added advantage of providing a place to put hot pans from a cooking surface.

Kitchen ventilation systems

Every kitchen needs a good ventilation system. Odors, smoke, excessive heat, and moisture are by-products of cooking. A properly installed ventilation system removes them before they fill up the kitchen and spread throughout the home. Perhaps a more serious consideration is that cooking produces more than 200 pounds of grease-laden moisture in the average home each year, according to the Home Ventilating Institute. Thus, good kitchen ventilation eliminates cleaning and maintenance problems in the kitchen and nearby rooms.

While wall or ceiling exhaust fans are still being made, hood-fan combinations are installed in nearly all kitchens today. The reason is because the hood-fan units are a far more efficient way to exhaust the odors and smoke than exhaust fans.

Types of hood fans

There are two basic types of hood ventilating systems available: ducted and nonducted. They come in various grades, prices, sizes, styles, and colors. In addition, some built-in cooking equipment, as discussed later in this chapter, comes complete with its own ventilating system.

With a ducted system, venting is achieved to the outside by the shortest and most direct route possible to keep the efficiency of the system at its highest. Most hoods offer two-way discharge, horizontal or vertical through standard-sized ducts. A hood may be installed directly on an outside wall or it may carry its discharge via ducts through the attic (to roof or eave) or through a soffit or false beam to the outside wall (Fig. 6.18). But remember when laying out the duct work that extra elbows drastically reduce the flow of air. Also, avoid changes in the duct size along the run, since duct-size changes waste fan power and create a place for grease to collect.

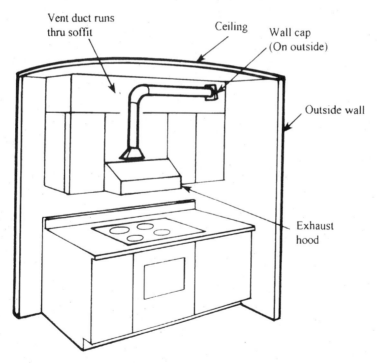

Figure 6.18a Exhaust hood vented through the soffit to the outside. (*Leon E. Korejwo*)

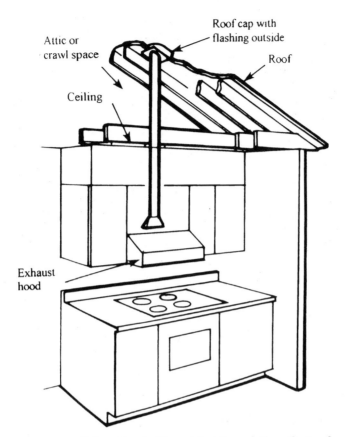

Figure 6.18b Exhaust hood with vent duct through the ceiling and roof. (*Leon E. Korejwo*)

Nonducted ventilating systems can be hung on the wall above a freestanding range or cooktop or can be suspended beneath the wall cabinets over the cooking surface. They are easy to install and ideal for any location where ducting to the outside is either difficult or not desirable (such as over an island installation or in apartment kitchens). They draw air through a filtering system and then return it to the room, filtering out the smoke, grease, and odors with special charcoal filters. While they do a fairly good job of removing smoke, grease, and odors, the ventless or nonducted hoods are not effective on heat and moisture.

Certain range models, as mentioned earlier in the chapter, come with a built-in vent system. The vent system uses channels behind the top oven to the cooktop platform. These built-in vent systems either require ducts or are ductless.

As for the hood fan, four types of blades are used:

1. The *propeller* type is a fan with three or more blades, much like an ordinary portable household fan. Its blades are pitched to deliver a maximum volume of air against relatively low resistance. It is often less expensive and

uses less power than others, but it is recommended where high air resistance is a factor, as with a duct having numerous elbows.

2. The *axial flow* type has blades resembling the contour of an airplane propeller. It is actually a refinement of the propeller principle. It works by driving air at a high velocity through a close-fitting tube. This type is used more often commercially than residentially.

3. The *centrifugal blower* is commonly called a "squirrel cage" fan. Its open revolving drum sucks in air and discharges it at right angles through blades (or vanes) set vertically around the circumference of the drum. This fan has a higher-powered motor than the propeller type.

4. The *mixed-flow impeller* combines the principles of squirrel cage and propeller types. On the intake parts, the blades are pitched like a propeller type. On the discharge ends, they are like blower blades. As a rule, this type takes less power than the squirrel-cage blower.

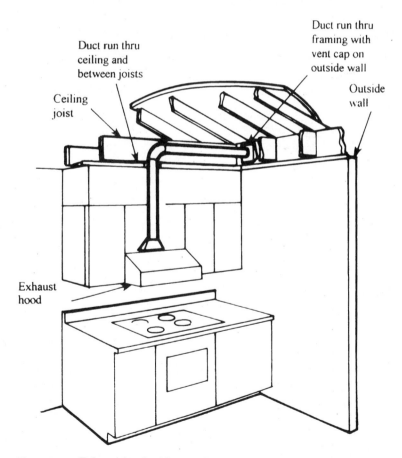

Figure 6.18c Exhaust hood with vent duct above the ceiling to the outside wall. (*Leon E. Korejwo*)

Today, most good hoods employ centrifugal blower fans. They seem to be the most effective. They overcome resistance most efficiently, thus giving quieter operation. The propeller fan is frequently used where there is little or no resistance in the system (no ductwork involved).

There are three basic ways in which fans are combined with range hoods:

- The fan is built right into the hood itself and does not take any of the cabinet space above it.

- The fan is not in the hood but is instead installed in the cabinet space directly above the hood.

- An exhaust fan is simply mounted in the wall above the cooking surface of the range and is vented directly to the outside.

When selecting a ventilating hood and fan, there are two important considerations: noise level and capacity. The latter is measured in cubic feet per minute (CFM) and refers to the volume of air the hood fan or blower removes from the room each minute of operation. Unlike a wall or ceiling exhaust fan, the hood fan's capacity is normally not based on the size of the kitchen; rather, it is determined by the size of the hood. Since the heat, odor, moisture, and smoke are trapped at their source, the fan capacity is independent of the kitchen size.

According to United States Department of Housing and Urban Development/Federal Housing Administration (HUD/FHA) and Home Ventilating Institute (HVI) standards, the minimal requirement for a range placed against a partition is a hood fan with a capacity of 40 CFM per lineal foot of hood, or about 120 CFM for a 36-inch (91-cm) hood. Island or peninsula locations for the same size call for 50 CFM per lineal foot, or 150 CFM for a 36-inch (91-cm) hood. The HUD/FHA minimums for range hoods apply when the natural ventilation of the kitchen is below minimum and when ventilation is installed optionally.

While these minimum standards may do the job, most kitchen planners, as well as the Small Homes Council/Building Research Council, recommend that the hood-fan capacity should be 100 CFM for each lineal foot of hood length, except for island and peninsula hood lengths. In other words, a 36-inch (91-cm) hood installed against a partition would require a fan with a capacity of 300 CFM, and the same size unit in an island or peninsula location should have a capacity of 360 CFM.

For top performance of any indoor barbecue, the HVI suggests hood fans that have at least a 600 CFM rating. The hood should extend at least 1 inch beyond each edge of the barbecue unit. In the case of an island or peninsula unit, where crossdrafts are a factor, it should extend at least 3 inches (8 cm) beyond each edge. The bottom of the hood should be from 16 to 24 inches (41 to 61 cm) above the barbecue, with 30 inches (76 cm) the absolute maximum.

In some grill arrangements, the vent is right on the cooking surface, eliminating the need for a hood. Instead, the surface vent pulls heat, odors, smoke, and grease downward, then ducts them outdoors. If the cooking top is on an outside wall, installation of a vent is simple. If not, ducts can be run under the floor to an outside wall. Incidentally, all CFM ranges on fans bearing the HVI label are certified after independent tests at Texas A & M University and therefore can be compared dependably with one another. CFM ratings can easily be identified either by the manufacturer's literature or by the label on the unit itself.

The kitchen wall or ceiling exhaust fan capacities are determined by the volume of the room and the number of times each hour that this quantity of air is removed from the room. The unit of measure is also cubic feet per minute (CFM). It is commonly recommended that the fan should be capable of exhausting the room 15 times per hour, equivalent to one air change every four minutes. Thus, in a kitchen that is 10 feet wide and 15 feet long with an 8-foot ceiling height, the capacity of the exhaust fan should be 300 CFM.

$$\text{Fan capacity} = \frac{\text{Volume (cubic feet in kitchen)}}{\text{Number of minutes for each air change}}$$

$$= \frac{10' \times 15' \times 8'}{4 \text{ minutes}}$$

$$= \frac{1200 \text{ cubic feet}}{4 \text{ minutes}}$$

$$= 300 \text{ CFM (cubic feet per minute)}$$

HVI members commonly translate this information as follows on the product tag or in literature for wall and ceiling exhaust fans:

This fan ventilates a residential:

Kitchen	80 square feet
Bathroom	150 square feet
Laundry	200 square feet

Many vent hoods have two-speed motor controls. Some have three set speeds, and a few come with solid-state controls that permit the user to dial an almost infinite number of speeds. Most also have built-in lighting (some halogen), and nearly all feature easy-to-clean, fire-retardant, full metal construction. Some have an indicator light that goes on after 30 hours of operation as a reminder to clean filters, which are usually dishwasher-safe aluminum. Cleaning is very important since the cleanliness of the system affects the efficiency of the entire unit.

As mentioned earlier, noise is a very important consideration. Some ratings, a recognized method of measuring loudness in numbers that correspond to the way people hear them, are given to vent hoods and exhaust fans. They help greatly to make accurate comparison of sound levels among different models. Fans rated at 4 sones (the level of conversation speech) make half the sound as those rates at 8 sones. All fans bearing the HVI label are certified for sones ratings as well as for CFM specifications. No certified kitchen exhaust fan or hood (up to 500 CFM capacity) can exceed 9 sones. Every certified fan must be rated for sones at maximum air delivery, but multispeed fans may also be rated for sones at lower speeds. Dual ratings permit selection of whatever combination of air delivery and quietness is desired in steps of 10 CFM and 0.5 sones. In many cases, the multispeed fan is a good solution, offering heavy-duty ventilation when needed and low-speed quietness at other times.

In addition, today's ventilation systems have touchpad control panels that feature a large LED operation light that makes it easier to select blower and light settings. Some have electronic sensors to automatically adjust the blower to the cooking conditions (i.e., to high speed when excess heat is detected), and automatically turns the blower to the last setting used. Many manufacturers have created systems with variable speeds and high-capacity centrifugal blowers with high-efficiency motors to produce an exhaust of 600 CFM at only 5 sones. They have downdraft ventilator systems for island and peninsula cooking where updraft venting is impractical. Downdraft ventilation eliminates the need for an overhead ventilating hood and fan.

Contoured range hoods feature distinctive curved hoods and glass visors to improve the cooktop view and contain cooking vapors until they are exhausted (Fig. 6.19). Hoods come in designer colors, durable baked enamel finish, or highly-polished stainless steel. Many hoods can slide in and out; some have wood finishes to match the cabinets, and others actually hide away into the cabinet wall.

Figure 6.19 A distinctive glass hood and stainless rail wrap around the body of the seamless-style range hood. (*Broan*)

Some ventilation systems are so advanced that, at the touch of a finger, they rise many inches above the cooktop to vent smoke, odors, and vapors. With the touch of a button, they can automatically lower flush with the cooktop. There are high-performance, restaurant-range, commercial-look systems on the market as well.

For best efficiency, distance from the cooking surface to the bottom of the hood should be 24 to 30 inches (61 to 76 cm). It is very important that installations be done by qualified installers who are knowledgeable in ventilation.

The microwave and ventilating hood may be combined into one appliance to vent away smoke from the range or cooktop below. This combination should only be done when the cook is tall enough to use them safely and conveniently.

7

Developing a design from concept to plans

The final step in planning a kitchen is making the plan or blueprint. First make a rough sketch of the room, usually on a grid or graph paper. Not all kitchen builders are responsible for preparing kitchen plans. However, all need to be able to read and understand them. If you are not responsible for the initial field measurements, be certain to verify any dimensions given to you before starting the project. Even if the original house plans for a remodeling project are still available, it is best for you to physically take measurements and check the room conditions. Changes in floor and wall levels may have taken place, and there may have been revisions that were not recorded on the plans. The verification of existing measurements and conditions in the kitchen is crucial to the success of the installation project. A video camera offers an excellent visual record of a kitchen layout.

Measuring the kitchen

To take a room's rough dimensions, a folding wood carpenter's rule is more accurate and easier to handle than a retractable tape measure. A tape requires someone to hold one end to keep it taut. It is, however, handier for taking full-length room dimensions. A clipboard with an 8-by-10 quadrille pad, pencil with an eraser end, a red pencil for special notations, and a 6-inch (15-cm) architect's rule are needed for recording the plan and dimensions. Also, a carpenter's level and a large steel square may be needed. If your budget allows, look into a laser device used to measure rooms.

The first step in making the rough sketch is to measure the room. Start at any corner above the counters at a comfortable height, measure the room, and mark all dimensions in a rough sketch. Take the overall width and length dimensions of the room, and draw its outline on the quadrille pad, using the blue lines as a guide. It is not necessary to do the drawing in scale at this time. Just approximate the location of doors and windows. If the dimensions are going to be turned over to a draftsman, redraw the sketch at the office, if there is any doubt that it is understandable.For accuracy, measure and show on your drawing all dimensions in inches or fractions of inches, never a mixture of feet and inches. It is easy to make mistakes when figuring in feet and inches. Remember that accuracy and correct information helps you, the manufacturer, and the client to save time and money. When using a calculator, the fractions must be converted to decimals of an inch. A conversion chart is handy.

After placing the room outline on the paper, take detailed measurements. Place the rule against the wall at a height of about 36 inches (91 cm). Begin at one corner and measure to the nearest door or window. Measure to the edge of the trim. Mark off this distance on the outline and also note the exact inches and fractions between arrow points as in the sketch. Also measure and note width of the trim and check the possibility, if needed, for cutting down the trim.

Next, measure the width of the door or window, including casing or trim. Mark off this distance on the outline and note the distance between arrow points. Note on the sketch the swing of each door and label them to show where they lead—such as to the dining room, hall, or outdoors. Proceed in this way around the room. Note all critical measurements and irregularities—chimneys, closets, radiators, or any other similar structures. When finished, add the detailed dimensions of each wall. Then take overall wall-to-wall dimensions. Make sure that these figures are equal.

Measure from the floor to bottom of the window stool (the underside of the inside window sill). This measurement is essential to show if units can go beneath windows and also to determine the height of the backsplash. If there is not sufficient clearance under the stool, check and note the possibility of cutting it down, if necessary. If impossible, use a window notch in the backsplash. Also measure from the bottom of the stool to the top of the window trim. Show these measurements in the space for the window.

Measure the height of the room from floor to ceiling. Enter this figure in a space below or to one side of the sketch. It is a good idea to take this measurement at two diagonally opposite points to see what allowance may have to be made for a floor that is out of level. This method won't indicate that the floor is out of level if the ceiling also slopes at the same angle. Therefore, as an added precaution, check the floor with a level.

Check also for any inequalities in the walls that must be considered in building the sink top or counter and in planning proper use of fillers. Check corners with a square and walls with a level. Measure in several spots in case the floor is uneven. Be sure to measure from the floor to the bottom of the window sill and from the top of the window trim to the ceiling. If there are discrepancies, be sure to record them for consideration when planning the cabinetry and counters.

It seldom happens this way, but in new construction the walls should be finished before you measure (sometimes this is referred to as clear dimensions). If you must measure before the interior is finished, measure the actual frame dimensions and add what the window and wall materials will be. Allow for the thickness, and then add a little extra.

Show the size and location of heat, air-conditioning, and ventilating outlets, electrical and plumbing outlets, pipes, chimneys, or any other obstructions that cannot be changed and may affect the plan. If you are familiar with architectural symbols, use them. Otherwise, make careful notations and add the symbols when redrawing the plan at the office.

When taking the measurements, try to determine what is inside the walls—gas, electric, or water pipes; ductwork; plumbing stacks; and chimneys—by reference to blueprints, if available, or by careful examination of the area all around the kitchen. That is, on remodeling jobs, it is necessary to check and take notes on the following:

1. Present drain location and distance from the plumbing stack or drain vent pipe
2. Air ducts concealed in walls, soil pipes, vent pipes, and water pipes
3. Exterior walls for construction type
4. Heating and location of ducts, radiators, pipes, etc.
5. Electrical system and size and type of service, condition of wires, and general condition
6. Any unusual situations
7. Which walls are bearing or nonbearing
8. Size and style of doors and windows to be used

On new construction, when finished dimensions are established, the kitchen planner is generally limited to the existing conditions, since any changes at this point would prove quite costly. In the case of a complete remodeling job, considerable latitude is usually allowed, and, consequently, more specific, detailed information is required. Remember, it is less expensive to plan around plumbing and heating systems and other obstructions than to relocate them.

Construction drawing sets

You may be working from a prepared set of plans or you may be preparing them yourself. In either case, a clear knowledge of how to read and understand the kitchen plan or blueprints is vital to the installation project. The building industry has an established set of symbols and material indications that are used consistently on building plans. You need to become familiar with the symbols and standards to successfully use the building plans. Usually a list of the symbols and material indications, as well as any abbreviations (i.e., different kinds of switches, lights, appliances, etc.) used on the plans, can be found on the title page (first sheet of the building plan). Before trying to read a set of plans or before beginning any work, be sure you understand the symbols used on the drawings. Take the time to become fully acquainted with the work on the plans.

Few people know how to understand and read blueprints or plans; therefore, fully explain layouts, etc., to your clients, even if they are unable to read the plans. In almost all instances, your clients will ask you about the drawings, and therefore, you need to be able to answer them. As well, it is extremely important that everyone involved with the project be given a set of drawings to review before starting work. Have them review the entire set against the work they will be performing for any possible conflicts, etc.

In many cases, plans of the project are required by the local building inspector or zoning officer before a permit is issued. A set of plans may be reviewed by the municipality to check for code compliance before issuing a building permit. The responsible party may be a zoning officer. If so, you must submit clear and complete drawings (in the instance of a remodeling project) of the existing floor plan as well as the new one. You may need to show areas adjacent to the kitchen so inspecting municipal officials can determine how the project will affect the rest of the house.

Drawing sets for kitchen projects with more limited scopes of work do not necessarily contain all of these items. However, if your project requires a building permit, many of these sheets are required regardless of project size.

Residential building projects ranging from new home construction to simple remodeling of an existing room are universally organized with drawings. The extent of a drawing or blueprint set is determined by municipal code compliance as well as the size and complexity of the job. A kitchen project for new building construction is only one part of the full construction drawing set. The designer of a kitchen needs certain drawings before he or she can start the design or proposal. The best way to demonstrate these drawings would be to discuss the largest project situation a builder must address. The reader may then decide what is needed for individual jobs as they occur. A well-developed set of plans provides you with all the information required to complete your installation work. It should generally include the following:

A building drawing set for complete new home construction includes:

1. *Title page*

2. *Site plan* (prepared by surveyor or engineer)

3. *Foundation plan* (prepared by an architect, builder, or designer)

4. *Floor plan* (prepared by an architect, builder, or designer)

5. *Elevations*, interior and exterior (prepared by an architect, builder, or designer)

6. *Structural framing plan* (prepared by an architect or builder)

7. *Construction details and sections* (prepared by an architect or builder; the kitchen drawing set is included here)

8. *Mechanical plans* (prepared by a subcontractor)
    a. Electrical (prepared by electrician or mechanical contractor)
    b. Plumbing (prepared by plumber or mechanical contractor)
    c. HVAC (prepared by an HVAC engineer or a mechanical contractor)

9. *Reflected ceiling plan (prepared by an architect, builder, or designer)*

10. *Specifications (prepared by an architect, builder, or designer)*

A kitchen drawing set includes:

1. *Title page with specifications* (optional)

2. *Kitchen floor plan* (Fig. 7.1)

3. *Kitchen soffit plan* (if highly detailed soffit work is required, include this drawing)

4. *Kitchen elevation drawings* (Fig. 7.2)

5. *Kitchen perspective drawings* (Fig. 7.3 through 7.5)

6. *Kitchen countertop plan*

7. *Kitchen mechanical drawings* (locations of electrical outlets, switches, appliance hookup, lighting, location and size of fixtures, plumbing, location of drain and pipe connection, low voltage, intercom, video, and security system, etc.)

A kitchen drawing set can be prepared by the designer, cabinet maker, supplier, etc. The drawings listed are an all-inclusive set for the full construction of a new residence. Remodeling may also require a demolition drawing, showing any walls, structure, or mechanical equipment to be removed before proceeding with new construction.

The kitchen designer's responsibilities may vary as to what information they provide. For each contract, as little as a floor plan may be adequate. But in some cases, a full set of design drawings may be agreed upon, including a kitchen floor plan; a kitchen mechanical plan (showing desired appliances, lighting, and receptacle and switch locations, with a legend for symbols and power requirements, location of water and drainage plumbing, as well as gas hookup locations), a kitchen soffit plan; and kitchen elevation drawings.

As reported by NKBA, by standardizing floor plans and presentation drawings, the kitchen remodeler:

- Limits errors caused by misinterpreting the floor plans

- Avoids misreading dimensions, which can result in costly errors

- Prevents cluttering floor plans and drawings with secondary information, which often make the documents difficult to interpret

- Creates a clear understanding of the scope of the project for all persons involved in the job

- Presents a professional image to the client

- Permits faster processing of orders

- Simplifies estimating and specification preparation

- Helps in the standardization of uniform nomenclature and symbols

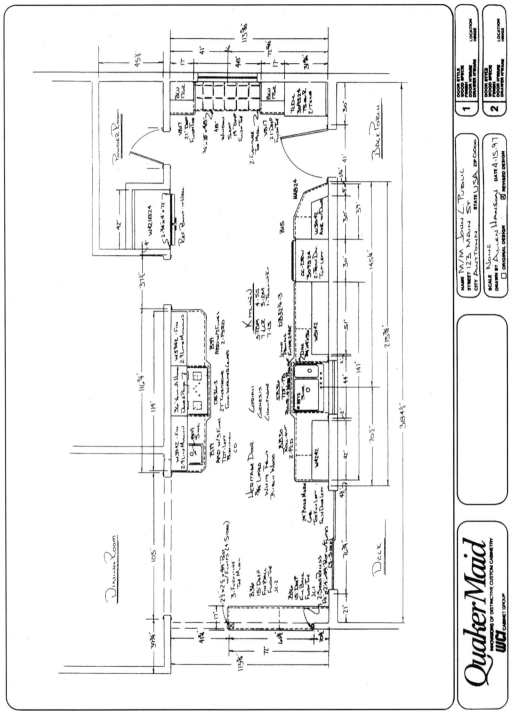

Figure 7.1 Kitchen floor plan. *(QuakerMaid Kitchens of Reading, Inc. / Allen V. Hanson, Designer)*

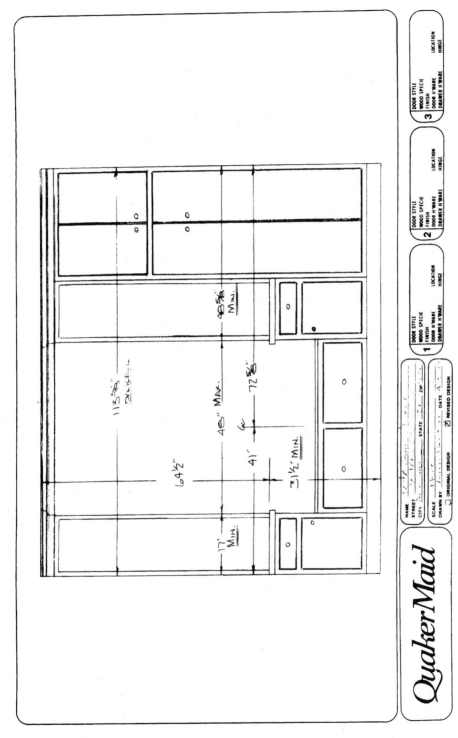

Figure 7.2 Kitchen wall elevation plan. (QuakerMaid Kitchens of Reading, Inc. / Allen V. Hanson, Designer)

Figure 7.3 Kitchen perspective. (*QuakerMaid Kitchens of Reading, Inc./Allen V. Hanson, Designer*)

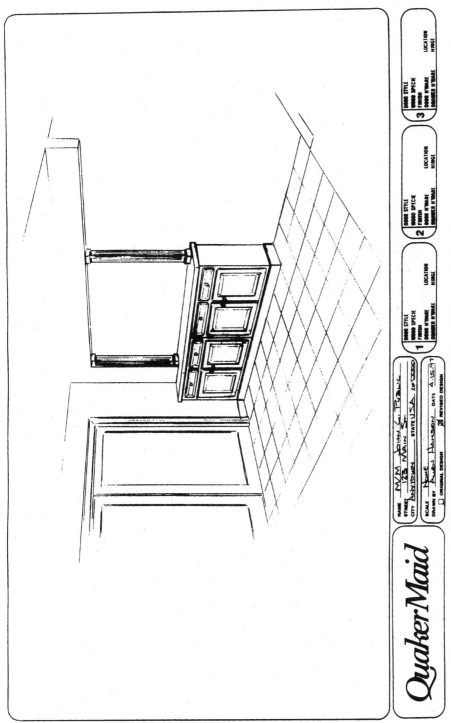

Figure 7.4 Kitchen perspective. (*QuakerMaid Kitchens of Reading, Inc./Allen V. Hanson, Designer*)

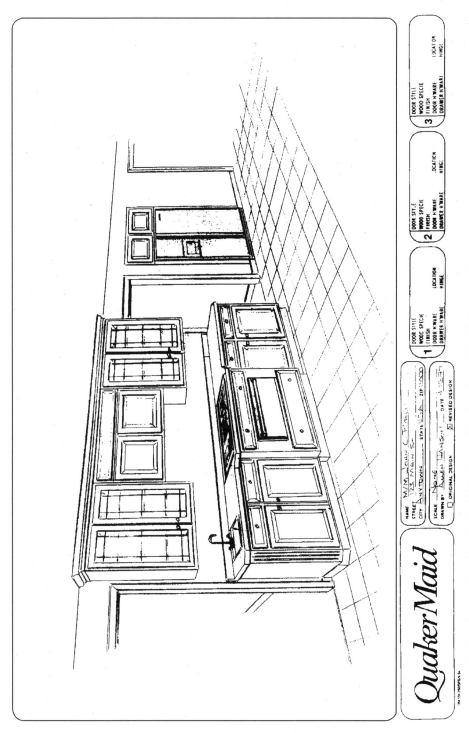

Figure 7.5 Kitchen perspective. *(QuakerMaid Kitchens of Reading, Inc. / Allen V. Hanson, Designer)*

Title page

The first page in a building plan set is the title page. It may include some or all of the following: the client's name; the location of the project; the name of the building or project; names of consulting engineers, designers, architects, etc.; a key to the symbols for materials, appliances, and cabinet information; a list of abbreviations; an index of the drawings contained in the whole building plan set; the number of sets distributed and to whom; a space for municipal approvals, one for each body responsible for review of plans; and local, planning commission, building inspector, county, or city stamps and signatures. The title page is optional and, therefore, all of this information may be included on the floor plan.

Site plan (or plot plan)

The site plan is an overhead view of the entire property around a building showing lot lines and layout of the land prior to construction. To the kitchen builder, one of the most important aspects of the site plan is the location of electric, water, sewer, gas services, and excavations. They are located by a surveyor with plotted points or dimensions and are subject to municipal approval if it is a new building or remodeling addition.

Foundation plan

This plan shows the extent of the building footings and foundations and generally describes how the building structure is attached to the foundation wall.

Floor plan

There are many types of floor plans, ranging from very simple sketches to completely dimensioned, detailed working drawings. A floor plan is drawn as if viewed from overhead, showing the outline and walls and partitions of a building as you would see them if the top of the house were cut off horizontally about 4 feet above the floor line (refer back to Fig. 7.1). Most people can envision the home as if they were looking down into the rooms and walking through them.

Prepare the floor plan layout of the kitchen to scale—either $\frac{3}{8}$ inch to the foot ($\frac{3}{8}$" = 1'0) or $\frac{1}{2}$ inch to the foot ($\frac{1}{2}$" = 1'0"). Metric dimensions for floor plans should be drawn to a scale of 1 to 20 (i.e., 1 cm = 20 cm) to the foot. This plan is the central reference point for all of the other construction drawings, and it is here that indications for details, sections, and schedules are referenced. The floor plan is usually the simplest of the drawings to read and understand. The rooms should be divided by "break lines" and should show all major structural elements (i.e., walls, door swings, door openings, windows, archways, stairs, equipment, and partitions) with adjoining areas indicated and labeled. It is also helpful to note the direction of joists, mark any bearing walls, and sketch in other features that might affect the remodeling plans, although these structural details may be shown on a separate plan to avoid crowding too much on the floor plan. When possible, it should depict the entire room. When the entire room cannot be depicted, it must show the area where cabinetry and appliances are permanently installed. Indicate all measurements, obstructions, and peculiarities.

When the kitchen is designed, all cabinets, appliances, and other equipment are drawn in place and labeled with proper nomenclature on the floor plan. Use a ruler or T-square to draw horizontal lines, a triangle to draw vertical lines and right angles to horizontal lines, and a compass for drawing the doors' directions of swing. Complete the floor plan using your sketch as a model. Drawing to scale reveals many problems on the board, saving many hours on the site later. When scaled figures do not agree, recheck the field sketch for errors.

Review all original survey notes and the householder's desires. Lightly lay in the three work center areas as per the basic layout. Lay in a dashed line (- - -) along the walls of the kitchen, 24 inches (61 cm) from the wall (remember that an unsquare room may lead to problems). This line indicates the base cabinet line, while countertops are depicted using a solid line. Lightly lay a line 12 inches (30 cm) from the walls; this line indicates the area for the wall cabinets.

Lay out the perimeter of the kitchen from the measurements made on the rough sketch (Fig. 7.6a). Be sure to transfer measurements accurately, and include all the details noted while measuring the room. Try several room arrangements by drawing on tracing paper or by making cutouts from templates and moving them around (Fig. 7.6b through 7.6e). Figure 7.7 shows sample templates of cabinets and general appliances. With the use of a photocopy machine, these templates may be increased or reduced to match a specific scale. As noted, these are only generic. The dimensions of these appliance templates—$\frac{3}{8}$-inch-to-1-foot scale—are typical. Be sure for the final drawing, however, to consult specifications in manufacturer's literature for cutout dimensions, required clearances, and exact size of appliances and cabinets that are going to be used. Any appliances that are presently in the kitchen and are going to be used in the new one should be measured accurately, with their indicated door swings.

According to NKBA's *Graphics and Presentation Standards for Kitchen Design*, the following dimensions should be shown on every floor plan:

- Overall length of wall areas to receive cabinets, countertops, fixtures, or any equipment occupying floor or wall space. This dimension should always be the outside line.

- Each wall opening (windows, arches, doors, and major appliances) and fixed structures (chimneys, wall protrusions, and partitions) must be individually dimensioned. Dimensions are shown from outside trim. Trim size must be noted in the specification list. Fixtures such as radiators remaining in place must be outlined on the floor plan. These crucial dimensions should be the first dimension line.

- Ceiling heights should appear on the floor plan. A separate plan for soffits is required when the soffit is a different depth than the wall or tall cabinet

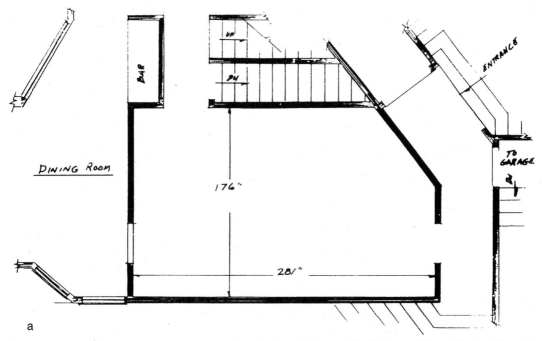

DINING ROOM

176"

281"

BAR

UP

DN

ENTRANCE

TO GARAGE

a

Figure 7.6a A rough sketch of the basic floor plan of a large kitchen. *(The House and Home Kitchen Planning Guide, Robert Scharff)*

below. A separate soffit plan is recommended when the soffit is to be installed prior to the wall or tall cabinet installation.

- Additional notes must be included for any deviation from standard height, width, and depth (cabinets, countertops, etc.) Be sure to indicate the thickness of the walls and the shape of the sink (square or angular). Use dotted lines to show appliances in open positions.

- The exact opening must be given in height, width, and depth for areas to be left open to receive equipment, cabinets, and appliances at a future date.

- Items such as island and peninsula cabinets must be shown with the overall dimensions given from countertop edge to opposite countertop edge, tall cabinet or wall. The exact location of the structure must be identified by dimensions that position it from two directions—from return walls or from the face of cabinets or equipment opposite the structure.

Cabinet type, shape, and size are identified by the manufacturer code or nomenclature (see Chap. 5), i.e., W 24 36, which is written within that cabinet's drawn outline on the kitchen plan. Any molding, trim, or finishing pieces are called out adjacent to the cabinet on which they are to be attached; an arrow points to their location.

Another method is to designate an encircled number to each item and locate that number on the floor plan and elevation. The information would

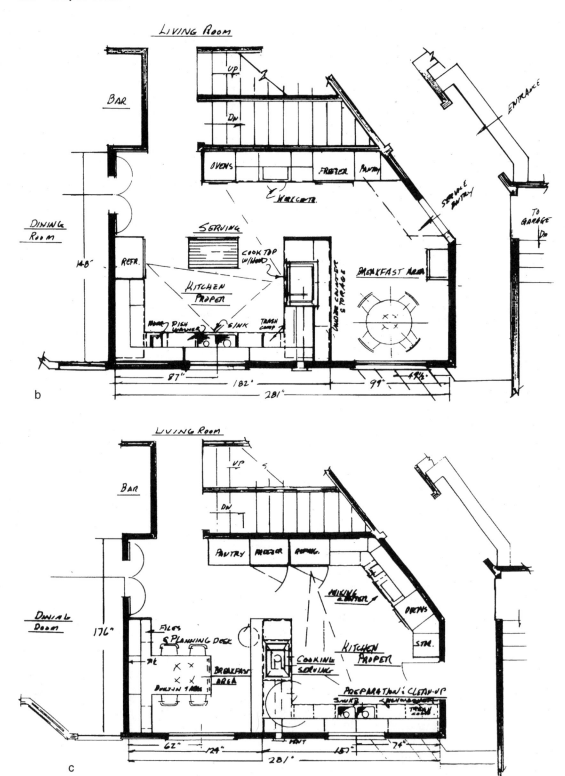

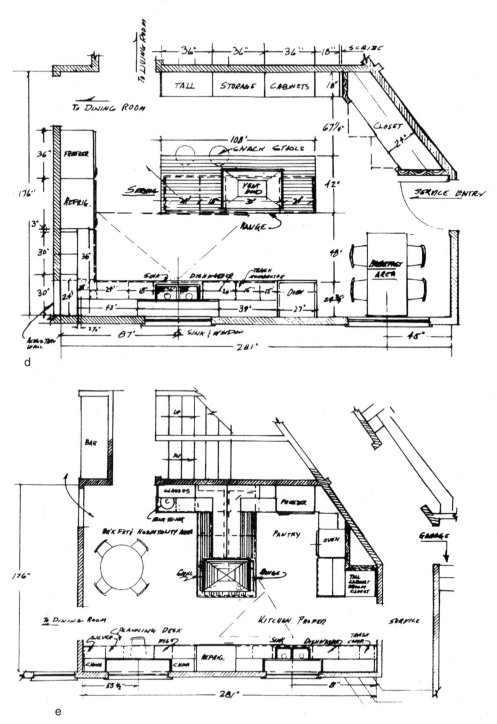

Figure 7.6b through 7.6e Rough sketch of the basic floor plan of a large kitchen: four different ways of laying out a kitchen in basically the same area. *(The House and Home Kitchen Planning Guide, Robert Scharff)*

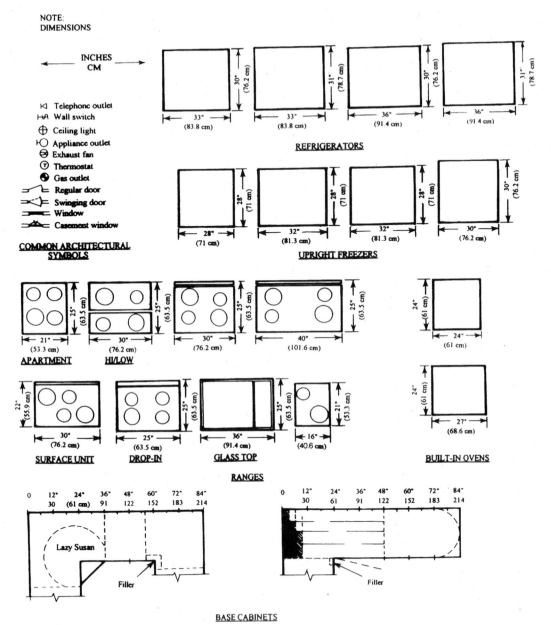

NOTE:
DIMENSIONS

← INCHES →
CM

⋈ Telephone outlet
⊢A Wall switch
⊕ Ceiling light
⊢○ Appliance outlet
⊖ Exhaust fan
Ⓣ Thermostat
● Gas outlet
⌐= Regular door
⨯⌐= Swinging door
≡ Window
⟐ Casement window

COMMON ARCHITECTURAL
SYMBOLS

REFRIGERATORS

UPRIGHT FREEZERS

APARTMENT   HI/LOW

SURFACE UNIT   DROP-IN   GLASS TOP

BUILT-IN OVENS

RANGES

Lazy Susan

Filler

Filler

BASE CABINETS

KITCHEN COMPONENT TEMPLATES
SIZES ARE ESTIMATES ONLY
ACTUAL DIMENSIONS VARY.
CONSULT MANUFACTURER SPECS.

Figure 7.7a Typical templates for kitchen components. Check the manufacturer's specifications for exact dimensions. Scale: $\frac{3}{8}$" = 1'. *(The House and Home Kitchen Planning Guide, Robert Scharff)*

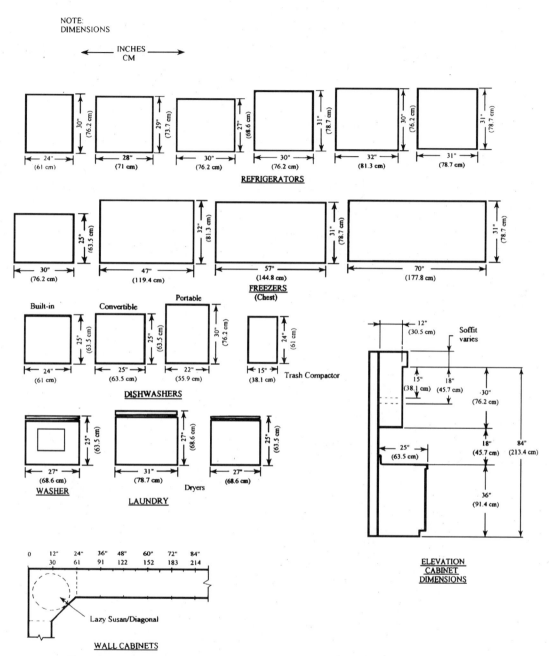

NOTE:
DIMENSIONS

INCHES
CM

REFRIGERATORS

FREEZERS
(Chest)

DISHWASHERS

Trash Compactor

WASHER

LAUNDRY

Dryers

ELEVATION
CABINET
DIMENSIONS

Lazy Susan/Diagonal

WALL CABINETS

Figure 7.7b Typical templates for kitchen components. Check the manufacturer's specifications for exact dimensions. Scale: $\frac{3}{8}$" = 1'. *(The House and Home Kitchen Planning Guide, Robert Scharff)*

then be listed in a table located on the same drawing sheet or on another sheet entirely. The table would organize size, style, and pertinent information without cluttering the floor plan. This method also aids in gathering the ordering data for purchasing.

Cabinet layout

The simplest method of laying out kitchen cabinets on the floor plan is the subtraction method. All the room measurements should be converted to inches (if not already done) and the size (width) of each cabinet or appliance used is deducted (subtracted) from the space available (wall dimension in inches). After deducting the width of all items, there should be zero inches of space remaining, and you will have the exact number of inches of cabinets needed to fill the wall space—neither too many nor too few.

To understand the subtraction method concept, study the illustrations in Fig. 7.8. For our purpose, let us assume that the U-shaped kitchen includes the following items:

- 33-inch (84-cm) double-bowl sink (requires a 36-inch cabinet)
- 36-inch (91-cm) refrigerator
- 25-inch (64-cm) drop-in range (can use any size base cabinet above 30 inches, or 76 cm, in width). The back wall is 141 inches (358 cm) with a window in the center. The right leg of the U is 87 inches (221 cm), while the left leg is 75 inches (191 cm).

As shown, the total space from the sink center (or window center) to the right wall measures 66 inches (168 cm). Deducting 18 inches (46 cm), or half of the 36-inch (91-cm) sink cabinet, leaves 48 inches (122 cm) of space to be filled. The first cabinet to be selected, once the sink cabinet has been determined, is the base corner cabinet. If a 33-inch (84-cm) rotary (lazy Susan or revolving) base corner cabinet is selected, this leaves 15 inches (38 cm), Space A, to fill.

An all-drawer unit should be next to the sink for convenient storage of the utensils used at this cleaning center: cutlery, dishtowels, etc. Since a 15-inch (38 cm) all-drawer cabinet is available, the back wall of the U-shaped kitchen plan is neatly filled from the center of the window to one corner.

To complete the back wall, let us plan another convenient 33-inch (84 cm) rotary corner cabinet. Space B in the illustration can be neatly filled with a 24-inch (61 cm) cabinet or dishwasher.

Now that the entire back wall is filled, we are ready to fill the 87-inch (221 cm) right leg of the U-shaped arrangement. The rotary corner cabinet previously selected occupies 33 of that 87 inches (221 cm), and the refrigerator occupies 36 inches (91 cm), so we have only 18 inches (46 cm), Space D, to fill. Since an 18-inch (46-cm) base cabinet is available, the right leg is complete.

The computation for the left leg is somewhat different. The illustration shows a 33-inch (84-cm) rotary corner cabinet plus a 36-inch (91 cm) base cabinet for the drop-in range, leaving Space C to be filled. Since there is no

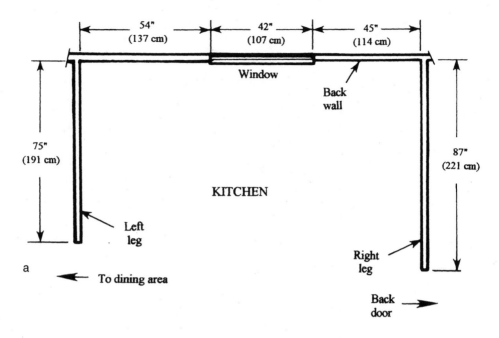

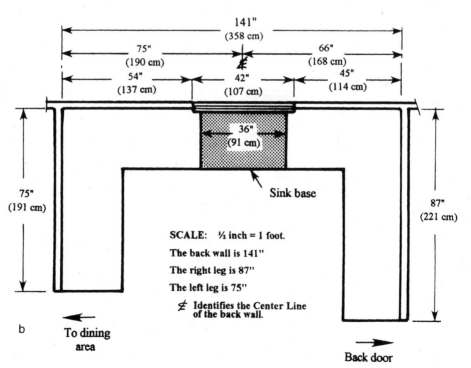

Figure 7.8 The subtraction method: laying out the base cabinets on the back wall. *(The House and Home Kitchen Planning Guide, Robert Scharff; Leon E. Korejwo, illustrator)*

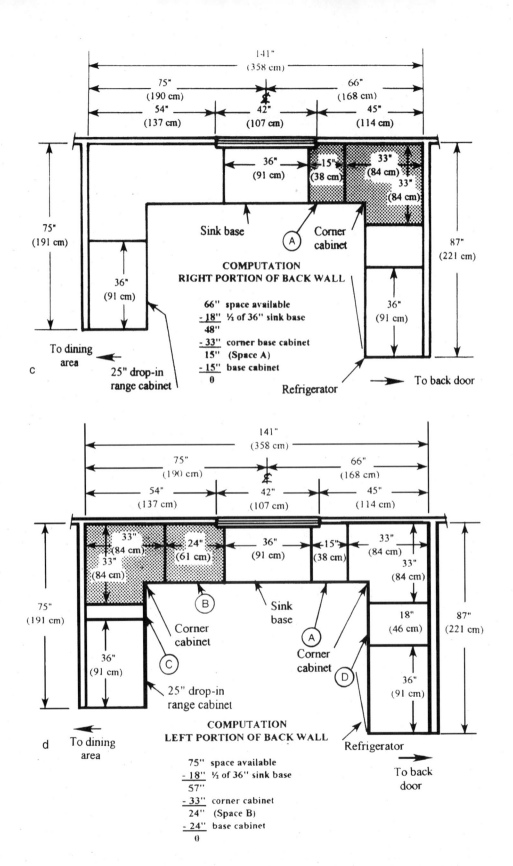

141"
(358 cm)

75"
(190 cm)

66"
(168 cm)

54"
(137 cm)

42"
(107 cm)

45"
(114 cm)

36"
(91 cm)

15"
(38 cm)

33"
(84 cm)

33"
(84 cm)

Sink base

Corner
cabinet

(A)

75"
(191 cm)

36"
(91 cm)

87"
(221 cm)

36"
(91 cm)

**COMPUTATION
RIGHT PORTION OF BACK WALL**

66" space available
- 18" ½ of 36" sink base
48"
- 33" corner base cabinet
15" (Space A)
- 15" base cabinet
0

To dining
area

c

25" drop-in
range cabinet

Refrigerator

To back door

141"
(358 cm)

75"
(190 cm)

66"
(168 cm)

54"
(137 cm)

42"
(107 cm)

45"
(114 cm)

33"
(84 cm)

24"
(61 cm)

36"
(91 cm)

15"
(38 cm)

33"
(84 cm)

33"
(84 cm)

33"
(84 cm)

(B)

Sink
base

(A)

18"
(46 cm)

75"
(191 cm)

Corner
cabinet

(C)

36"
(91 cm)

Corner
cabinet

(D)

36"
(91 cm)

87"
(221 cm)

d

To dining
area

25" drop-in
range cabinet

**COMPUTATION
LEFT PORTION OF BACK WALL**

Refrigerator

To back
door

75" space available
- 18" ½ of 36" sink base
57"
- 33" corner cabinet
24" (Space B)
- 24" base cabinet
0

6-inch (15 cm) base cabinet (it would be uneconomical) and since fillers should be limited to a maximum of 3 inches (8 cm) to minimize waste storage space, the easiest way to fill this 6-inch (15-cm) space would be to increase the size of the range cabinet by 6 inches (15 cm), substituting a 42-inch (107-cm) cabinet for the 36-inch (91-cm) cabinet previously selected. This completely eliminates Space C (Fig. 7.8e).

To complete the U-shaped kitchen layout, cabinets for the upper wall are needed (Fig. 7.8f). For maximum storage and beauty, the width of each wall cabinet should correspond to the width of the appliance or base cabinet beneath it, except at the end of the arrangement, when open shelves are frequently more attractive and economical. The subtraction method can be used in planning any shape kitchen.

Drawing floor plans and details

To make the final floor plans and details, proceed as follows:

1. Working from the cleanup center, place the sink cabinet, using the centerline of the window as a guide. The sink is generally placed beneath the window; however, this is not a firm and fast rule. If design, function, or aesthetics require it to be elsewhere, place it freely in that area and then concentrate on making that area the focal point of the kitchen. Design around it.

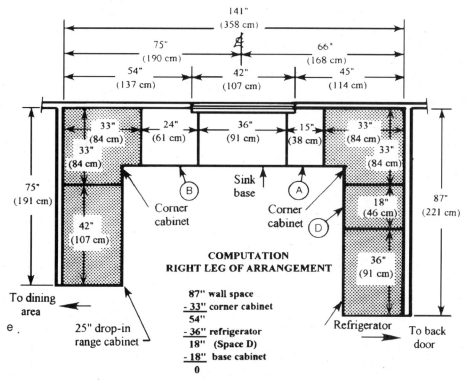

Figure 7.8 *(Continued)* The subtraction method: laying out the base cabinets on the back wall. *(The House and Home Kitchen Planning Guide, Robert Scharff; Leon E. Korejwo, illustrator)*

2. Place the dishwasher next, if one is to be incorporated into the plan. Remember that for better function, if the homemaker is right-handed, it should be placed on the left of the sink; if he or she is left-handed, it should be placed on the right side. This placement should be the order of first preference; however, if a compromise is necessary due to space limitations, place it as near to the sink as possible.

3. Next, work toward the nearest corner, subtracting the size of the dishwasher and half of the sink cabinet size from the measurement of the wall to the centerline of the window.

4. Decide what method and size of the cabinet that is going to be used in turning the corner. Place this cabinet in the appropriate area, noting visually the appearance.

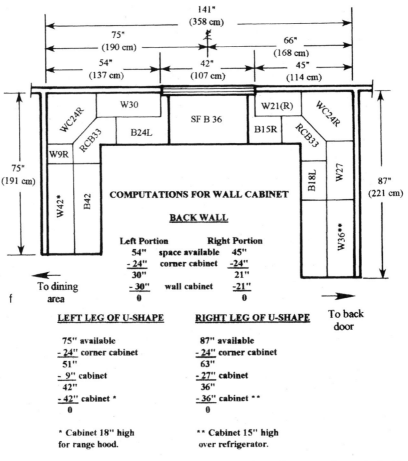

Figure 7.8 *(Continued)* The subtraction method: laying out the base cabinets on the back wall. *(The House and Home Kitchen Planning Guide, Robert Scharff; Leon E. Korejwo, illustrator)*

5. Work back toward the sink and dishwasher area and adjust the location of the sink and dishwasher, if necessary and within acceptable limitations. Place base cabinets, if space permits, between the dishwasher and sink cabinet and the corner cabinet, noting every cabinet that is placed, according to function and type of storage required.

6. Place the range in the area originally designated and adjust to accommodate necessary cabinetry on both sides of the range. The normal flow for right-handed people is from left to right; the opposite is true for left-handed people. When laying out a kitchen plan, work toward this end with this rule in mind.

7. Check wall cabinet function and operation in this area and develop the swing of the single-door cabinets to guarantee full-door opening and accessibility. Place the balance of cabinetry on this leg, again reviewing function and type of storage required. When working between two walls, incorporate a filler somewhere in the layout of this leg to compensate for walls that are out of plumb.

8. After the left leg is completed, return again to the base leg (the leg with sink and dishwasher) and work from the right side of the sink and dishwasher toward the right-hand corner or right leg, whichever the case may be. (We are assuming that the corner nearest the window was the left corner, or left leg.) If another corner cabinet is required, determine the type of cabinet according to its intended use and the size of the cabinet according to the space availability. Place it in the appropriate space required.

9. Place all other necessary cabinetry in line, using the subtraction method, and end the run or right leg with the refrigerator and possibly the utility or broom cabinet.

10. After all of the base cabinets have been placed and the measurements have been checked, proceed to place the wall cabinets in their proper sequence. Use the subtraction method again to prove out wall measurements. Contrary to popular belief, it is not necessary for the stiles of the wall cabinets to align themselves with the stiles of the base cabinets; however, where possible, it tends to balance the design of the kitchen and layout.

A floor plan does not show the customer how the room really looks or how it would actually appear to them as they stand in it. For this effect, we add interpretive drawings that include elevations and perspective drawings. A perspective drawing shows a section or entire kitchen in perspective, from a viewpoint at eye level. You also might add sketches or other types of drawings to help the customer visualize the kitchen. Always draw to a consistent scale, and clearly include the dimensions of any feature and its location. Do not omit dimensions for others to guess or estimate. Someone else's interpretation may be different. Prints or computer printouts are not always accurate enough to measure with a scale. If you have designed correctly, it saves others from wasting time recalculating.

Soffit plan

A soffit plan is not used very much today. However, if highly detailed soffit work is required, include this drawing. If soffits must be built before hanging wall cabinets, show side views and dimensions of cabinets at the walls. Interpretative or perspectives may also include this information. Keep in mind any lights that are to be installed in the soffit.

Construction plans for remodeling

When building remodeling projects, the floor plan or construction may only encompass the area of a home that is to be remodeled and those affected by the remodeling. The purpose of the construction plan is to show the relationship of the existing space with that of the new design. The construction plan is detailed separately so that it does not clutter the floor plan. However, if construction changes are minimal, it is acceptable to combine the construction plan with either the floor plan or mechanical plan. Existing walls are shown with solid lines or hollowed out lines at their full thickness. Wall sections to be removed are shown with an outline of broken lines. New walls show the material symbols applicable to the type of construction or use a symbol that is identified in the legend to distinguish them from existing partitions.

Interpretive drawings

Drawings (orthographic projections) and perspective renderings are considered interpretive drawings. They are used as an explanatory means of understanding the floor plans. The interpretive drawings should never be used as a substitute for floor plans. In the case of a dispute, the floor plans are the legally binding documents. Because perspective drawings are not drawn to scale, many kitchen specialists include a disclaimer on their renderings:

> This drawing is an artistic interpretation of the general appearance of the floor plan. It is not meant to be an exact rendition.

Elevations

Elevations help with the installation, making it easier to visualize what the finished kitchen will look like (refer back to Fig. 7.2). These drawings can take the form of a front view of all wall areas receiving cabinets and equipment as shown on the floor plan. Drawn to scale, these drawings also provide the designer with a problem-solving tool. Elevations should be designated on the floor plans with arrows in alphabetical order, leading from left to right. The elevation should then be drawn in the same order. According to NKBA, an elevation drawing should illustrate a front view of all cabinets, showing fixtures and equipment. The following features are most effectively shown on elevations:

- Cabinets with toekick and finished height.
- A portion of the cabinet doors and drawer fronts should indicate style and, when applicable, placement of handles or pulls.
- Countertops indicate thickness and show backsplash. Also, note heights of the range hood, light fixtures, window trim, and any valances.
- All doors, windows, or other openings in walls that will receive equipment. The window/door casing or trim is listed within the overall opening dimensions.
- All permanent fixtures, such as radiators, etc.
- All main structural elements and protrusions, such as chimneys, partitions, etc.
- Centerlines for all mechanical equipment.

Perspective drawings

Designers have the option of preparing a one-point or two-point perspective, with or without the use of a grid (refer back to Figs. 7.3 through 7.5). The minimum requirements for perspectives shall be the reasonably correct representation of the longest cabinet or fixture run or the most important area in terms of usage. They need not show the entire kitchen. Separate sectional views of significant areas are acceptable.

Anyone can learn to draw reasonable perspective views even without previous art training. Systems have been worked out that are entirely mechanical. Only two things are needed—patience and practice. While one learns to draw floor plans and elevations, visualization (spatial thinking) should be practiced. In other words, while drawing layouts, constantly practice seeing in your mind how the kitchen area and furnishing would look in three dimensions. This is called "seeing in the round." Perspectives are *not* drawn to scale, but proportion is the key to successful drawings.

Perspective charts

Two mechanical methods for drawing perspective views are available. Printed charts (often called grids or screens) are available at drafting supply houses or art stores. Practicing with a grid sheet helps. The grid is divided for you, and the squares decrease proportionally as they get further away from the viewer.

Projection method

One-point perspective view is simpler than *two- point*, but if a room is long and narrow, the view tapers too much. The advantage of this method is that three walls can be shown. With *two-point* perspective, two or three views may be required, but the result is more photographic. With these methods, many variations are possible, just as if a camera was held high, low, or at one end of the room or other. Refer to Fig. 7.9 for the steps (steps 1 through 8) generally necessary in making a two-point perspective drawing.

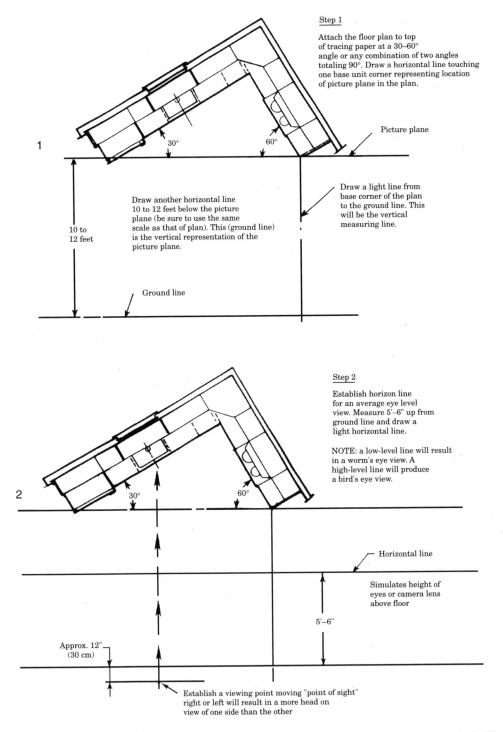

Step 1

Attach the floor plan to top
of tracing paper at a 30–60°
angle or any combination of two angles
totaling 90°. Draw a horizontal line touching
one base unit corner representing location
of picture plane in the plan.

Picture plane

30°   60°

Draw a light line from
base corner of the plan
to the ground line. This
will be the vertical
measuring line.

Draw another horizontal line
10 to 12 feet below the picture
plane (be sure to use the same
scale as that of plan). This (ground line)
is the vertical representation of the
picture plane.

10 to
12 feet

Ground line

Step 2

Establish horizon line
for an average eye level
view. Measure 5'–6" up from
ground line and draw a
light horizontal line.

NOTE: a low-level line will result
in a worm's eye view. A
high-level line will produce
a bird's eye view.

30°   60°

Horizontal line

Simulates height of
eyes or camera lens
above floor

5'–6"

Approx. 12"
(30 cm)

Establish a viewing point moving "point of sight"
right or left will result in a more head on
view of one side than the other

Figure 7.9 Steps in making a two-point perspective kitchen plan by using the projection method. *(The House
and Home Kitchen Planning Guide, Robert Scharff; Leon E. Korejwo, illustrator)*

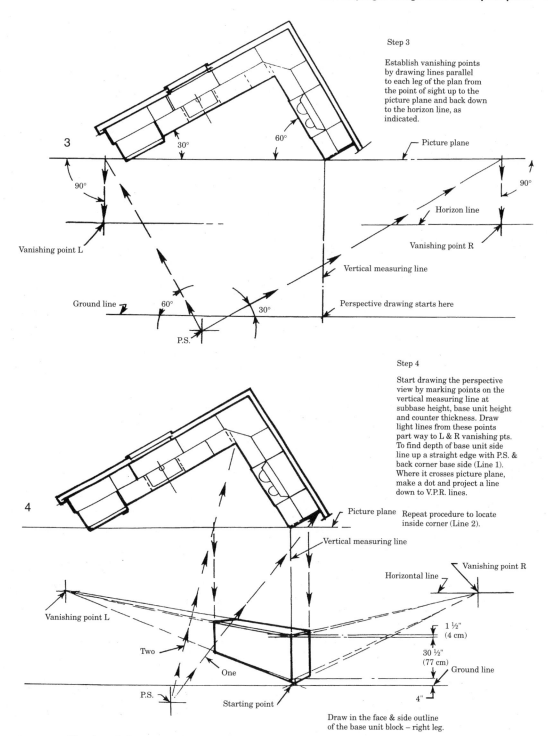

Step 3

Establish vanishing points by drawing lines parallel to each leg of the plan from the point of sight up to the picture plane and back down to the horizon line, as indicated.

3

30°   60°

Picture plane

90°   90°

Horizon line

Vanishing point L

Vanishing point R

Vertical measuring line

Ground line   60°   30°   Perspective drawing starts here

P.S.

Step 4

Start drawing the perspective view by marking points on the vertical measuring line at subbase height, base unit height and counter thickness. Draw light lines from these points part way to L & R vanishing pts. To find depth of base unit side line up a straight edge with P.S. & back corner base side (Line 1). Where it crosses picture plane, make a dot and project a line down to V.P.R. lines.

4

Picture plane   Repeat procedure to locate inside corner (Line 2).

Vertical measuring line

Vanishing point R

Horizontal line

Vanishing point L

1 ½"
(4 cm)

Two

30 ½"
(77 cm)

One   Ground line

P.S.

Starting point   4"

Draw in the face & side outline of the base unit block – right leg.

Figure 7.9   (Continued) Steps in making a two-point perspective kitchen plan by using the projection method. *(The House and Home Kitchen Planning Guide, Robert Scharff; Leon E. Korejwo, illustrator)*

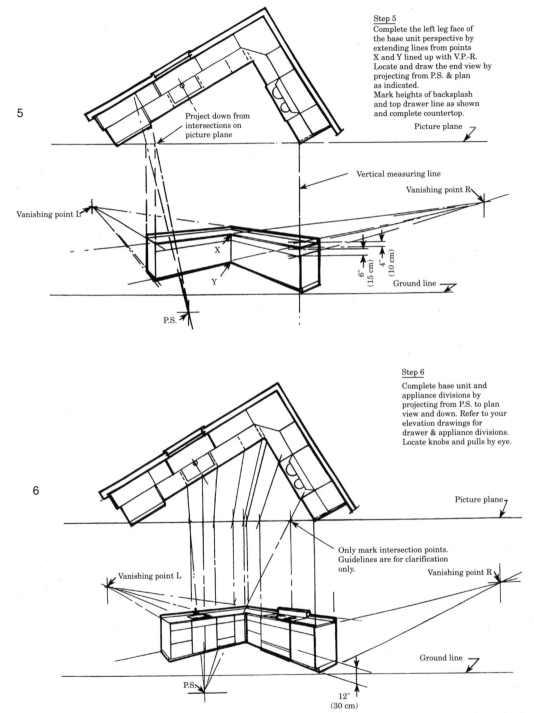

5

Step 5
Complete the left leg face of
the base unit perspective by
extending lines from points
X and Y lined up with V.P.-R.
Locate and draw the end view by
projecting from P.S. & plan
as indicated.
Mark heights of backsplash
and top drawer line as shown
and complete countertop.

Picture plane

Project down from
intersections on
picture plane

Vertical measuring line

Vanishing point R

Vanishing point L

X

Y

6" (15 cm)

4" (10 cm)

Ground line

P.S.

6

Step 6

Complete base unit and
appliance divisions by
projecting from P.S. to plan
view and down. Refer to your
elevation drawings for
drawer & appliance divisions.
Locate knobs and pulls by eye.

Picture plane

Only mark intersection points.
Guidelines are for clarification
only.

Vanishing point L

Vanishing point R

Ground line

P.S.

12"
(30 cm)

Figure 7.9 (Continued) Steps in making a two-point perspective kitchen plan by using the projection method.
(The House and Home Kitchen Planning Guide, Robert Scharff; Leon E. Korejwo, illustrator)

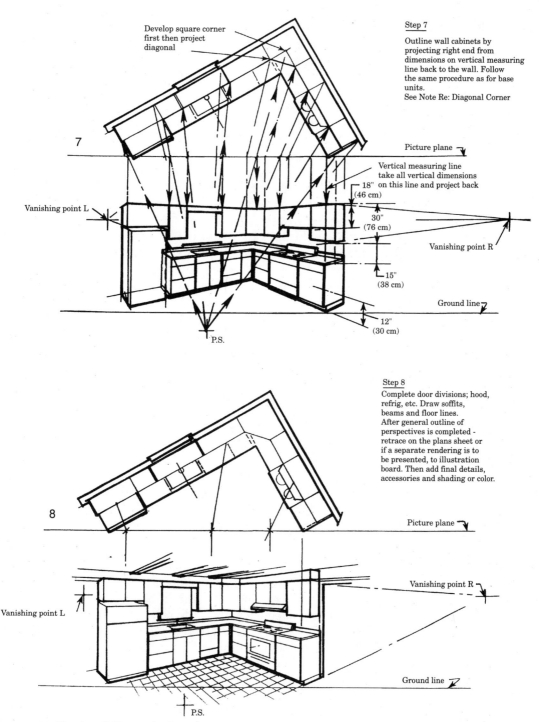

Develop square corner
first then project
diagonal

Step 7

Outline wall cabinets by
projecting right end from
dimensions on vertical measuring
line back to the wall. Follow
the same procedure as for base
units.
See Note Re: Diagonal Corner

Picture plane

Vertical measuring line
take all vertical dimensions
18" on this line and project back
(46 cm)

30"
(76 cm)

Vanishing point L

Vanishing point R

15"
(38 cm)

Ground line

12"
(30 cm)

P.S.

8

Step 8
Complete door divisions; hood,
refrig, etc. Draw soffits,
beams and floor lines.
After general outline of
perspectives is completed -
retrace on the plans sheet or
if a separate rendering is to
be presented, to illustration
board. Then add final details,
accessories and shading or color.

Picture plane

Vanishing point R

Vanishing point L

Ground line

P.S.

Figure 7.9 *(Continued)* Steps in making a two-point perspective kitchen plan by using the projection method.
(The House and Home Kitchen Planning Guide, Robert Scharff; Leon E. Korejwo, illustrator)

Mechanical plan

The mechanical plan, usually found at the back of the plan set, consists of the electrical/lighting, plumbing, heating, air conditioning, and ventilation systems. If any minor wall or door construction changes are part of the plan, they should also be detailed on the mechanical plan. Indicate where plumbing, gas, and electric lines enter the room and how they will reach the appliances. Also, mark the tentative locations of electrical outlets, switches, and lighting fixtures.

The various mechanical drawings can be difficult to understand. Each trade has its own subset of symbols, and these symbols are usually included as a part of the mechanical drawings. The mechanical legend should be prepared on the plan. This legend is used to describe the meaning of each symbol for special-purpose outlets, fixtures, or equipment.

The mechanical plan should show an outline of the cabinets, countertops, and fixtures without cabinet nomenclature. It should include only the information and proper symbols for that. The location of equipment should be noted. The overall room dimensions should be listed. Be sure to note the location of the sink plumbing and gas pipes (where they are and where you plan to relocate them). You must have a clear understanding of all mechanical work that affects your installation. You need to read features (i.e., heating supply vent located in the toe space of a base cabinet) on these drawings and coordinate for their installation, such as providing openings.

Centerline dimensions must be given for all equipment in two directions when possible. The mechanicals requiring centerlines include cooktops, refrigerators, dishwashers, compactors, sinks, wall ovens, microwave ovens, fan units, light fixtures, heating and air conditioning ducts, and radiators. Centerline dimensions should be pulled from return walls or from the face of cabinets or equipment opposite the mechanical element. Any differences from the plan that are discovered as a result of your installation work should be reported to the appropriate mechanical trade promptly for resolution. If a room is very irregular in shape, or walls are uneven, it is possible to locate the known centerline of the room and mark dimensions from it.

Reflected ceiling plans

This plan indicates the location of all ceiling features, including skylights, mechanical vents, light fixtures, soffits, steps in the ceiling construction, etc. If the ceiling is a suspended acoustical tile ceiling, the reflected ceiling plans indicate the size and layout of the tile grid. This plan is extremely important for you to be sure that equipment and cabinetry extending up to the ceiling will not interfere with soffits or any other ceiling features. If soffits are to be constructed above the cabinets, the depth of the soffits needs to be carefully coordinated with the cabinet depth. A reflected ceiling plan may not be provided in some projects. Separate plans for the mechanical, construction, etc., help to clearly identify such work without cluttering the kitchen floor plan.

Specifications

Specifications are written instructions describing the basic requirements for construction of a building. They should clearly describe sizes, types, and quality of all building materials and work affected by the job (either directly or indirectly). The methods of construction, fabrication, or installation and the expected quality of work to be produced are also spelled out explicitly. They define the area of responsibility between you and the purchaser. They must clearly indicate which individual has the ultimate responsibility for all or part of the work. In addition, information that cannot be conveniently included in the drawings, such as the legal responsibilities, methods of purchasing materials, and insurance requirements, is included in the specifications. If there is a difference between the drawings and the specifications, the information contained in the specifications is to be followed unless there is a note in the drawings or specifications.

If you are hiring a subcontractor, specifications tell that person: "These are the materials you must use, this is how you must use them, and these are the conditions under which you undertake this job." In other words, specifications help to guarantee you that the contractor delivers the job as specified. They also help ensure that project is done according to standards that the building laws require. Specifications may be listed on a separate form, may be part of the working drawings, or may be a combination of both.

As reported by the NKBA, the following "Delegation of Responsibilities" shall apply: Kitchen specialists are responsible for the accuracy of the dimensioned floor plans and the selections and designations of all cabinets, appliances and equipment, if made or approved by them.

- Any equipment directly purchased by the kitchen specialist for resale should be the responsibility of the kitchen specialist. Further, they must be responsible for supplying product installation instructions to the owner or the owner's agent.

- Any labor furnished by the kitchen specialist, whether by their own employees or through subcontractors paid directly by them and working under their direction, should be the kitchen specialist's responsibility. There should not be a *delegation of total responsibility to the subcontractor working under these conditions.*

- Any equipment purchased directly by the owner or the owner's agent from an outside source should be the responsibility of the owner or the owner's agent. The same applies to any subcontractor, building contractor, or other labor directly hired and/or paid by the owner or the owner's agent.

- Specifications should contain descriptive references to all areas of work.

- All specification categories must be completed. If the job does not cover any given area, the words *Not Applicable, N/A,* or *None* should be inserted.

- In each area, the responsibility of either the kitchen specialist or the owner or the owner's agent must be assigned.

In all cases, the owner and the owner's agent must receive a completed copy of the project documents *prior* to the commencement of any work.

One of the most useful ways to increase productivity in the kitchen installation business is through the use of standard master specifications. If specifications must be drawn up from scratch for each project, especially when unfamiliar products are being used, a great deal of time is required to complete this task. While it is true that no two jobs are the same, a large percentage of the work for kitchen installations can be standardized.

There is no question that developing standard specifications is a time-consuming project, and many builders are not willing to commit the time necessary to undertake the task. Still, the benefits are tremendous. Using standard specifications saves a lot of time and research for the builder or retailer/dealer/designer who is writing the specifications. These standard specifications can be carefully written to eliminate vagueness or gray areas that are often found in hastily written specifications. Standard specifications such as those designed by the NKBA also allow for fine-tuning of the estimating system for increased accuracy. As an operation is repeated time and time again, using the same method and products, closer attention can be given to labor and materials costs for that particular item, and the unit price can be adjusted accordingly.

Standard specifications are easily adaptable to a computer system. A project specification can be quickly and accurately assembled from the specifications stored on the computer. New specifications can continually be added so that they can be reused in future projects. Standard specifications that call out the use of the same products on a regular basis mean that builders learn the individual characteristics of each product and increase their efficiency at installation of the product. This familiarity leads to a reduction in mistakes and an increase in productivity. And finally, it is much easier to keep abreast of price changes when the standard specifications are written around a limited number of standard products.

To set up standard specifications, the builder should review projects over the past year and try to identify operations that were repeated more than five times. One of the ways to classify the elements of a job is to use a 16-category system standard developed by Construction Specifications Institute (CSI). It presents the material specifications in an order very close to the order in which the products will be installed. Or the builder can organize specifications in the categories of work established by a unit-cost estimating manual. In some cases, standard specifications need to give the customer some selection in style or color. In the case of flooring, for example, the usual method is to select a brand name with a low-, medium-, or high-priced selection, which is almost always enough choice for a customer. On rare occasions, it may turn out that a customer wants a particular color that is only available from another manufacturer. When it is necessary to have more style choices, say for floor tile, it is still best to stay with one manufacturer or supplier.

When selections have been made, standard specifications can be customized for a specific job by including specific model numbers. This should routinely be done at least for windows, skylights, flooring, and plumbing fittings. For items that are subcontracted, meet with subcontractors to develop standard specifications. For example, the electrician and the builder can decide that a 200-amp circuit breaker box by a particular manufacturer, with 24 circuits and a master cutoff, will be used on all jobs. That will be the standard specification, and the electrician will always know that this box is expected.

Builders can also agree on certain minimum requirements with their subcontractors, such as using only copper romex cable for branch wiring. With this sort of arrangement, builders and their subcontractors always know what to expect, and customers can be assured of the quality and safety that they can expect.

When builders have developed master specifications and put them into a book or on the computer, the time required during estimating to write up specifications can be enormously reduced. Master specifications can even be referred to by number, which helps identify them easily in the computer. Of course, after master specifications are developed, builders must continue the same ongoing process of review that was used to create the original specifications. Projects must be reviewed periodically to see if new procedures or techniques are being adopted that can be added to the master specifications. In this way, the master specifications can continue to grow with the company.

Specifications are generally written by an architect or by a specifications writer. You, as well as any of your subcontractors, should review the drawings and specifications to verify that all codes and project expectations have been met.

Building plans and specifications are actually legal documents that are a part of your installation contract. By accepting the terms and conditions of your contract, you are agreeing to provide the materials and labor necessary to produce all work indicated on the drawings, unless otherwise stated. In the event that the designer or architect fails to include some portion of work that might be necessary to complete the project, be sure to report the possible omission and request a change order to complete this additional work.

Software/computer drawings

With the proper software, kitchen specialists can produce suitable reproduction drawings of kitchen designs and layouts (Fig. 7.10). Some computer program product drawings completely eliminate manual boardwork. They can help you and your clients turn ideas into working plans accurate enough to be used in construction. In addition to CAD floor plans and perspectives, many of these programs provide a "bill of materials." But the program cannot replace a skilled designer's knowledge or experience.

There are several software/computer programs available. Each operates slightly differently, but each can save you time in presenting a visual kitchen plan. As with manual drawings, practice hones the skills. Verification of figures and dimensions is often seen as a waste of time, but a minor oversight can cause a cabinet or appliance misfit that may require a work stoppage and costly material reorder.

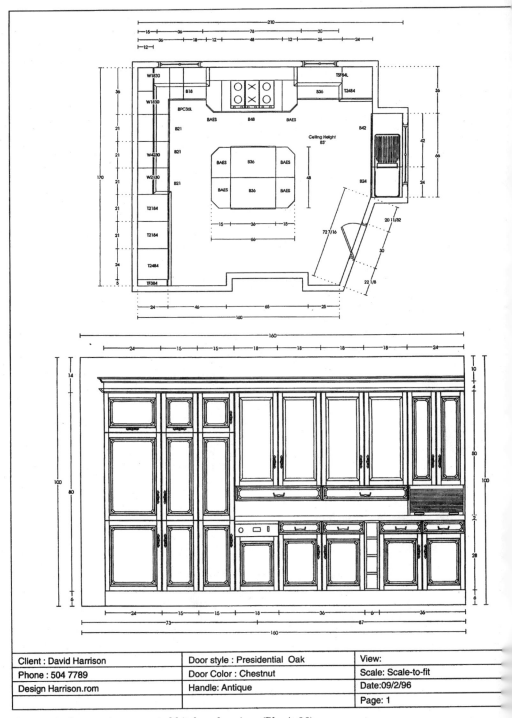

Figure 7.10 A computer-generated kitchen drawing. *(Planit 96)*

8

Getting the kitchen ready

It is time to get the kitchen ready for the installation. The cabinets, appliances, and other components are all on order, and the plans are in place. It is time to tear out the old kitchen and prepare for the new one.

As mentioned in Chap. 2, be certain that you and the client are clear on what existing materials will be removed and discarded. If items are to be saved—i.e., plumbing fixtures, cabinets, etc.—be sure that the client provides an area where these items can be stored. In addition, decide who is responsible for the removal of any salvaged items.

More in the kitchen than in any other room in the house, the kitchen remodeler must be on the lookout for design and structural traps. For example, work put in by amateurs is often overengineered (and hard to remove) and often ignores codes (which makes it harder to get the job accepted) and standard practices (so studs, pipes, and wires may not be where one would expect them). The main thing to remember is to work carefully. The goal is to remove the old kitchen with as little damage to the surrounding areas as possible, to minimize unintended reconstruction work.

Removing appliances and fixtures

The kitchen is dismantled in the opposite order in which it is assembled. The first items to remove are the appliances and fixtures. Before beginning, locate the water and gas shutoffs, as well as the electric service panel. Locate the circuit breakers for each of the appliances and turn them off. Be sure to test each individual appliance to verify that the electricity is off before you begin working on it. Shut off the gas at the fixture supply or at the main, and shut off the water to the sink and dishwasher, either at the shutoffs under the sink or at the main shutoff valve. Verify that the water and gas are off before continuing. Shutting off a supplied service may require notifying the utility providing that service, such as a gas company. Not informing the utility may be a violation.

Removing the appliances

Most appliances are installed in roughly the same manner, although there may be some variation from brand to brand. In most cases, the installation procedures can be reversed to remove the unit.

Ovens   Ovens are attached with four or more flathead screws that are driven through the front of the oven case on either side of the oven cavity into the surrounding wood of the cabinets. Remove the screws, and slide the unit out to gain access to the power cable or gas line. Disconnect the line, then remove the oven.

Drop-in cooktop   A drop-in cooktop is usually held in place with two or more thumbscrews. If the unit has a lid that lifts up, these thumbscrews are usually located inside, at the two side edges. If there is no lid, look inside the cabinet, against the bottom of the countertop. Loosen the screws until they swing out of the way, or remove them completely. Disconnect the power cable or gas line, and remove the unit.

Drop-in range   To remove a typical drop-in range, remove any screws that go through the oven into the cabinet. Lift the lid around the burners, and remove any screws or thumbscrews that are securing the unit against the counter. Slide the unit out of the opening and place it on the floor. Remove the gas line or flexible metal electrical cable.

Freestanding range   If the range is freestanding, simply slide the unit away from the wall, then locate and remove the plug from the large wall socket.

Removing the plumbing

The first step is to remove the plumbing under the sink, either by disconnecting the fittings or cutting the lines. You need to disconnect the hot and cold water lines and the drain lines and trap. Do not worry about removing the faucet—it comes off with the sink. The sink itself is held in by one of four methods:

1. *Stainless steel sinks* have a channel welded to the underside of the sink rim, with clips and screws that hold the sink tight against the counter. Loosen the screws and remove or swing the clips aside, then pry up the sink.

2. Many *tiled-in sinks* are set on the countertop decking and then held in place with tile. You need to chip off the old tile to completely expose the rim of the sink; then pry the sink up.

3. *Sinks with sink rims (or flush mounts)* are similar to stainless-steel sinks, except that the rim is a separate piece. The screws and clips hold the sink, the rim, and the counter all sandwiched together. Simply remove the screws and clips.

4. *Self-rimming sinks* have a smooth, rounded edge and no sink rim. Nothing holds them in place except a bead of sealant between the underside of the sink rim and the counter, plus their own weight. To remove them, just pry up.

Removing cabinets and countertops

Once they are empty, and all doors and drawers have been removed, cabinets and countertops can be removed fairly easily. Be certain that the water was turned off at the shutoff valves. The appliances and plumbing fixtures should have been disconnected and removed. Usually it is most efficient to remove the upper cabinets first. If the cabinets are modular, you will find they are screwed together through the edges of the face frame. These should be the first screws to be removed. Next, any screws holding the cabinets to the wall must be removed. In some cases, nails are used instead of screws. Use a nail puller to get under the head and raise the nail; then remove the nail with a hammer. If salvaging them, be careful to avoid damage to the cabinets. Always take caution to avoid damage to the surface of the wall.

Once all of the upper cabinets have been removed, the countertops are next. In some cases you may find that screws have been driven up through cleats on the top of the base cabinets into the underside of the counters. Simply remove the screws. Remove any screws or brackets holding the countertops to the cabinets, and unscrew the take-up bolts on mitered countertops. A utility knife can be used to cut caulk beads along the edge of the countertop and backsplash. Remove trim moldings at the edges and tops of cabinets with a flat pry bar or putty knife. Use a flat pry bar to lift the countertop away from the base cabinets, as well as to remove base shoe moldings and baseboards. If the countertop cannot be pried up, use a reciprocating saw or jigsaw with a coarse wood-cutting blade to cut the countertop into manageable pieces for removal. Be careful not to cut into the base cabinets.

To remove ceramic tile, chisel the tile away from the base with a ball peen hammer and masonry chisel. A tile countertop that has a mortar bed can be cut into pieces with a circular saw and abrasive masonry-cutting blade. It is imperative to always remember to wear eye protection and heavy gloves when breaking tile.

When the counters have been removed, the base cabinets can be taken out. Unscrew them as you did with the upper cabinets, remove the doors, take out the screws from the face frame, and then remove the screws from the wall.

Some older styles of cabinets were built on site when the house was constructed. These cabinets, in many cases, have no backs, and there is little to be done to salvage them. Begin by unscrewing and removing the doors. Using a hammer from the inside of the cabinet, knock the face frame free of the shelves. Depending on how the cabinets were originally built, either remove the sides to completely free up the shelves or pry the entire cabinet off the wall or off the cleats. Beware that these types of cabinets were assembled with a number of finishing nails.

Built-in cabinets should be cut into pieces and discarded. Old cabinets can be salvaged if they are modular units that were installed with screws. Remove trim moldings at the edges and tops of cabinets with a flat pry bar or putty knife. Remove vinyl base trim. Work a pry bar or putty knife underneath to peel off the vinyl, as well as to remove base shoe moldings and baseboards. During this removal process, wall surfaces can be protected with scraps of wood. Keep in mind while removing valances that some are attached to cabinets or soffits with screws, while others are nailed and must be pried loose. Any screws holding the cabinets to the wall must be removed. Cabinets can be taken apart piece by piece with a hammer and pry bar or cut into manageable pieces with a reciprocating saw. Once the old counters and cabinets are removed, remove them from the work area.

Removing flooring

With any new kitchen, replacing the flooring is almost inevitable. If the old flooring is in good condition and adheres well to the floor, you may be able to lay the new flooring right over it (Tables 8.1, 8.2, and 8.3). In most cases, a better overall finish is achieved by removing the old flooring.

If you find an instance in which the cabinets were installed first, followed by underlayment, and then flooring, you will find large gaps in the underlayment when the cabinets are removed. The best thing to do in this situation is to simply remove the underlayment, flooring, and all.

To remove resilient sheet goods, slit them with a utility knife, work a spade underneath, and then peel up the flooring. Old resilient floor coverings that are embossed or cushioned should be removed or covered with plywood underlayment before installing new flooring (see "Preparing for resilient flooring" at end of this chapter). Be sure to remove the felt backing entirely by moistening the glued felt with soapy water and with a floor scraper or drywall knife. Ceramic tile that is damaged or loose must be removed by breaking the tiles with a heavy hammer or sledgehammer and prying up the pieces with a cold chisel. Remove baseboards and base shoe with a drywall knife (protects wall surfaces) and pry bar. Kitchen carpeting is almost always glued down. It must be scraped off completely to prepare for the new flooring.

When all of the appliances, cabinets, countertops, and flooring have been removed, study the proposed layout to prepare for what else needs to be removed. If the walls need to be opened up (for example, to provide for access for new plumbing or wiring) do as much of this as possible now. Finally, completely clean the kitchen, sweep up all of the debris, and remove all of the old items to their proper designated area.

The structural system

When a kitchen is being assembled in a new house, structural problems should not arise. However, when planning to remodel an old kitchen, problems commonly arise. Even if the home is only a few years old, it always seems like the windows are never quite where they ought to be, or the chimney flue forms a jutting corner right where your client wants you to add a cabinet. And, just about as common are problems that are already in the wall. That is, plumbing lines, wiring, and doors never seem to be where they should be. Electrical lines and outlets can be moved or added with little trouble, but reworking existing piping and ductwork involves more complicated solutions (see Chap. 9). As discussed in Chap. 2, be certain that you had a preliminary inspection. As a result of this inspection, you should have discussed with the client the cost of the desired changes. Inform your client of the expense of taking out and putting in new fixtures. Replacing fixtures also means renovating the electrical, plumbing, and heating lines that serve the kitchen.

Before you begin work, it is extremely important to analyze the house structural system. Do not assume the architect or dealer or designer has provided all this information for you. You need to be able to identify bearing and non-bearing walls, study the floor and ceiling framing, and determine in which direction the members are oriented (important in determining locations for plumbing and ductwork runs and recessed lighting fixtures). Architectural drawings generally show only schematic layouts, with locations and, hopefully, sizes of pipe to be used. The routing is left to be done "in the field."

If the design does not provide the required structural support (for a masonry floor, for example), immediately notify the architect, dealer, or designer so he or she can modify the plans and the client can be made aware of the additional working costs.

Extensive framing

Before you begin any framing work, have your plan worked out, to scale, on paper. Since most of the other interior work depends on the framing, it is crucial that you get the planning and framing work done properly.

Existing floor and ceiling structural members can present problems. Floor joists and subflooring could be rotted. Repairing damaged framing generally is not a big deal, but it is extremely important that the existing joists be strong enough to handle the live load of the planned space.

The term "wall framing" includes primarily the vertical studs and horizontal members of interior and exterior walls that support ceilings, upper floors, and the roof. The wall framing also serves as a nailing base for wall-covering materials (see Chap. 11).

The wall framing members used in conventional construction are generally nominal 2-by-4-inch (5-by-10 cm) studs spaced 16 inches (41 cm) on center (Fig. 8.1). Depending on the thickness of the covering material and the purpose of the wall, 24-inch (61-cm) spacing is sometimes used. Top plates and sole plates are also nominal 2 by 4 inches (5 by 10 cm) in size. Headers over doors or windows in load-bearing walls consist of doubled 2-by-6-inch (5-by-15-cm) and deeper members, depending on the span of the opening. Ceiling height for the first floor is 8 feet (244 cm) under most conditions. It is common practice to rough-frame the wall (subfloor to top of upper plate) to a height of 8 feet $1\frac{1}{2}$ inches (248 cm).In most construction, precut studs are often supplied to a length of 7 feet $8\frac{5}{8}$ inches (236 cm) for plate thickness of $1\frac{5}{8}$ inches (4.13 cm).

When dimension material is 1½ inches (3.8 cm) thick, precut studs would be 7 feet 9 inches (236 cm) long. This height allows the use of 8-foot (236-cm)-high drywall sheets and still provides clearance for floor and ceiling finish or for plaster grounds at the floor line.

You are likely to encounter three types of framing systems: platform-frame construction, post-and-beam construction, and balloon-frame construction. Each system has its own set of unique characteristics that makes it generally easy to identify. While the platform method is more often used because of its simplicity, you may encounter any one of these systems when remodeling a home.

Platform-frame construction   The wall framing in platform construction is erected above the subfloor, which extends to all edges of the structure (Fig. 8.2). A combination of platform construction for the first-floor sidewalls and full-length studs for end walls extending to end rafters of the gable ends is commonly used in single-story houses.

One common method of framing is the horizontal assembly (on the subfloor) or "tilt-up" of wall sections. This system involves laying out precut studs, window and door headers, cripple studs (short-length studs), and window sills. Top plates and sole plates are then nailed to all vertical members and adjoining studs to headers and sills with 16-penny (16d) nails. Let-in corner bracing

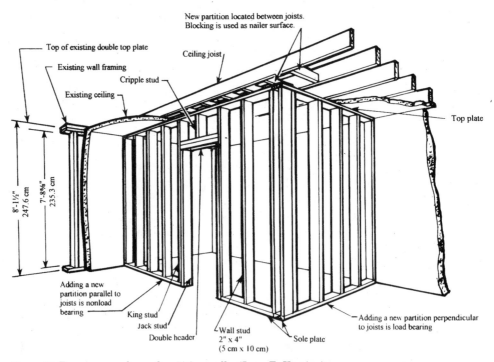

Figure 8.1 Framing members of partition walls. (*Leon E. Korejwo*)

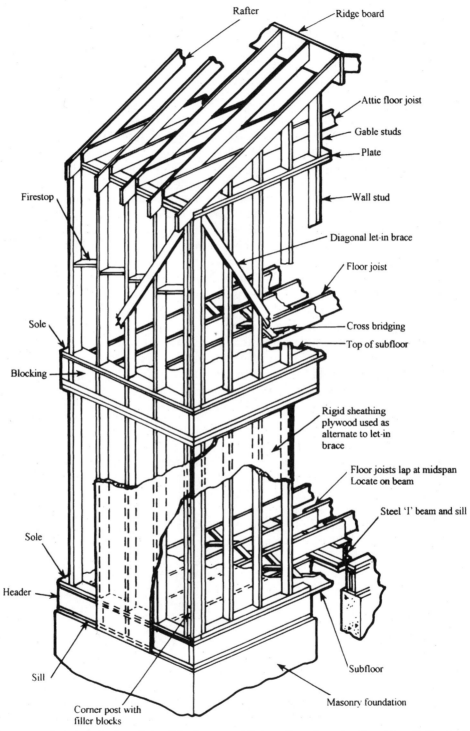

Figure 8.2 A typical platform-frame (also known as Western-frame) construction. (*Leon E. Korejwo*)

should be provided when required. The entire section is then erected, plumbed, and braced.

Post-and-beam construction   The post-and-beam form of building can be seen in historical structures dating back to the earliest European influence. Gaining popularity in current rural structures, it presents a different set of circumstances for the remodeler. Massive wood members carry the loads, leaving open spaces in the wall normally taken up by the wall studs. These open areas between the beams and posts, referred to as glazed areas, are free to be remodeled without weakening the wall, although side loads (created by wind, called *shear* or *racking*) require diagonal bracing. Braces may be diagonal wood members. Bracing may also be achieved by using sheathed framed walls or solid masonry in the glazed areas. These braces must not be removed. Areas below interior beams within the house can remain open or can be closed in. In post-and-beam houses, all of the building's loads are carried on heavy lumber posts.

This type of construction, while not adaptable to many styles of architecture, is simple and straightforward.

Today, specialists in timber framing can combine post-and-beam construction with panels filled with insulation. Many panels have drywall on the inside and siding on the outside. The insulating values of these nonbearing panels is very good. Without making any structural changes to the house, these panels can be removed. Historical buildings used pegs or other wooden attachments to join the members. Current modern methods of steel angle and straps and plate systems ease assembly or reconstruction for today's carpenters.

Balloon-frame construction   The main difference between platform and balloon framing is at the floor lines (Fig. 8.3). The balloon wall studs extend from the sill of the first floor to the top plate or end rafter of the second floor, whereas the platform-framed wall is complete for each floor. Balloon-frame construction uses wood studs rather than posts. Because the studs extend continuously from the top of the foundation to the roof, the stud cavity makes it easy to pull new wiring or reroute plumbing supply lines. The use of firestops, however, creates blockages that may be in areas difficult to see and penetrate for new wiring or plumbing.

In most areas, building codes require that firestops be used in balloon framing to prevent the spread of fire through the open wall passages. These firestops are ordinarily 2-by-4-inch (5-by-10-cm) blocking placed between the studs or as required by local regulations.

In balloon-frame construction, both the wall studs and the floor joists rest on the anchored sill. The studs and joists are toenailed to the sill with 8d nails and nailed to each other with at least three 10d nails. The ends of the second-floor joists bear on a 2-by-4-inch (5-by-10-cm) ribbon that has been let into the studs. In addition, the joists are nailed with four 10d nails to the studs at these connections. The end joists parallel to the exterior on both the first and second

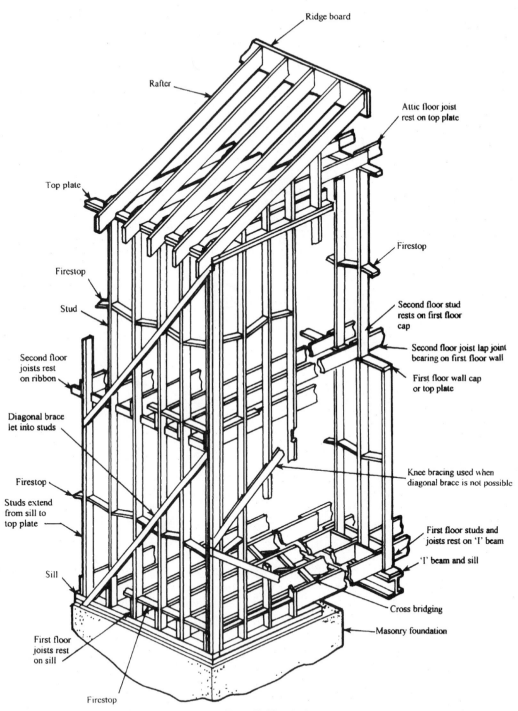

Ridge board

Rafter

Attic floor joist
rest on top plate

Top plate

Firestop

Firestop

Stud

Second floor stud
rests on first floor
cap

Second floor
joists rest
on ribbon

Second floor joist lap joint
bearing on first floor wall

First floor wall cap
or top plate

Diagonal brace
let into studs

Knee bracing used when
diagonal brace is not possible

Firestop

Studs extend
from sill to
top plate

First floor studs and
joists rest on 'I' beam

'I' beam and sill

Sill

First floor
joists rest
on sill

Cross bridging

Masonry foundation

Firestop

Figure 8.3 Typical balloon-frame construction. (*Leon E. Korejwo*)

connections. The end joists parallel to the exterior on both the first and second floors are also nailed to each stud.

Walls

Frequently, you'll run into the problem of a wall not being where it should be in the new plan. Before starting demolition, check the wall carefully. It may be a loadbearing one. Loadbearing walls carry the weight of the roof or upper floor down to the foundation. The entire stability of the house is dependent on these bearing walls. Any bearing wall must be replaced with a beam at ceiling level supported at the side walls or alternative support system. Nonbearing walls are independent of the structural system and some serve merely as partitions separating interior spaces. They are relatively easy and inexpensive to remove, provided they don't carry drain and vent lines.

About the only visible difference between a bearing and nonbearing wall is that the bearing wall has a double top plate. It can be identified by drilling a small hole in the wall a few inches from the point where it meets the ceiling. Even better, go to the basement or attic and note the direction the floor joists run. Nonbearing walls usually run parallel to joists; bearing walls always run perpendicular to joists. Double joists or extra members built into a floor under a wall indicate that it is load-bearing.

Any openings to exterior walls (i.e., doors and windows) that are loadbearing, have structural framing (headers) above them. Headers distribute weight from above to the vertical framing at either side of these openings.

In some cases, a wall may need to be added. Doing so is quite simple and can be accomplished by following standard construction procedures.

Removing existing walls

In a remodeling job, a few walls might need to be torn out. In most cases tearout is not too difficult, but some circumstances can make the removal of walls a bit more challenging. Finding animals, such as a colony of honey bees, snakes, or rats can give you major complications when tearing out walls.

Beware of electrical wires, plumbing, or heating piping, which are usually located in interior walls. It is always best to open walls with a hammer rather than a saw. If you rip into a wall with a reciprocating saw, you might wind up shocking yourself. Turn off the circuit breaker for that area and test the outlets and lights. Other areas may be affected by that circuit. Hitting a wire with the face of a hammer is not likely to electrocute you.

Plumbing pipes can also be hidden in existing walls. Cutting through a water pipe can really cause severe problems. If a drain or vent pipe runs through a wall you want to remove, the pipe must be relocated. As with electrical wires, using a hammer to open existing walls is a lot safer than a saw when working around plumbing pipes. Heating, ventilation, and air-conditioning (HVAC) ducts can also present a challenge when remodeling a kitchen. Electrical wires are fairly easy to relocate. Plumbing is tougher to move than electrical wires, but it is easier to relocate plumbing than the heating and air-conditioning ducts. When involved with a large remodeling project, you are probably going to be rearranging most of the mechanical systems (see Chap. 9).

Removing drywall or plaster You can usually tell the difference between drywall and plaster by knocking on the wall. Drywall has a hollow sound between the studs; plaster is more solid. Removing plaster requires a slightly different procedure than removing drywall. Tearing out either type of wall is a messy job. Be sure to move any furniture, protect the floor by covering it with drop cloths, and hang plastic or damp sheets in all doorways. Always wear a filter mask, goggles, gloves, and head protection when you take out a wall. Cut out the drywall between the studs with a circular saw, reciprocating saw, or handsaw. Expect to make lots of dust at this point. Cut the studs in half and wrench out the pieces. For plaster over metal lath, use a metal-cutting blade or simply batter the plaster with a crowbar or sledge hammer until it falls off the wall in chunks, then pry away the lath. Use caution in removing metal lath and nails. It is usually corroded, and sharp edges result from cutting them.

Removing interior partitions Removing interior partitions is a basic procedure. Once you get around the mechanical obstacles, ripping out nonload-bearing walls is easy. A reciprocating saw, a hammer, and a nail puller make the job simple.

Removing load-bearing walls Load-bearing walls present more complications than simple partition walls. Structural walls can be removed, but some concessions must be made. You might have to install an I-beam or some other form of support. Instead of removing a wall, consider opening it up using a beam and using portions of the existing wall to support the load.

When opening a load-bearing wall, you must reinforce it to support the weight bearing on it (Fig. 8.4a). A temporary support on both sides of the wall is erected with studs and top and bottom plates. It should be slightly higher than the existing wall and wedged up to raise the joists and relieve the load. Shut off the circuit breaker that supplies the wiring. In addition, shut off the water supply valves. Plan ahead to rewire or change piping while this wall is open. Have all supplies on hand. Cut away the existing wall and ceiling surface, being careful not to damage wiring or piping inside the framework (for clarity, the temporary bracing is only partially shown in Fig. 8.4b). Remove the unwanted studs by cutting them at midspan and twisting the halves while using a pry bar. The top plate must remain intact. Rewire or move pipe. Figure 8.4c shows a new beam properly sized to support the load for the span. The beam members should be sized by an architect to sufficiently support the load. Double studs at the corners are cut to fit and nailed to the existing sole plate. Cut away the unused soleplate. If the span is wide, support the beam with extra pairs of studs by extending the beam further into the wall. Finish the wall with drywall.

Before you start removing support walls, talk with an engineer, architect, or contractor. Experienced carpenters can do a good job designing support sys-

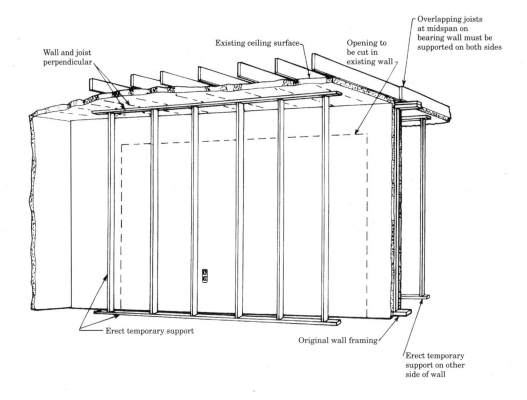

Wall and joist
perpendicular

Existing ceiling surface

Opening to
be cut in
existing wall

Overlapping joists
at midspan on
bearing wall must be
supported on both sides

Erect temporary support

Original wall framing

Erect temporary
support on other
side of wall

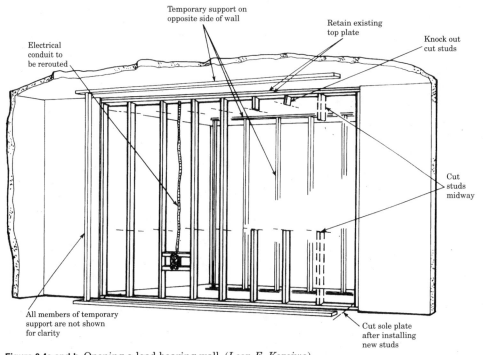

Temporary support on
opposite side of wall

Retain existing
top plate

Knock out
cut studs

Electrical
conduit to
be rerouted

Cut
studs
midway

All members of temporary
support are not shown
for clarity

Cut sole plate
after installing
new studs

Figure 8.4a and b Opening a load-bearing wall. (*Leon E. Korejwo*)

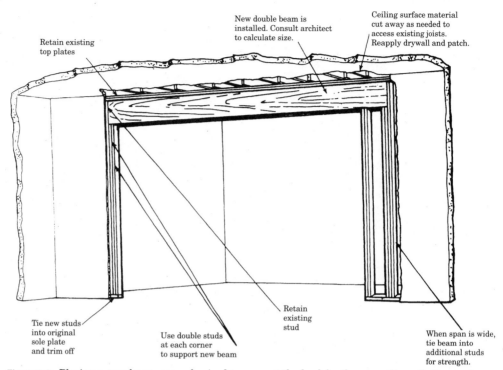

Retain existing
top plates

New double beam is
installed. Consult architect
to calculate size.

Ceiling surface material
cut away as needed to
access existing joists.
Reapply drywall and patch.

Tie new studs
into original
sole plate
and trim off

Use double studs
at each corner
to support new beam

Retain
existing
stud

When span is wide,
tie beam into
additional studs
for strength.

Figure 8.4c Placing a new beam properly sized to support the load for the span. (*Leon E. Korejwo*)

tems, but this approach may be risky. If a problem shows up later, the individual who made the decision on what to use as a replacement for bearing walls can be held responsible. Therefore, it is most important to consult an engineer or architect.

Wall preparation

If the kitchen installation includes the removal and replacement of wall finishes, check the insulation in the walls. Because the designer or architect has not had the opportunity to look inside the walls, you should report your findings so the designer or architect can determine if additional insulation is warranted.

Wall insulation is manufactured in several R-values, or insulating values, corresponding to the thickness of the wall studs. Walls with 2-by-4-inch (5-by-10-cm) studs can accommodate R-11 batt insulation and, in some cases, R-13 or R-15, while 2-by-6-inch (5-by-15-cm) stud walls can accommodate R-19 batt insulation.

Figure 8.5 shows a room with a variety of options that apply to construction and insulation types—how a variety of insulations may be applied to different wall conditions. Blanket insulation is sold in rolls of widths made to fit between wall studs of either 16 or 24 inches (41 or 61 cm) between centers

(Fig. 8.5a). The rolls are fitted with allowance for the studs. The installer only cuts it in lengths to fit from floor to ceiling. It comes in thicknesses of 1½ inches (4 cm), 2 inches (5 cm), or 3 inches (8 cm) for 2-by-4-inch studded walls. Thicker widths are available for 2-by-6-inch (5-by-15-cm) studded walls, although attempting to compress thicker insulation of higher "R" value into a 2-by-4-inch (5-by-10-cm) wall does not yield that higher "R" value. Figure 8.6 shows the proper method to install insulation backed with kraft paper.

The rolls usually have instructions printed on the backing to ease installation. Blanket facings have a paper "flange" that when scored, folded, and stapled, fits into the studs with a dead air space and vapor barrier.

Batt insulation is sized to fill the space between studs spaced 16 or 24 inches (41 or 61 cm) on center and can be purchased faced or unfaced. It is sold in precut lengths of 4 or 8 feet. Faced insulation has a vapor barrier installed on the warm side of the wall, and it can be stapled to the face of the studs (Fig. 8.5b). Unfaced insulation is stuffed between the studs, but a separate vapor barrier, such as a sheet of polyethylene, must be installed over this insulation on the warm side of the wall (Fig. 8.5c).

Insulation should be installed in all new wall construction. Even when not required for structural reasons, many homes are now being constructed with 2-by-6-inch (5-by-15-cm) studs to accommodate R-19 wall insulation to meet more-stringent energy codes. When reviewing installation plans, check the wall sections to verify the size of the studs being used in the exterior walls.

In existing homes with no insulation, blown-in loose fill insulation is an option if the wall finishes are not being removed (Fig. 8.5d). However, it is expensive if a limited wall area is being insulated. It is best done as a part of a complete house insulation project.

Rigid-foam or Styrofoam insulation can be used to insulate the interior side of a concrete or concrete block wall. The sheets are mechanically fastened to the wall, and studs or furring strips provide a means to install a fireproof layer of drywall over the insulation (Fig. 8.5e).

Rigid-foam insulation sheets may be used on frame walls; some have a foil facing. This facing, placed adjacent to a dead-air space towards the living area, provides further insulation (Fig. 8.5f). Rigid insulation is available as an exterior sheathing material and adds 'R' value to a wall, but because it has little structural value it must be supplemented by diagonal bracing at the corners of walls. Diagonal braces at the corners of the walls must be cut and nailed into the studs, or a sheet of plywood is used on each side of the corner instead of the insulation sheet (Fig. 8.5g). Often plywood is used in place of the insulating panels at the corners to brace a wall.

Doors, windows, and skylights

Doors, windows and skylights can turn a dull kitchen into a bright, cheerful place to live, work, and play. Kitchens almost always benefit from more natural light. Surveys have indicated that homebuyers are favorably impressed by rooms that have an abundance of natural light. To some

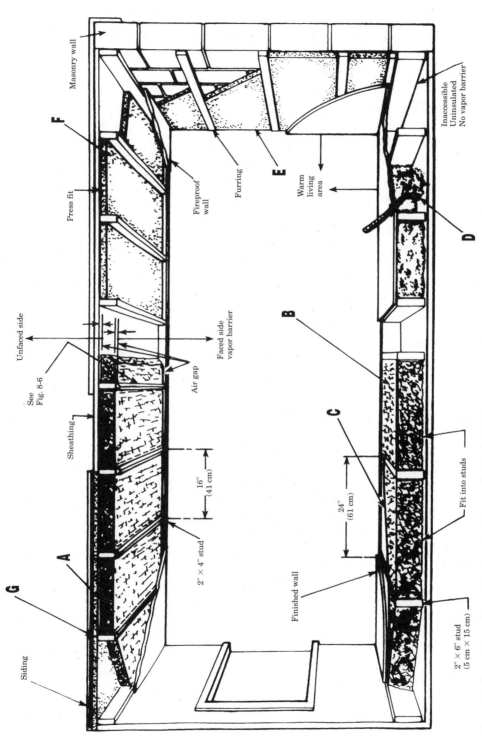

Masonry wall

F

Press fit

Unfaced side

See
Fig. 8-6

Sheathing

A

G

Siding

Fireproof
wall

Furring

E

Warm
living
area

Faced side
vapor barrier

Air gap

16"
(41 cm)

2" × 4" stud

B

C

24"
(61 cm)

Finished wall

Fit into studs

2" × 6" stud
(5 cm × 15 cm)

D

Inaccessible
Uninsulated
No vapor barrier

Figure 8.5 Several possible methods to insulate walls. (*Leon E. Korejwo*)

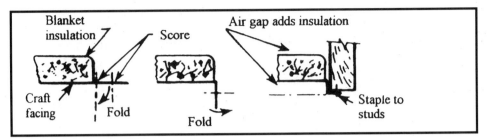

Figure 8.6 The proper method to install kraft-paper insulation. (*Leon E. Korejwo*)

extent, the existing construction of the kitchen can affect the options, but most buildings can be adapted to accept a wide variety of windows, doors, skylights, and roof windows.

Adding a lot of natural light to a small kitchen can make it appear larger and more appealing. Unfortunately, a tall window cannot be installed in an average kitchen, and wide windows won't fit between wall cabinets, but there are alternative solutions. If wall space is at a premium because of the kitchen cabinets or counters, and there is attic space or just a roof above the kitchen, skylights can be installed. Another option to suggest to your client is to consider a garden window. These big windows are shaped in a way that allows them to fit nicely in most kitchens without consuming unnecessary wall space. Another way to get extra light into a kitchen might be the use of a nine-light glazed door (a door with nine panes of glass in the top half), a terrace door, or a sliding-glass door.

Doors

When installing a new kitchen or remodeling an existing one, additional interior doors, exterior doors, or moving or closing up an existing door may be required. Almost every modern door has a vertical stile and horizontal rail framework. This construction helps counteract wood's tendency to shrink, swell, and warp with humidity changes.

Solid panel doors With a solid panel door, you can see the framing. Spaces between frame members can be paneled with wood, louvered slats, or glass (Fig. 8.7a).

Flush doors Flush doors (hollow-core or solid-core), which have no panels, hide their framing beneath two or three layers of veneer. Alternating the direction of the veneer minimizes warping.

Hollow-core doors A hollow-core door, usually $1\frac{3}{8}$ inches (3 cm) thick, may be filled with a lighter material, such as corrugated cardboard (Fig. 8.7b).

Solid-core doors A solid-core flush door, usually $1\frac{3}{4}$ inches (4 cm) thick, has a dense center of hardwood blocks or particleboard glued together within the internal stile and rail framework that is hidden under the veneer (Fig. 8.7c). Bypass doors come in pairs. Panel or flush, solid or hollow-core, they roll along an overhead track and are guided by metal or nylon angles screwed to the floor and header. Folding doors are hinged together. One slides along a track, and the other pivots on fixed pins. If two doors are installed together, the unit is referred to as French doors. With French doors or other pairs of doors, one door

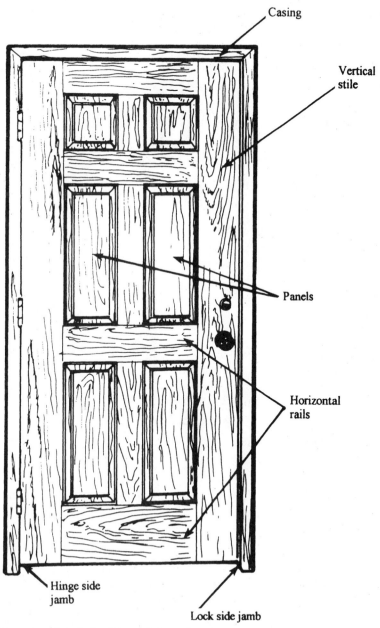

Figure 8.7a A typical solid panel door. (*Leon E. Korejwo*)

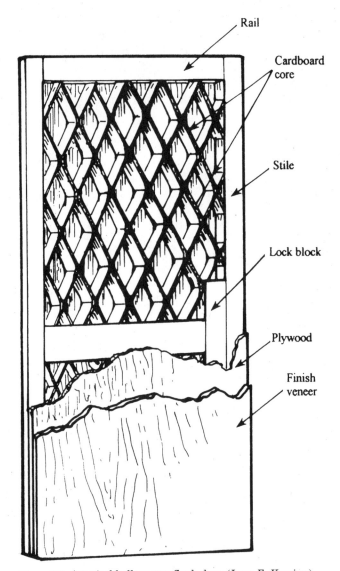

Rail

Cardboard core

Stile

Lock block

Plywood

Finish veneer

Figure 8.7b A typical hollow-core flush door. (*Leon E. Korejwo*)

is classified as inactive. Be sure you know which panel is to be the inactive door before the door is ordered.

Interior doors   Interior doors are manufactured in wood, flat panel, six-panel composite, and French doors. They do not need to be as sturdy as exterior doors. Hollow-core flush interior doors are common. In areas where space is limited, the design may call for you to install a pocket door. A pocket door can be either a panel door or a flush door. It is made to slide into a pocket in the wall (Fig. 8.7d). A number of door manufacturers make pocket door units that can be framed right into the wall during construction. Have the framing, door track, and door as an assembly. In this case, be careful to keep plumbing and wiring clear of the area of the wall where the pocket will be located. Unless plywood or furring has been provided under the drywall, cabinets cannot be hung on a wall containing a pocket door.

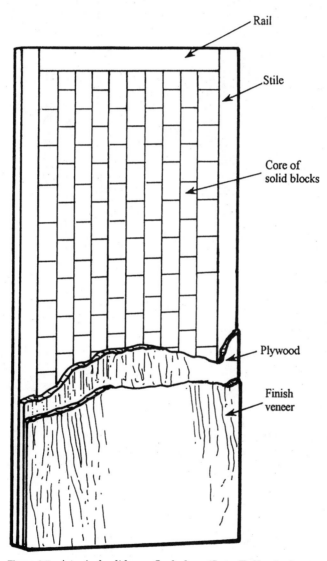

Figure 8.7c A typical solid-core flush door. (*Leon E. Korejwo*)

Figure 8.7d A pocket door with frame as one unit. (*Leon E. Korejwo*)

Exterior doors   Exterior doors are manufactured in wood, fiberglass, metal-insulated, sliding-glass, and terrace doors. They provide weather protection and security while allowing people to move from the inside to the outside. While wood doors are still most prevalent, steel and fiberglass doors are gaining popularity. Fiberglass and steel doors seal tightly, in most cases are filled with a high R-value foam insulation, and generally do not warp. A wood door may have an R-value of 1 to 5 compared to R-6 to R-12 of a steel door. To keep out cold air and rain, doors must seal tightly. Over time, old-fashioned wood panel doors can become leaky because they expand and contract with changes in weather. This problem can be remedied by providing weather stripping.

Prehung doors   Installing prehung doors is highly recommended. Prehung units do not cost much more than slab doors and the components needed to make them operational, but they can be installed in a fraction of the time needed to fabricate a complete door unit.

A prehung door usually comes complete with a split jamb and prefabricated trim (Fig. 8.8). The door can have either lower-cost fingerjoint trim or solid trim. Either trim is fine if it is to be painted, but fingerjoint trim looks out of place when it is stained.

Door installaion

When framing rough openings for door installations, you must allow room for shimming the door (Fig. 8.9). Rough openings should generally be 2 to 2½ inches (5 to 6 cm) larger than the width of the door to allow enough room for the jamb. The header is normally framed 3 inches (8 cm) higher than the door height. This height must consider the type and thickness of floor material

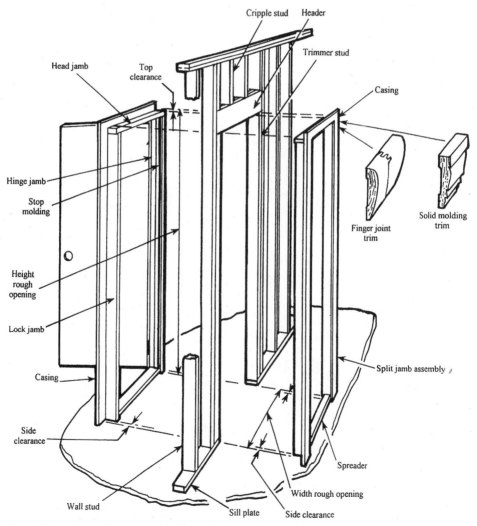

Figure 8.8 A prehung door assembly. (*Leon E. Korejwo*)

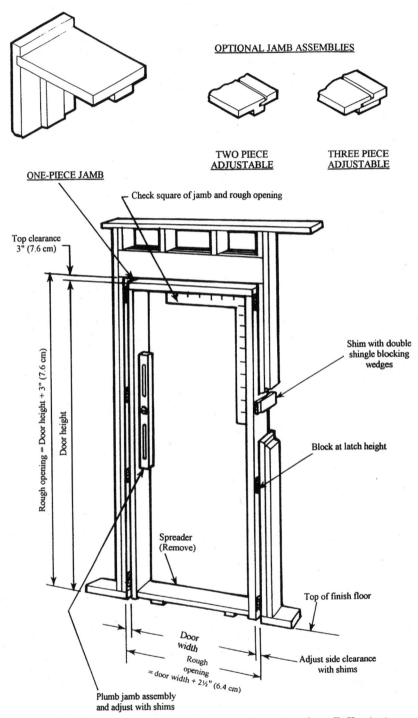

OPTIONAL JAMB ASSEMBLIES

TWO PIECE
ADJUSTABLE

THREE PIECE
ADJUSTABLE

ONE-PIECE JAMB

Check square of jamb and rough opening

Top clearance
3" (7.6 cm)

Rough opening = Door height + 3" (7.6 cm)

Door height

Shim with double
shingle blocking
wedges

Block at latch height

Spreader
(Remove)

Top of finish floor

Door
width

Rough
opening
= door width + 2½" (6.4 cm)

Adjust side clearance
with shims

Plumb jamb assembly
and adjust with shims

Figure 8.9 Installation of a prehung door in a rough opening. (*Leon E. Korejwo*)

being used. The door must be hung straight and plumb to operate properly. Different styles of prehung door assemblies are shown in Fig. 8.9. Interior door frames are made up of two side jambs and a head jamb and include stop moldings upon which the door closes. The most common of these jambs is the one-piece type. The two- and three-piece adjustable jambs are also standard types. Their primary advantage is in being adaptable to a variety of wall thicknesses.

Doors are either right-handed or left-handed. Your client can determine the handedness by facing the door as it is swinging toward them. A door that is left-handed has the latch on the left, while a right-handed door has the latch on the right. The swing of the door can be crucial in kitchen projects. It is possible to solve a swinging door problem with bifold doors, swinging cafe doors, or pocket doors.

Windows

Sometimes it is necessary, or at least desirable, to replace existing windows. Adding windows is a job that anyone with basic carpentry experience can usually handle. However, the process can get difficult at times. Ductwork, a plumbing vent, or a chaseway for electrical wires may be located in the same place you want to put a new window. Local code requirements might insist that new windows be added as the use of space is changed.

There are a wide variety of window styles from which to choose. Double-hung windows are the most common type; single-hung windows are not used very often. Casement windows, awning windows, bay windows, fixed glass, and bow windows are some additional types.

Window installation

Cutting in the rough opening for a new window can be a little tricky. However, once the wall is opened up, framing the window opening is pretty simple. Examine the structure of the wall before including major wall changes.

Some windows are held in place with a nailing flange. The flange is set against the exterior wall sheathing and screwed in place. The wall studs of balloon-framed houses carry the load of the second floor and roof; cutting into them may be best handled by a subcontractor. Siding is installed over the flange. Not all windows have nailing flanges. Some windows are made so that they are nailed into place through their sides. Follow the manufacturer's recommendations when installing any window.

When preparing to cut in a new window, it is important to remember to start on the inside of the building. If you work from the inside and discover an obstacle that could prevent the window installation, you need to only patch the interior wall, but if you start the work from the outside, you have to repair the siding and sheathing, which is a much larger project.

Once the window hole has been cut out on the inside wall, you can see if there are going to be any problems installing the windows in the desired location. If there is no reason to change plans, you can continue with the process.

This process entails the alteration of existing framing to accommodate a header, jack studs, and cripple studs. Some of the exterior siding and wall sheathing also must be cut away (Fig. 8.10).

The space for the new window is termed *rough opening*. Its dimensions are supplied by the manufacturer of the window. These are usually inches larger to provide for thermal expansion and contractions, as well as building settling, shrinkage, or binding, which can distort the new frame. This space is shimmed with tapered slivers of wood shingles to adjust, level, and fasten the window inside the rough opening.

The last step, when the rough opening is completed, is installing the window. This step usually goes smoothly. It is the early stages of this job that present you with the most problems. Remember, to keep the risk to a minimum, work from the inside out.

Double-hung windows include heavy sash weights concealed behind the frame's side jambs (Fig. 8.11a). Connected via a rope-and-pulley system, the weights provide a counterbalance that not only makes the sashes easier to open, but also holds them in any vertical position you choose.

A series of stops fitted to the jambs provides channels in which the sashes slide. Check the top view and note that though the outside bind stop is more or less permanently affixed, the parting stop and inside stop can be pried loose if you want to remove the sashes.

Newer double-hung windows replace the weight-and-pulley mechanisms with a pair of spring-lift devices. With both types, the lower sash comes to rest

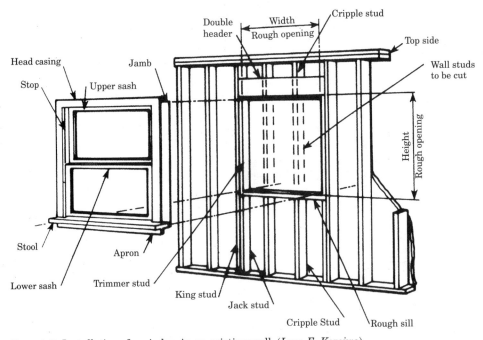

Figure 8.10 Installation of a window in an existing wall. (*Leon E. Korejwo*)

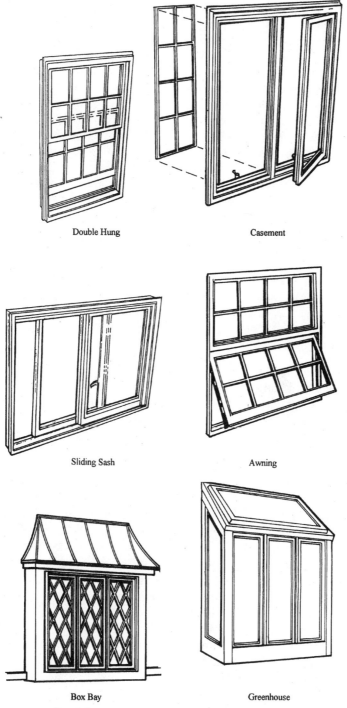

Double Hung   Casement

Sliding Sash   Awning

Box Bay   Greenhouse

Figure 8.11 Typical windows: double-hung (a); casement (b); sliding sash (c); awning (d); box bay (e); and greenhouse (f). (*Leon E. Korejwo*)

behind a flat stool; its outside counterpart, the sill, slopes so water can run off. Trim—called *casing* at the sides and top, and an *apron* below—covers any gaps between the jambs and the wall material. Casement windows open and close door-fashion, usually with the help of a crank-type operator (Fig. 8.11b). With some double-glazed casements, though, the muntins (or grilles) snap to the inside of the window to facilitate cleaning or are absent altogether.

As with double-hungs, sliding sashes open up only 50 percent of the total window area for ventilation (Fig. 8.11c). Some have one fixed and one sliding sash; with others, both sashes slide along continuous tracks. Sliding windows may have wood or metal construction.

Awning sashes tilt outward, under the direction of a scissors- or hinge-type cranking system (Fig. 8.11d). Some awnings slide downward as they tilt, so they can be opened to an almost horizontal position for maximum air flow.

Jalousie windows also let in lots of air; each turn of the crank pivots a series of glass slats for maximum flow control. The frames here consist of short metal channels at either end of the slats. Those glass-to-glass joints tend to leak air, so jalousies are usually found only in breezeways, porches, and other zones not-normally heated. Bay windows, such as the box bay (Fig. 8.11e) or the glass-enclosed greenhouse (Fig. 8.11f) open an area without major structural remodeling.

Installing a skylight

Cutting a skylight into an existing roof is not a big problem. However, the location of a skylight may be restricted by barriers such as electrical lines, heating ducts, or structural framing. Moving the location is less costly than rerouting utility lines or making structural changes. Skylights can be purchased in a size that allows them to fit between existing rafters. A rule of thumb concerning size is to buy a skylight with an area of 10 to 15 percent of the area of the floor in the room. To install them, you simply cut a hole in the roof, set the skylight in place, and seal around it. In vaulted ceilings, the job is not complicated at all.

If there is attic space between the roof and living space, the task takes on a few twists—a light box must be built (Fig. 8.12). Building a light box is really quite simple. Once you have located a place for the skylight, a plumb bob is used to find an appropriate spot on the ceiling below (done in the attic). A hole is then cut in the roof for the skylight, and the ceiling is also cut out, providing you with a hole in the roof and a corresponding hole in the ceiling below. Lumber is then used to frame the light box. The framing can be attached to the rafters on either side of the holes. The inside of the light box should be framed so that the skylight can give maximum light. Once the framing is complete, drywall can be hung and finished.

Larger skylights may span two or more rafters. To install these, rafters must be cut and new framing installed to maintain the roof's structural integrity.

Skylights are available in many different shapes and sizes. Some are operable—they can be opened—and others are fixed. Operable and fixed skylights look much the same. They both allow the same amount of light to

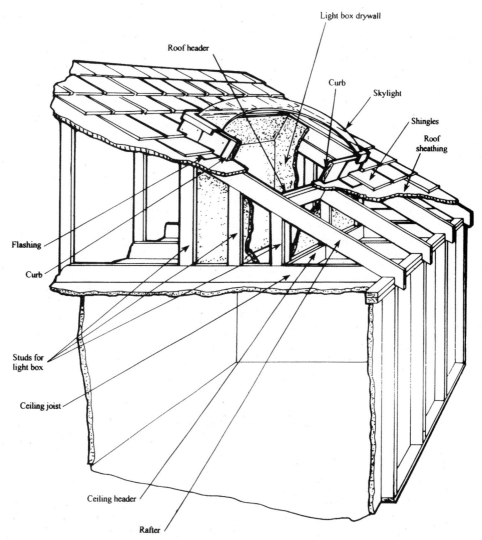

Light box drywall

Roof header

Curb

Skylight

Shingles

Roof
sheathing

Flashing

Curb

Studs for
light box

Ceiling joist

Ceiling header

Rafter

Figure 8.12 A skylight with a light box through the kitchen ceiling. (*Leon E. Korejwo*)

enter a home, but operable skylights also provide ventilation. Improvements in glass permit the use of large areas of glass without mullions, which are the support strips used between the panes of glass. Glazing options in good-quality glass skylights provide insulation, shield rooms from the sun's ultraviolet rays, and create a buffer against outside noise.

Floors

The choice of flooring style affects several construction factors that must be dealt with early on. Figure 8.13a shows a floor and its supporting structure for a typical platform-framed house. Wood floor systems and concrete slab floors are the two most common types of floor systems. Concrete slab floors are used when there is no basement or crawlspace. Wood floor systems are used when a house has a basement or crawlspace.

Wood floors

A *wood floor system* can be thought of as a wood platform on which the house is built. The platform is usually built of structural wood or engineered joists. Both types of joists are covered by subflooring. Subflooring is a panel material, such as plywood. The structure of a wood floor system generally does not need to be altered unless you are installing a new piece of equipment that would exceed the bearing or weight capacity of the existing wood platform.

Before ordering materials or beginning the preliminary work of preparing the subfloor for new flooring, check the basic floor and supporting structure. In older homes, you may find that the wood flooring system has sagged due to settling and shrinkage, creating a dip in the flooring surface. This settling does not necessarily mean that a structural weakness exists. A number of nonstructural changes may be required. Thoroughly inspect the existing flooring for indications of problems. If the existing floor is being removed as a part of the project, you can examine the subfloor. If the finish flooring is to remain, you need to inspect the subflooring from the underside by going into the crawlspace or basement. From this vantage point, you should be able to examine the floor joists, foundation, and subfloor, and determine the extent of any water damage and estimate how much subflooring needs to be replaced. Especially look for signs of a spongy or deteriorating subfloor, which indicates that there is or has been water leaking on the subfloor. Spongy or deteriorated subfloor may indicate more extensive damage due to leaks in plumbing or walls and roof.

Insects, such as termites and carpenter ants, as well as animal nesting, is unseen from above the floor. Termites build tunnels across masonry to reach the wooden framework. The surface appears solid while they consume the centers of the wood members. Professional exterminators and structural repairs affect time and costs. During this inspection, check the plumbing and electrical runs for condition or modification problems. If you find any indication of structural damage, notify the client of the damage and the anticipated cost to correct the problem.

The masonry foundation should be sound, with no crumbling mortar or loose or missing bricks or stones. Block or poured concrete foundations are more obvious, and cracking in mortar, while common, can worsen due to frost and moisture penetration. Repointing is advisable. Large separations or cracking may indicate a problem in the footer. These conditions may be outside the scope of a cabinetmaker, but they must be brought to the attention of the client. Changing the cabinet work or adding a partition over an uneven floor makes installation more time-consuming. Proper structural repairs may actually be more cost-effective, requiring fewer cosmetic coverups.

If the finish flooring is sheet goods, the entire existing finish floor has to be removed and replaced. If the floor is ceramic or vinyl tile, it may be possible to remove only the affected portion and replace it once the subfloor has been repaired. Once the floor surface has been leveled, you can install a new finish floor.

If you are installing a new finish floor material, it is generally best to remove the existing flooring materials rather than overlaying the existing flooring with new flooring. Many installers install new underlayment whether they remove the old flooring or not. Existing imperfections in the old finish flooring usually end up showing through the new flooring material at some point. The advantages of laying new flooring over old are that you bypass the messy job of removing old flooring, and you gain some soundproofing and insulation. Some disadvantages are that you are unable to inspect the subflooring and make corrections, and there may not be enough space left above to install appliances, such as the dishwasher. Check with the flooring manufacturer for specific instructions and requirements for flooring underlayment. Keep in mind that the failure to follow specific recommendations can void some manufacturers' warranties.

An extremely important matter to be aware of is that a variety of old flooring materials were manufactured with asbestos in them. Asbestos is no longer used in flooring materials, but if you are remodeling a house, it may already exist in the flooring. If you have reason to believe that the existing flooring contains asbestos (which is a recognized health hazard), leave the flooring in place. Do not sand it. If you have any doubts about the composition of the existing flooring, consult with a flooring expert before removing the flooring. In addition, check with the local state health department or local Environmental Protection Agency office. It requires an approved method of disposal, and undoubtedly there will be additional asbestos removal costs.

Concrete floor slab

In a house with a concrete floor slab, the problems are somewhat different. Concrete slabs are prone to cracking due to stress on the slab or soil movement beneath the slab. If you are to install a new finish floor on an existing concrete slab, an existing crack might cause a tile floor to crack along the line of the existing crack. If you are installing sheet goods, the crack might telegraph through the new finish flooring. A crack can be masked by using a cleavage membrane material manufactured for this purpose.

Whether or not there are problems with the concrete slab, completely remove the existing finish floor before installing a new floor. Old adhesive from the existing flooring needs to be completely removed so there is a good smooth surface for the new flooring.

Because plumbing, piping, and sometimes heating ductwork or piping is installed before pouring the concrete slab, any additional or relocated plumbing rough-in will probably require cutting or breaking out part of the existing floor slab. Obviously, this work needs to be done before any new finish floor work begins. When the concrete slab is patched, feather the joint between the existing slab and the patch so that the joint does not telegraph through the new finish floor.

Finish flooring must be compatible with the materials beneath it. Not all types are suitable for all kinds of conditions. Almost any kind of flooring material can be used over a wood floor system, including ceramic tile (use a thick-set mortar base or a backerboard), vinyl, and hardwoods (in strip or parquet patterns). Wood is affected by moisture, so only laminated products should be considered for rooms below grade.

The best finish floor materials over a concrete slab are sheet goods or vinyl, ceramic tile installed with an organic adhesive, and hardwood parquet flooring (but only if installed with a moisture barrier). If installing wood strips over a concrete slab, the best approach is to place the floor on sleepers or screeds. Sleepers are treated 2-by-4s laid flat on the slab. The sleepers must be protected from moisture by a vapor barrier. If the slab has no existing vapor barrier beneath it, ground moisture will penetrate to the sleepers. A layer of mastic, covered by 15-pound, asphalt-saturated felt paper, and finally another coat of mastic is needed. A layer of poly sheet between the sleepers and the final flooring is advised. For air circulation, a gap $\frac{1}{2}$ inch (1 cm) between last board and nail works well. Rows of staggered sleepers 12 to 16 inches (30 to 41 cm) on center are placed at right angles to the flooring.

Putting the floor on sleepers allows a strip wood flooring system, but the floor will be at least 2 inches (5 cm) higher than in adjacent rooms. Either use finish floor materials that can be installed directly on the concrete slab or find a transition between adjacent flooring materials.

Do not make any final flooring decisions until you know the kind and condition of the subfloor and underlayment the new floor will cover. With proper preparation, a concrete subfloor, because it is rigid, can serve as a base for almost any type of flooring. Other subfloors are more flexible and are not suitable for rigid materials, such as masonry and ceramic tile, unless they are built-up with extra underlayment or floor framing. Keep in mind that too many layers underneath can make the kitchen floor awkwardly higher than surrounding rooms.

The finished floor is only as good as the material over which it is laid, so it is essential that the subfloor is properly prepared to guarantee a smooth surface. Assuming the substructures are sound and sturdy, the subfloor must be adequately prepared to receive the new flooring.

Types of subflooring

Subflooring is used over the floor joists to form a working platform and base for finish flooring. It usually consists of square-edge or tongue-and-groove boards no wider than 8 inches (20 cm) and not less than $\frac{3}{4}$ inch (2 cm) thick or plywood that is $\frac{1}{2}$ (1 cm) to $\frac{3}{4}$ inch (2 cm) thick, depending on species, type of finish floor, and spacing of joists.

Subflooring boards may be applied either diagonally (most common) or at right angles to the joists. When subflooring is placed at right angles to the joists, the finish floor should be laid at right angles to the subflooring. Diagonal subflooring permits finish flooring or butt joints to be laid either parallel or at right angles (most common) to the joists (Fig. 8.13a). End joints of the boards should always be located directly over the center of the joists. Plywood subfloor is nailed to each joist with two 8d nails for widths under 8 inches (20 cm) and three 8d nails for 8-inch (20 cm) widths.

The joist spacing should not exceed 16 inches (41 cm) on center when finish flooring is laid parallel to the joists or where parquet finish flooring is used, nor should it exceed 24 inches (61 cm) on center when finish flooring at least $\frac{25}{32}$ (2 cm) inch thick is at right angles to the joists. Where balloon framing is used, blocking should be installed between ends of joists at the wall for nailing the ends of diagonal subfloor boards (Fig. 8.13b).

Plywood should be installed with the grain direction of the outer plies at right angles to the joists and be staggered so that end joints in adjacent panels break over different joists (Fig. 8.14). Plywood should be nailed to the joist at each bearing with 8d common or 7d threaded nails for plywood that is $\frac{1}{2}$ (1 cm) to $\frac{3}{4}$ inch (2 cm) thick. Space nails 6 inches (15 cm) apart along all edges and 10 inches (25 cm) along intermediate members. When plywood serves as both subfloor and underlayment, nails may be spaced 6 to 7 inches (15 to 18 cm) apart at all joists and blocking. Use 8d or 9d common nails or 7d or 8d threaded nails.

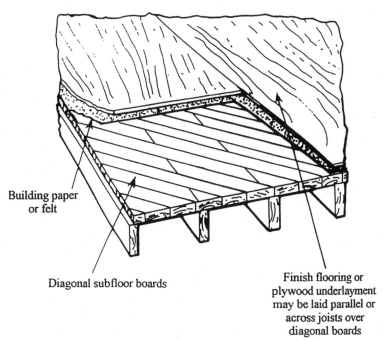

Diagonal subfloor permits underlayment or finish floor to be laid across or parallel to joists

Building paper or felt

Diagonal subfloor boards

Finish flooring or plywood underlayment may be laid parallel or across joists over diagonal boards

Figure 8.13a A diagonal subfloor permits the underlayment or the finish floor to be laid across or parallel to joists. *(Leon E. Korejwo)*

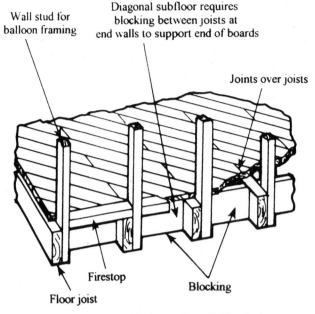

Wall stud for
balloon framing

Diagonal subfloor requires
blocking between joists at
end walls to support end of boards

Joints over joists

Firestop

Blocking

Floor joist

Figure 8.13b Balloon-frame blocking. (*Leon E. Korejwo*)

If necessary, lay an underlayment of untempered hardboard or plywood directly over the subfloor to raise the surface of new flooring to the same height as the old. Be sure to ensure a level transition between the new floor and adjacent rooms and make it unnecessary to trim door casings, doors, or moldings. In some instances, the addition of new flooring and plywood underlayment may interfere with the replacement of appliances. In these cases, be sure to remove the old flooring. Carefully examine the old flooring to determine what preparation, if any, must be completed before the new flooring is installed.

Underlayment

Correctly installed, underlayment provides a flat, uniform surface, eliminates irregularities in the subfloor, strengthens the floor, and creates a more secure and stable surface for laying finish flooring. It comes in 4-by-8-foot sheets. For underlayment, choose either $\frac{1}{4}$-inch (0.6-cm) plywood or hardboard or up to $\frac{1}{2}$-inch (1-cm) particle board. Do not install underlayment in exceedingly moist or humid weather or at times when the atmosphere is unusually dry. Techniques for installing underlayment vary, depending on whether the subfloor consists of plywood sheets or individual boards. Check the subfloor for uneven areas with a long, straight 2-by-4 laid on edge. Correct the low spots by shimming from below; protruding joints or fasteners will be revealed later on a new glossy surface. Loose joints will later crack through.

A new type of underlayment on the market that is gaining popularity is a fiber-reinforced underlayment, which provides an excellent base for a variety of floor coverings. This type of underlayment is smoother, more durable, and more moisture-tolerant than conventional wood-based underlayments. It is made by blending cellulose fiber and gypsum throughout the panel, giving it a uniformly smooth surface. Its high-density surface provides superior indentation resistance compared to traditional plywood underlayment. In addition, it is fire-resistant. It makes a perfect base for reapplying resilient floor covering, vinyl tiles, carpet, ceramic tile, quarry tile, and wood flooring.

Preparing for resilient flooring

The preparation work required before installing sheet vinyl is crucial to success (Table 8.1). Most resilient flooring is so pliable it conforms to irregularities in the subfloor. For this reason, it is essential that care be taken to prepare the subfloor properly to guarantee a smooth surface. If the surface that is to be covered with new flooring is rough, the finished job is not going to look good. When preparing an old floor for new vinyl, always install new underlayment. Hardboard or plywood that is $\frac{1}{4}$ inch (0.6 cm) or thicker works well. It provides a clean, porous base for the adhesive that secures the vinyl in place. It is best to have the old floor removed before adding the underlayment.

It might be necessary, even when new underlayment is installed, to fill cracks that form between the sheets. It is also important to make sure that nails and screw heads are not protruding from the underlayment. Any that stick up are going to show in the new vinyl.

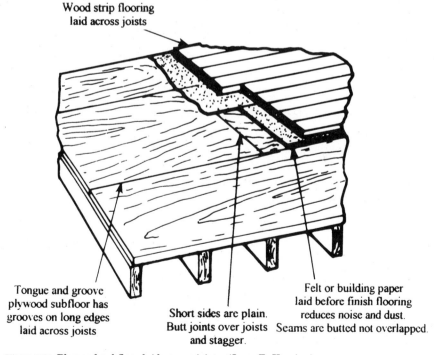

Wood strip flooring laid across joists

Tongue and groove plywood subfloor has grooves on long edges laid across joists

Short sides are plain. Butt joints over joists and stagger.

Felt or building paper laid before finish flooring reduces noise and dust. Seams are butted not overlapped.

Figure 8.14 Plywood subfloor laid across joists. (*Leon E. Korejwo*)

Be sure that you are clear about the manufacturer's installation recommendations and any restrictions that may apply to the flooring selected. Most of today's resilient floorings can be installed on any grade. A few, though, should not be laid on concrete in contact with soil. Refer to Table 8.1 for the information necessary to properly prepare a concrete slab, a wood subfloor, or an existing floor for new resilient flooring.

Table 8-1 Preparation guidelines for resilient flooring

Surface applied over	Preparation steps
Concrete slab (on or below grade)	☐ Be sure the slab is smooth, level, and clean. ☐ Newly poured concrete floors need at least 90 days to dry. ☐ Check that it is completely dry and take steps to ensure it stays dry. Any moisture coming through the concrete will eventually cause the flooring to loosen. ☐ Be sure to remove grease, oil, and other foreign material. ☐ Remove old paint or sealer. ☐ Fill any low spots, cracks, or joints in the surface with latex underlayment compound. ☐ Cover the surface with a sealer or other moisture barrier. ☐ If the slab is too uneven to be completely leveled by patching, lay a new plywood subfloor (on screeds) or pour a new thin concrete slab over the old.
Plywood subfloors	☐ If the subfloor is new, make sure the panels are securely attached with ring-shank nails, cement-coated nails, or screws and that the heads are flush with the surface. ☐ Remove every bit of old surface flooring, including cushioning, felt backing, and grout. A layer of underlayment is advised. ☐ If old flooring is too difficult to remove, cover the floor with a new underlayment of plywood or untempered hardboard (at least 1/4 inch thick). ☐ Leave a 1/8-inch (0.3-cm) gap between panels to allow for later expansion.
Woodboard subfloors	☐ It is extremely difficult to make woodboard subfloors smooth and level to properly install a resilient floor covering. ☐ Cover with 1/4-inch (0.6-cm) minimum-thickness underlayment-grade plywood or untempered hardboard. ☐ New resilient flooring can only be applied over an old floor surface that has been properly prepared to provide a clean, level base.
Old resilient flooring	☐ If the old flooring, cushioning, or underlayment is damaged, remove it. ☐ If the old floor is completely smooth, solid (not cushioned), and firmly secured, new resilient flooring can be installed directly over old resilient flooring. ☐ Clean the surface of the old flooring thoroughly, removing old wax or finish. ☐ Do not sand the old floorcovering. It may contain asbestos fibers (see text). ☐ Solid vinyl tile should not be laid directly over existing resilient flooring; install plywood underlayment first.
Old wood flooring	☐ New flooring can be laid directly over an old hardwood floor only if it is completely level, smooth, and in good condition. ☐ If old wood flooring is in poor condition, install plywood or hardboard underlayment. Otherwise, joints may show through. ☐ A wood floor makes a good base for vinyl tile only if there is an adequate ventilation space or crawl space beneath it, at least 24 inches (61 cm) above the ground level.
Old ceramic tile, slate, or masonry flooring	☐ If possible, remove old flooring. ☐ If the old floor is level, it can be covered with a plywood subfloor laid on screens, a latex underlayment compound, or a cement-base leveling compound. ☐ Resilient flooring should never be installed directly over old ceramic tile, slate, or masonry flooring with an uneven surface. ☐ Solid masonry-type floors can be covered with a new concrete slab.

Preparing for ceramic tile

Unless it is installed over a perfectly sound subfloor, the most carefully laid ceramic tile floor will eventually reveal flaws, and cracks will appear. Inspect the subfloor from above and below for any necessary repairs. Inspect the floor structure and joists, look for loose boards or a random low spot in a subfloor, which indicate defects. If the subfloor and structure are sound, you can lay ceramic tile over subfloors made of plywood panels or individual boards (Fig. 8.15a).

Use an underlayment suitable to your selected adhesive, such as fiber-reinforced concrete. Traditionally, ceramic tile was installed on floors using the mud-set method—the tile was set in a bed of mortar. The mud-set method allows the ceramic tile to be set level even if there are dips or imperfections in the subfloor when using plywood. Fiberglass-reinforced board is currently recommended. When using these underlayments, a thin-set adhesive method may be recommended by the tile manufacturers. Since you will be raising the level of the floor itself, check the bottoms of the doors for adequate clearance. If necessary, trim the door. A wood base is suitable if the boards or panels are securely fastened to the joists. Baseboard trim, heating

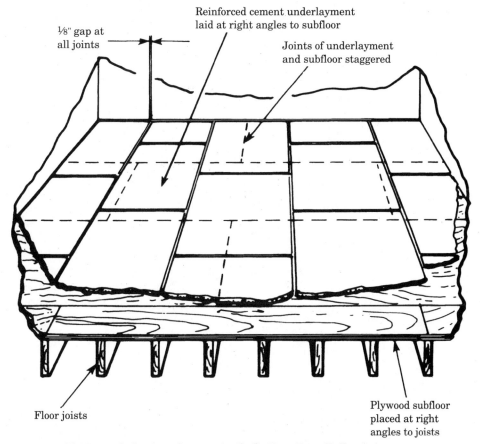

Figure 8.15a Placing underlayment for ceramic tile flooring. (*Leon E. Korejwo*)

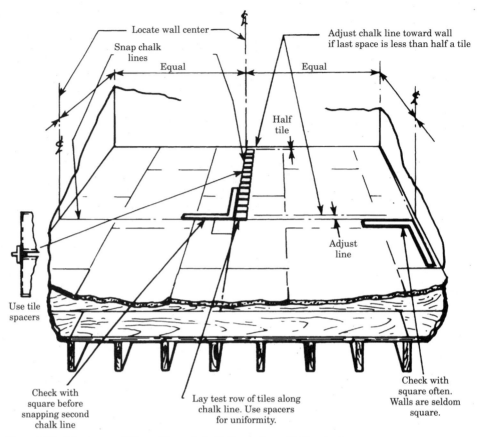

Figure 8.15b A test row of tile and layout chalk lines. (*Leon E. Korejwo*)

elements, and existing cabinetry will also be affected. Figures 8.15b and 8.15c briefly illustrate a method of installing ceramic floor tiles.

Ceramic tile can now be installed much more simply than the mud-set method using a thin layer of adhesive applied with a trowel. It requires a very flat, rigid surface, meaning that board floors or damaged resilient tile must be covered with plywood or underlayment such as the fiber-reinforced concrete before the tile can be laid. Adhesives have been developed that permit installation of ceramic tile in an area with some moisture, making it an appropriate choice to cover a below-grade concrete slab. Refer to Table 8.2 for the information necessary to prepare a concrete slab, a wood subfloor, or an existing floor for new ceramic tile flooring.

Note: Ceramic tile should be installed over subfloors no less than $1\frac{1}{8}$ inches (3 cm) total thickness. Thinner subflooring may flex, causing the tiles to break or the grout to crack.

Preparing for wood flooring

Preparing a proper base for any type of wood flooring can be more demanding than putting in the new flooring itself. Wood floors are seldom installed in rooms below grade or in areas subject to dampness. It is important to check an on-grade concrete slab carefully for moisture before installing wood flooring; be certain it will stay dry over the years. Similarly, the space below a standard floor supported by joists and beams should be properly ventilated and protected from moisture for wood flooring to be laid over it.

If there is already an old floor in place, the best approach is to expose the subfloor. Leaving the old floor surface increases the weight of the system and takes away from the headroom. If you want to leave an old floor in place, cover it with asphalt felt, as you would a wood subfloor.

Strip flooring is laid lengthwise in a room and normally at right angles to the floor joists. A subfloor of diagonal boards or plywood is normally used under the finish floor (Fig. 8.16). Strip flooring of this type is tongue-and-grooved and end-matched (Fig. 8.17). Strips are random lengths and may vary from 2 to 16 feet or more. The tongue fits tightly into the groove to prevent movement and floor squeaks. All of these details are designed to provide beautiful finished floors that require a minimum of maintenance. Blind-nail boards without damaging the finish. Drive nails into the tongue with a nail set. Use a block and bar to tightly force the board tongue into the groove without damaging it (Fig. 8.18).

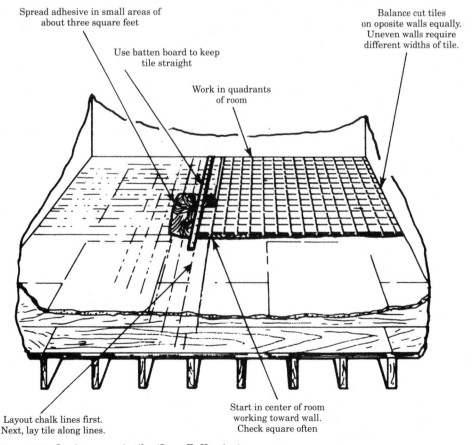

Spread adhesive in small areas of about three square feet

Use batten board to keep tile straight

Work in quadrants of room

Balance cut tiles on oposite walls equally. Uneven walls require different widths of tile.

Layout chalk lines first. Next, lay tile along lines.

Start in center of room working toward wall. Check square often

Figure 8.15c Laying ceramic tile. (*Leon E. Korejwo*)

Table 8-2 Preparation guidelines for ceramic-tile flooring

Surface applied over	Preparation steps
	Ceramic tile should be installed over subfloors no less than 1 1/8-inch (2.9-cm) total thickness.
Concrete slab	☐ A concrete slab makes the best possible base for the tile.
	☐ New or old, the concrete must be completely dry, clean, and level before you can begin preparing it for tile installation.
	☐ Once dry, make sure it is clean and free from grease, oil, and old paint of other finishes.
	☐ Fill any holes, low areas, or cracks in the slab with a concrete patching compound or mastic compound.
Plywood subfloor	☐ For best results, an underlayment of reinforced panels is recommended.
	☐ Be certain plywood panels are securely attached to joists with no protruding nails.
	☐ If plywood panels move when walking on them, reinforce the floor with a second layer of plywood or underlayment.
	☐ Plywood underlayment panels should be at least 3/8 inch (0.9 cm) thick if using mastic, 5/8 inch (1.6 cm) thick for epoxy adhesive.
	☐ Stagger the second layer of plywood panels or underlayment to be sure joints do not fall directly over those in the layer below.
	☐ Leave 1/8-inch (0.3-cm) gaps between panels.
	☐ Drive nails through panels into the joists before applying the second layer.
Woodboard subfloors	☐ If subfloor is made of individual 4- or 6-inch (10-or-15-cm) boards, be sure that each board is securely attached.
	☐ To prevent warping, cover the subfloor with underlayment before installing ceramic tile.
	☐ Strip, plank, or parquet floors in good condition can be covered with ceramic tile using underlayment.
	☐ If old floor is sound and level, you only need to give it a rough sanding to remove the old finish and smooth the rough areas.
	Whenever possible, old flooring should be removed before installing new ceramic tile flooring. Not only is it easier to examine the subfloor, and make necessary repairs, but also the new floor will be level with the floors in adjacent rooms. Underlayment for ceramic tile floors must resist moisture and be rigid enough to prevent flexing, which cracks tile and grout. Several types are available, including 1/4-inch (0.6-cm) exterior-grade plywood, reinforced concrete panels, and reinforced gypsum board specially designed for tile. Check manufacturer recommendations.
Old resilient flooring	☐ Well-bonded resilient flooring, if level and in good condition, can be successfully covered with tile. Underlayment is recommended.
	☐ Resilient flooring that is badly damaged should be removed or covered with a layer of underlayment.
	☐ Cushioned resilient flooring (sheet or tile) is too springy to be used as a base for ceramic tile and must be removed.
	☐ Old resilient flooring may contain asbestos—use caution! Consult local codes for approved removal and disposal methods.
Old wood flooring	☐ Wood floors, wood strip, and parquet flooring are not smooth enough to serve as backing for ceramic tile.
	☐ Cover with a layer of underlayment.
Old ceramic tile, slate, or masonry flooring	☐ Tile can be applied over old ceramic tile if the old tile is in good condition, clean, and well-bonded; otherwise, remove it. It raises the floor level excessively.
	☐ If there is evidence of water damage, the tiles and backing may have to be removed and the moisture problem corrected.
	☐ Clean the old tile with a degreasing agent.
	☐ An underlayment layer, or new concrete laid over the old, is best. The subfloor must be level for a good bond.

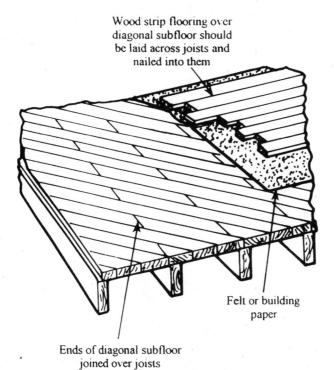

Figure 8.16 Wood strip flooring installed over diagonal subflooring. (*Leon E. Korejwo*)

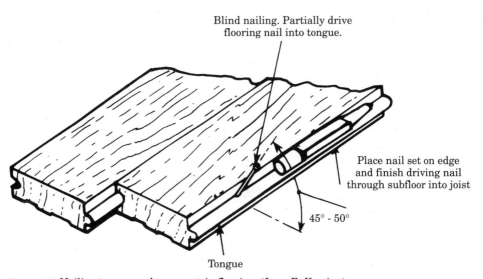

Figure 8.17 Nailing tongue-and-groove strip flooring. (*Leon E. Korejwo*)

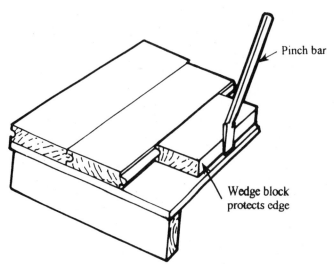

Pinch bar

Wedge block
protects edge

Figure 8.18 Lay each succeeding course by fitting the groove edges of
flooring pieces into the tongue edges of the preceding course. Use a
block and bar to force the board tongue into the groove without
damaging the flooring. (*Leon E. Korejwo*)

Parquet (block) flooring is made in a number of patterns. Blocks may vary in
size from 4-by-4 inches (10-by-10 cm) to 9-by-9 inches (23-by-23 cm) and larger.
Solid wood tile is often made up of narrow strips of wood splined or keyed
together in a number of ways. Wedges of the thicker tile are tongue-and-
grooved, but thinner sections of wood are usually square-edged. Plywood blocks
may be $\frac{3}{8}$ (1 cm) inch and thicker and are usually tongue-and-grooved (Fig.
8.19). Many block floors are factory-finished and require only waxing after installation.

Whichever of the three types of wood flooring your client has chosen, preparing a reliable base requires the same steps. Wood floors are typically laid over
a concrete slab, over a wood subfloor supported by joists and beams, or, in
some cases, over an existing floor, depending on the old floor's composition and
condition and on the kind of wood flooring you plan to install. Strips or planks
can be fastened to a wood floor or subfloor or to plywood or 2-by-4-inch sleepers
(or screeds) installed over a dry, level concrete slab (Fig. 8.20). See Table 8.3. If
the concrete is uneven, it is best to build a subfloor suspended over the concrete
base. Parquet flooring requires a solid, smooth, continuous subfloor, whether of
boards or plywood panels, concrete, or even resilient flooring. Laminated wood
tiles can be laid directly in mastic on a thoroughly dry concrete slab that has
a waterproofing membrane below or above to keep it dry, but blocks of solid
wood pieces are best installed over a plywood base built over the slab.

Unlike parquet tiles, strip-and-plank flooring can't be cemented directly to concrete. If you don't want to build a subfloor, lay a polyethylene vapor barrier, fasten down 2-by-4-inch (5-by-10-cm) sleepers, and then nail the flooring to the
sleepers. Be sure the concrete is properly sealed against moisture. If the floor is

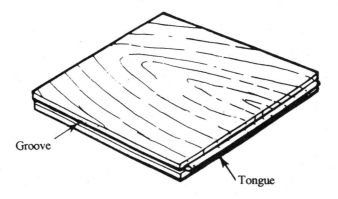

Groove

Tongue

Parquet wood block

Figure 8.19 Tongue-and-groove wood block flooring. (*Leon E. Korejwo*)

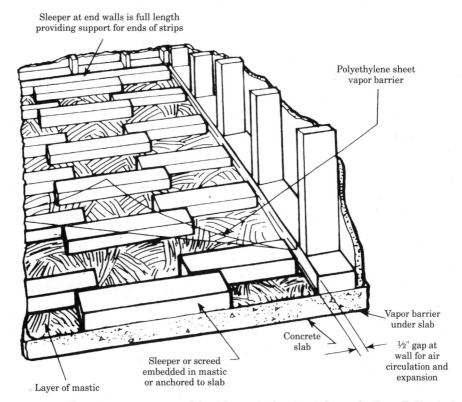

Sleeper at end walls is full length
providing support for ends of strips

Polyethylene sheet
vapor barrier

Vapor barrier
under slab

Concrete
slab

½" gap at
wall for air
circulation and
expansion

Sleeper or screed
embedded in mastic
or anchored to slab

Layer of mastic

Figure 8.20 Sleepers over a concrete slab with a vapor barrier underneath. (*Leon E. Korejwo*)

cold, lay rigid foam insulation between the sleepers. Allow some space above the foam for air circulation. A single layer of polyethylene film laid under new wood flooring is usually considered adequate moisture protection for a dry slab floor (Fig. 8.21).

A wood subfloor, either plywood or boards, should be covered with a plywood underlayment. The combined thickness of subfloor and underlayment should be $1\frac{1}{4}$ inches (3 cm) with underlayment at least $\frac{3}{8}$ inch (1 cm). If the combination is not thick enough, cover it with a layer of $\frac{3}{8}$-inch (1 cm) exterior-grade plywood.

The laying of wood strip flooring should be completed after the other interior wall and ceiling finish is completed (see Chap. 11), windows and exterior doors are in place, and most of the interior trim, except base, casing, and jambs, is applied, so that the flooring will not be damaged by construction activity. Board subfloors should be clean and level and covered with a deadening 15-pound asphalt-saturated felt. This felt stops a certain amount of dust, somewhat deadens sound, and, where a crawlspace is used, increases the warmth of the floor by preventing air infiltration. Due to expansion and contraction, always allow a gap at the wall, finishing with molding to conceal the gap.

In some cases, tongue-and-groove flooring needs no additional subflooring and can be nailed directly to the joists. When installing over this type of existing floor, eliminate loose or squeaky joints before installing a new surface.

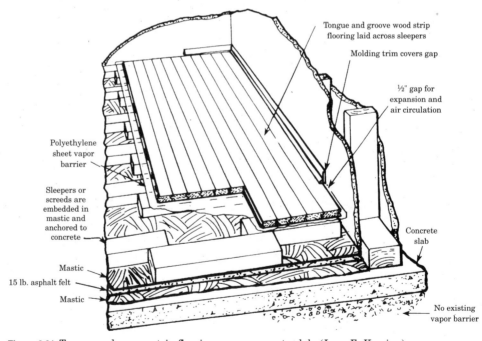

Figure 8.21 Tongue-and-groove strip flooring over a concrete slab. (*Leon E. Korejwo*)

Table 8-3 **Preparation guidelines for wood flooring**

Surface applied over	Preparation steps
Concrete slab	☐ Be sure the slab is smooth, level, and clean. ☐ Check that it is completely dry and take steps to ensure it stays dry. Any moisture coming through the concrete will eventually cause the flooring to loosen, rot, or mildew. ☐ Sweep the slab clean to remove dust and dirt. Do not clean the floor with water. Use a chemical cleaner that removes grease and oil. ☐ Even if a below-grade or on-grade slab appears dry, apply a good vapor barrier (polyethylene) under the floor to safeguard against future moisture problems. ☐ Except for glue-down parquet or laminated planks, wood flooring cannot be secured directly to a concrete slab. ☐ Lay sleepers (strips of wood) with a plywood subfloor of 3/4-inch (1.9-cm) tongue-and-groove exterior plywood (required for wood block floors) and a vapor barrier to allow for air circulation, and prevent moisture buildup.
Wood subfloor supported by joists and beams	☐ Plywood common pine or fir 1-by-4 or 1-by-6 boards provide a suitable base for wood flooring. ☐ Plywood that is 3/4 inch (1.9-cm) thick is considered the best subfloor material. ☐ Lay subfloor boards or planks diagonally across the joists and nail them in place using a 1/16-inch (.16 cm) gap between them for expansion. ☐ Plywood subflooring with interlocking tongue-and-groove joints have those edges on the 8-foot sides only. They are laid across the joists, while the 4-foot lengths are laid over the centerline of the joists to reduce panel separation. ☐ It is possible to lay wood flooring over an old floor that is in good condition. For the most reliable base, remove the old flooring to get down to the subfloor and make any necessary repairs, or install underlayment.
Old resilient flooring	☐ If in good condition, flat, and securely fastened down, old resilient flooring can serve as a base for new wood block, but not strips or planks. ☐ If damaged, the flooring should be removed or covered before installing a new wood block floor. ☐ If too difficult to remove old vinyl flooring, cover with 1/4-inch (0.6-cm) underlayment plywood or hardboard.
Old wood flooring	☐ Old wood flooring must be structurally sound and perfectly level. A wood floor makes a good subfloor if it has no seriously damaged or loose boards. ☐ Examine the old floor and support structure carefully for damaged members or moisture damage. ☐ If installing new wood block flooring, remove all wax, varnish, or other finish from the old floor.

Refer to Table 8.3 for information necessary to prepare a concrete slab, a wood subfloor, or an existing floor for new wood flooring.

Preparing for masonry flooring

Masonry flooring can be laid over a ground-level concrete slab or a suspended wood floor. However, putting masonry over a suspended floor in a frame home can be complicated because of the weight. If setting masonry over a wood floor, the floor must be protected by a moisture barrier of asphalt felt.

Masonry exerts tremendous weight and should be laid over concrete. A concrete slab provides the best-possible subfloor. Make sure the slab is clean so that mortar will bond to it. Slate or flagstone may be set in a bed of mortar on a concrete slab or in mastic or adhesive over a wood subfloor. Preparing a base for masonry flooring is simple if the flooring is to be laid over a concrete slab at ground level. If masonry flooring was planned as part of the original building, allowances have been made for the extra weight, and the support system was planned accordingly. But if brick or stone flooring is being laid in an older home as part of a remodeling project, the old floor, in most cases, will need reinforcing. Always consult a building design professional before putting masonry flooring over any subfloor other than a ground-level concrete slab.

Stone is available in rough-hewn pieces (irregular thickness and shape) or in uniform tiles. You need a thick mortar base over a concrete slab to lay a flooring of rough-hewn stone. The irregularity of the material makes it necessary to provide a cushion of mortar that compensates for the differences in thickness from piece to piece.

You can use thin-set mortar for stones that can be cut into pieces of uniform thicknesses, such as slate and marble. Keep in mind that the thickness of these stones, plus the mortar bed, raises the height of the floor considerably. Though most bricks can be used for both walls and floors, *pavers* are intended specifically for floors. Pavers are available in regular or slightly less than regular thicknesses, or as splits, which are half as thick as regular bricks. Full pavers can be laid on a concrete slab; splits require a concrete slab topped with a mortar bed or spread with thin-set adhesive. Wooden subfloors require the lighter-weight splits. Because this precut material is a lighter floor, it requires less structural support than rough-hewn stone in a full bed of mortar. The subfloor requirements are the same as those for ceramic tile. If installing a masonry floor, particularly one of natural stone, be sure to check the local building code for subfloor requirements.

Mechanical systems

A kitchen builder must be familiar with mechanical systems, such as plumbing; gas; electrical; heating, ventilation, and air-conditioning (HVAC); and cooking ventilation. Generally, you use trained subcontractors for much of this work, but, as the builder, you need to have a clear understanding to accurately plan, coordinate, and install a kitchen.

Kitchen plumbing

Because most of a plumbing system is hidden inside walls and floors, it may seem complex and overwhelming; however, kitchen plumbing needs are fairly simple and straightforward. The plumbing system of the home provides a means of bringing water to an outlet and a means of taking used water away. In today's kitchen, water is needed at many places, no longer just the kitchen sink. Water must be supplied to the dishwasher, food disposer, water purifier, ice maker, and special preparation or service areas, such as a bar or vegetable sink. It is also necessary to provide a means of draining wastewater away from each of these sinks and appliances (with the exception of ice makers).

Once installed, the plumbing system is relatively permanent. Changes to existing systems are usually expensive because they require walls to be opened and holes to be cut in floors and ceilings. The effect changes may have on rooms located below or above the kitchen, or both, can also increase costs considerably.

The total plumbing system includes all pipes, fixtures, and fittings used to convey water into and out of homes. It can be divided into three basic areas:

1. The water-supply system

2. The drainage system

3. The fixtures and appliances

The object of it all, of course, is to make water available where it is wanted in the home, and to get rid of water, plus wastes, after it has served its purpose.

Water-supply system

In every plumbing system, there must be a source of water and pipes to carry it to the fixtures (Fig. 9.1). This water-supply system must be adequate to ensure the homeowner has pure water for drinking, to supply a sufficient quantity of water at any outlet in the system (at the correct operating pressure), and to furnish the homeowner with hot or cold water, as required.

The main supply line (regulated by a main shutoff valve) coming into a house carries cold water. This source is provided by either a municipal water company or a private underground well. Figure 9.2 shows a typical jet-pump type private underground well. If the source is a municipal supplier, the water passes through a meter that registers the amount of water used. From this main supply line (through one pipe), a branch splits off and is joined to a hot-water heater. From this point, branches lead to the various fixtures and faucets in the home. The water supply to fixtures and appliances is controlled with faucets and valves. The supply mains should be graded to one low point in the basement so that a drain cock permits complete drainage of the entire supply system. Any portion of the piping that cannot be so drained must be equipped with a separate drain cock. As a rule, a pitch of $\frac{1}{4}$ inch (0.6 cm) to each foot of pipe is sufficient to permit proper drainage.

The water-supply system should deliver water to the sink and appliances in the quantity and rate needed. The size of pipe, number of fittings, the length of pipe from the main, the pressure available, and the number of other fixtures in use at the time determine the rate at which water flows from the faucet. In remodeling projects, if new plumbing fixtures are added, the pipe size of risers and main arteries (or branches) might need to be increased.

The service from the main or well should be underground to avoid freezing or mechanical damage. A valve to control the water (the main shutoff) should be located inside the house near the entrance of the water-service pipe. The water meter (if installed) is usually located at this point.

The size of the water piping from the main to the individual fixtures is determined by the number and type of fixtures served by the pipe. The following relationship can be used as a general "rule-of-thumb:"

▫ For service to three or more fixtures, use $\frac{3}{4}$-inch (2-cm) pipe.

▫ For service to two or fewer fixtures, use $\frac{1}{2}$-inch (1.3-cm) pipe.

Each fixture is supplied by a fixture-supply pipe. Table 9.1 shows the minimum sizes of supply pipes for specific fixtures. This table is a good starting point. For most fixtures, a $\frac{1}{2}$-inch (1.3-cm) pipe should be run to the fixture's cutoff valve. Under most codes, the water service to the house might be a $\frac{3}{4}$-inch (2-cm) pipe, but a 1-inch (3-cm) water service pipe is a better choice. The larger water pipe

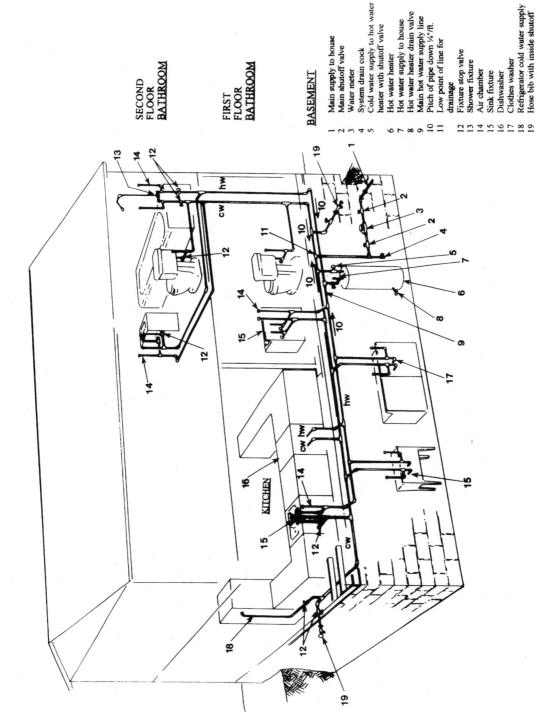

SECOND
FLOOR
BATHROOM

FIRST
FLOOR
BATHROOM

BASEMENT

1   Main supply to house
2   Main shutoff valve
3   Water meter
4   System drain cock
5   Cold water supply to hot water
    heater with shutoff valve
6   Hot water heater
7   Hot water supply to house
8   Hot water heater drain valve
9   Main hot water supply line
10 Pitch of pipe down ¼"/ft.
11 Low point of line for
    drainage
12 Fixture stop valve
13 Shower fixture
14 Air chamber
15 Sink fixture
16 Dishwasher
17 Clothes washer
18 Refrigerator cold water supply
19 Hose bib with inside shutoff

KITCHEN

Figure 9.1 A typical water supply system. (*Leon E. Korejwo*)

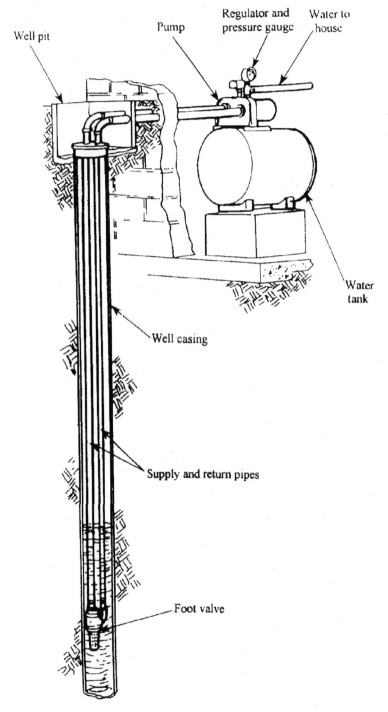

Figure 9.2 Jet-pump type well system. (*Leon E. Korejwo*)

TABLE 9.1 Minimum sizes for fixture water-supply lines

Fixture	Supply pipe size (in inches)
Clothes dryer	$\frac{3}{4}$ inch (2 cm)
Clothes washer	$\frac{3}{4}$ inch (2 cm)
Dishwasher	$\frac{1}{2}$ inch (13 mm)
Hose bib	$\frac{1}{2}$ inch (13 mm)
Hot-water heater	$\frac{3}{4}$ inch (2 cm)
Ice maker	$\frac{1}{8}$ inch (3 mm)
Kitchen sink	$\frac{1}{2}$ inch (13 mm)
Laundry tub	$\frac{1}{2}$ inch (13 mm)
Sink risers	$\frac{1}{4}$ inch (6.3 mm)

allows for future expansion and provides a higher volume of water to the interior system. Each supply pipe should be equipped with a valve or "stop" so that repairs can be made to an individual fixture without interrupting service to other fixtures in the house.

Water heaters Hot water is obtained by routing cold water through a water furnace. This heater may be part of the central heating plant or a separate unit. When part of a central system, a separate hot-water storage tank is generally provided to hold the heated water. On the other hand, when a separate heater is used, the water is stored within the unit (Fig. 9.3).

A separate heater unit may be electric, oil, or gas-fired, but all are automatically controlled by a preset thermostat. Each style of heater comes in a wide variety of sizes. All automatic heaters have the necessary internal piping already installed, and the only connections required are the hot- and cold-water and fuel lines. Oil or gas-fired water heaters also require flues to vent the products of combustion.

A new hot-water heater might be necessary when new plumbing is added to the present system. Even if the present water heater is functioning properly (not scaled up or rusted out), there may not be sufficient hot water available for the family. Actually, the size of the hot-water storage tank needed in the house depends on the number of persons in the family, the volume of hot water that may be needed during peak-use periods, and the recovery rate of the heating unit. A good rule to follow when estimating the capacity of the tank required is 10 gallons per hour for each member of the family. For a family of four, for example, the hot-water demand is 40 gallons of hot water per hour. This does not mean that the system operates continuously at that capacity, but it must be capable of producing that amount of hot water to keep up with normal usage. If any unusual demands are anticipated, a larger capacity should be provided.

The recovery rate of water heaters varies with the type and capacity of the heating element. Temperature and pressure-relief valves are on all hot-water heaters and hot-water storage tanks. Their function is to relieve pressure in the tank and water pipes should any other piece of control equipment in the system fail and the water temperature reach a point high enough to cause a

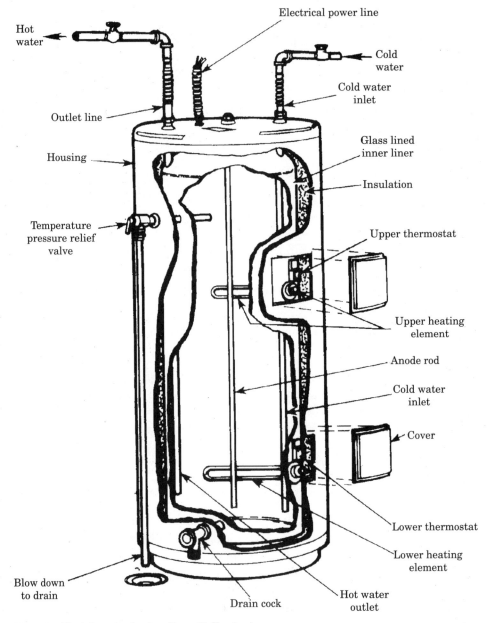

Hot water

Electrical power line

Cold water

Cold water inlet

Outlet line

Housing

Glass lined inner liner

Insulation

Temperature pressure relief valve

Upper thermostat

Upper heating element

Anode rod

Cold water inlet

Cover

Lower thermostat

Lower heating element

Blow down to drain

Drain cock

Hot water outlet

Figure 9.3 Electric water heater. (*Leon E. Korejwo*)

dangerous pressure that would rupture the tank and pipes. Another important device on the heater is the drain cock or valve. Located at the bottom of the storage tank, the drain cock or valve allows for the draining of the tank. A shutoff is also located on the cold-water intake pipe.

New dishwashers and laundries add to the hot-water draw. Make sure the equipment has the capacity to handle them.

In some houses, the distance from the water heater to the point of use may be quite far and may result in long waiting periods for hot water. A circulation line can be installed so that hot water is always available at each fixture. As an alternative to the circulating system, a separate water heater should be considered for the remote area. Or, in some instances, it may be convenient to install a small hot-water booster in the kitchen or laundry if the primary source of hot water is too small in storage or recovery capacity to provide sufficient hot water.

Water softeners In some areas of the country, it is necessary to make hard water soft by piping the domestic water supply through a device called a water softener. Most water softeners have few moving parts and consume little power (Fig. 9.4). The water is treated as it flows through a special chemical that removes the objectionable minerals that make the water hard. Depending on the hardness of the water, the rate of consumption, and the unit's capacity, there comes a time when the chemical must be regenerated or the equipment must be cleaned and renewed. Different types of chemicals and equipment may be required to treat a specific water-hardness problem. For this reason, an analysis of the water should be made. This test may be performed by various local agencies or with a kit that can be obtained from a plumbing supplier.

Drainage system (drain-waste-vent system)

Drainage (strictly controlled by code in most localities) is the complete and final disposal of the wastewater and the sewage it contains. A drainage system, therefore, consists of

1. The pipes that carry sewage away from the fixtures (Fig. 9.5).
2. The place where the sewage is deposited (Fig. 9.8).

The concept of plumbing centers on two physical principles—pressure and gravity. When water is delivered to an outlet, some provision must be made to drain away waste or excess water. The drainage system differs from the water supply in one very important respect.

In supply lines, water flows under pressure. In the drainage lines, flow is entirely by gravity, and the pipes must be designed and installed carefully to ensure flow at a velocity adequate to keep the pipes clean. Because they are small in diameter and do not depend on gravity, supply lines are easily rerouted. The drainage system is also more complex than the supply system in that it consists of three parts, all of which are needed in every installation, even if only one fixture is served. The flow of wastewater starts at the fixture trap, the device that stops sewer gases from entering the house. It flows through the fixture branches to the soil stack. It continues through the house drain and the house sewer and finally reaches the city sewer system, or, in a

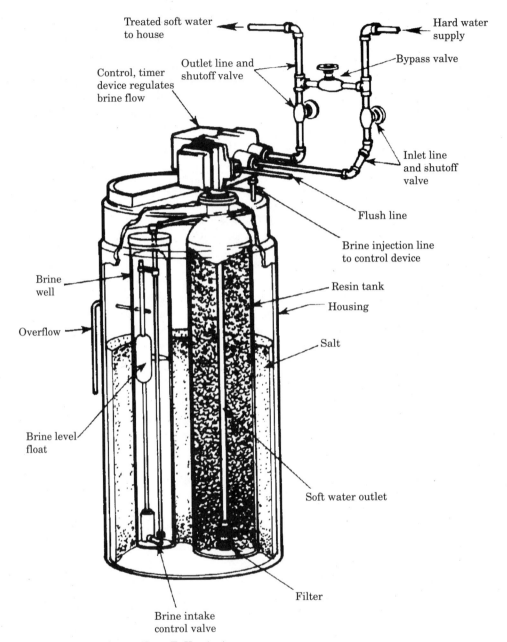

Figure 9.4 Water softener. (*Leon E. Korejwo*)

private system, a septic tank. Waste stacks carry only water waste. These parts are as follows:

1. *Traps* are water seals that carry the wastewater to the drain lines and prevent the backflow of air or sewer gas into the house. Figure 9.6a shows a trap as its liquid is being siphoned out by suction in the drain line. Other

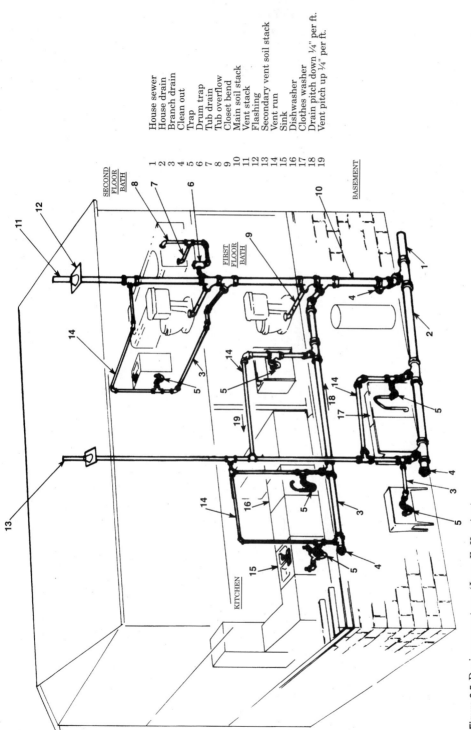

1   House sewer
2   House drain
3   Branch drain
4   Clean out
5   Trap
6   Drum trap
7   Tub drain
8   Tub overflow
9   Closet bend
10 Main soil stack
11 Vent stack
12 Flashing
13 Secondary vent soil stack
14 Vent run
15 Sink
16 Dishwasher
17 Clothes washer
18 Drain pitch down ¼" per ft.
19 Vent pitch up ¼" per ft.

SECOND FLOOR BATH

FIRST FLOOR BATH

KITCHEN

BASEMENT

Figure 9.5 Drainage system. (*Leon E. Korejwo*)

drains downstream can also draw the liquid when they are drained. Figure 9.6b shows how gases pass over the low water. Figure 9.6c illustrates how adding a vertical vent pipe breaks the vacuum in the horizontal drain, thereby allowing liquid to remain in the curve of the trap. The liquid seal blocks rising sewer gases from entering the room. Traps should be accessible and as close as possible to each fixture. Sometimes they are part of the fixture. Plumbing codes restrict the distance a sink's trap can be located from its drain/vent line. The trap never empties and retains enough water to maintain the seal. Trap styles are: S-trap. deep seal P-trap, and running-trap (Fig. 9.6d), P-trap (Fig. 9.6e), or P-traps using a J-bend fitting (Fig. 9-6f) which connect the fixture and the drain pipe. When the fixture drains, water flows through the trap and into the drain pipe. Most traps used today are P-traps, which drain into the wall by way of a 90-degree trap arm. On existing installations, you might also find S-traps, which drain into the floor via a 180-degree trap arm. (This type is no longer allowed by code in many areas.)

2. *Drainage lines* are pipes, either vertical, called *stacks*, or horizontal, called *branches*, that carry the discharge from the fixtures to the house sewer (see Fig. 9.1). They are usually concealed in walls or floors. The pipes receiving the discharge from the water closet are known as soil lines. Those receiving the discharge from other fixtures are known as waste lines. Some plumbing codes require a grease trap in the waste line from the kitchen sink. Others specifically prohibit them. This trap, which is designed to prevent grease and oil from entering the sewer, must be cleaned frequently for efficient operation.

3. *Vent lines* are pipes extending upward through the roof. They allow air to flow into or out of the drainage pipes, thus equalizing air pressure in the drainage system and protecting the water seal in the traps. Without a vent pipe, the rush of water down a drain could cause a siphoning action that would pull the water seal out of the trap and let sewer gases enter the home. The vent pipe is generally located right behind the sink.

The relationship between the trap, the drainage line, and the vent line is very important and has a great influence on the location of plumbing fixtures, particularly in remodeling work where the relocation of fixtures is desired. Actually, the required maximum distance from a trap to a vent does not usually present any problems in residences, except when island sinks are used. This problem can usually be solved by using a drainage line large enough for the distance to the nearest vent. The actual span varies from one locality to another, but generally speaking, a sink cannot be moved more than 3 to 4 feet without either adding or extending branch lines; in some situations, an entirely new run of piping from basement to roof may be needed. Remember that some plumbing code regulations do not permit sinks or other water-using appliances anywhere other than along a wall.

All vents must terminate outside the house. The vent terminal must be carried through the roof at full size. It must be at least 3 inches (8 cm) in

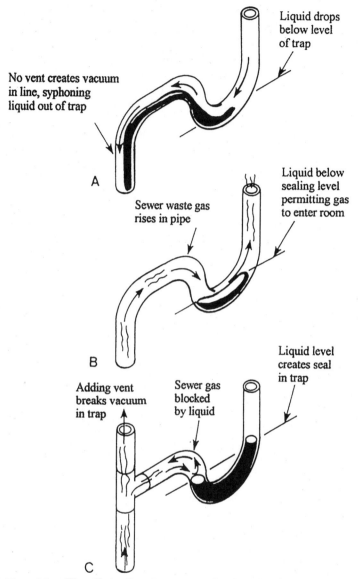

No vent creates vacuum in line, syphoning liquid out of trap

Liquid drops below level of trap

Sewer waste gas rises in pipe

Liquid below sealing level permitting gas to enter room

Adding vent breaks vacuum in trap

Sewer gas blocked by liquid

Liquid level creates seal in trap

A

B

C

Figure 9.6a-c The effects of vents on trap seals. (*Leon E. Korejwo*)

diameter to prevent clogging by frost. The minimum extension above the roof is 6 inches (15 cm). *Individual vents* serve one fixture with a trap (Fig. 9.7a). A *common vent* serves two fixtures (Fig. 9.7b). When a fixture discharges through its drain, it is called a *wet vent* (Fig. 9.7c). A *dry vent* does not act as a drain (Fig. 9.7d).

Each trap installed in a drainage system must be vented. Traps may be provided with individual vents (sometimes referred to as back-vents or continuous vents) or a common vent. The latter is useful when sinks are located side by side or when a new fixture is to be added to an existing drainage system. A new fixture could be added below or above an existing fixture as long as the lower fixture is the one with the greater flow.

The drain/waste lines from each fixture or group of fixtures are connected to the building drain. The building drain takes the waste material to the sewer system or septic tank. The building drain must be installed with the proper slope—$\frac{1}{4}$ inch (0.6 cm) per foot (Fig. 9.7e). Steeper slopes provide higher velocities, increasing the carrying capacity of the pipe, and tend to keep the drain pipe clean.

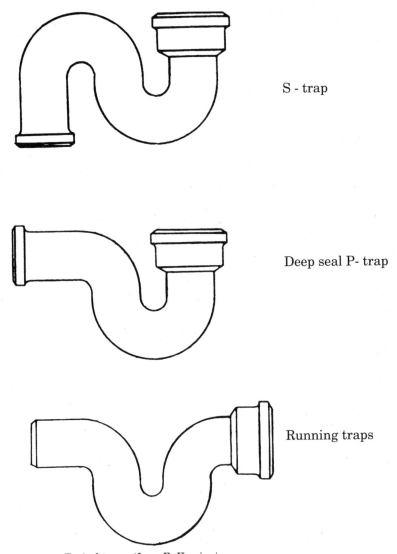

S - trap

Deep seal P- trap

Running traps

Figure 9.6d Typical traps. (*Leon E. Korejwo*)

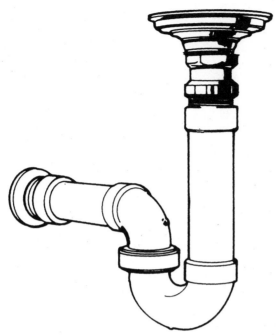

Figure 9.6e Typical P trap. (*Leon E. Korejwo)*

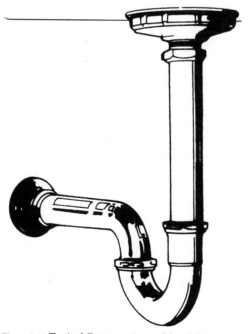

Figure9.6f Typical P-trap using a J-bend fitting. (*Leon E. Korejwo)*

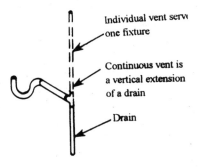

Individual vent serve
one fixture

Continuous vent is
a vertical extension
of a drain

Drain

Figure 9.7a Individual vent. (*Leon E. Korejwo*)

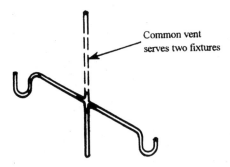

Common vent
serves two fixtures

Figure 9.7b Common vent. (*Leon E. Korejwo*)

A septic tank treats waste right on the property in an enclosed tank. They are watertight receptacles that receive the discharge of the drainage system. The liquids are discharged into the soil outside of the tank in what is referred to as a tile field (or drain field); the solids in the waste biodegrade in the septic tank (Fig. 9.8).

The city sewer system carries waste away from the property. If adding new fixtures to the kitchen, double-check that the capacity of the septic system can handle the additional fixtures. If the residence is hooked up to a city sewer system, be sure the waste line is properly sized to handle the total fixtures.

To be safe, the drainage system has to meet five basic requirements:

1. All pipes in this system must be pitched (slanted) down toward the main disposal so that the weight of the waste causes it to flow toward the main disposal system and away from the house. Because of gravity flow, the waste lines must be larger than the water-supply lines, in which there is pressure.

2. Pipes must be fitted and sealed so that sewer gases cannot leak out.

3. The system must contain vents to carry off the sewer gases to where they can do no harm. Vents also help to equalize the air pressure in the drainage system.

4. Each fixture that has a drain should be provided with a suitable water trap, so that water standing in the trap seals the drain pipe and prevents the

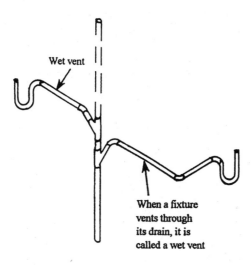

Wet vent

When a fixture vents through its drain, it is called a wet vent

Figure 9.7c Wet vent. (*Leon E. Korejwo*)

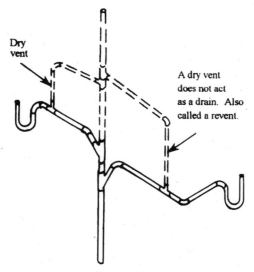

Dry vent

A dry vent does not act as a drain. Also called a revent.

Figure 9.7d Dry vent. (*Leon E. Korejwo*)

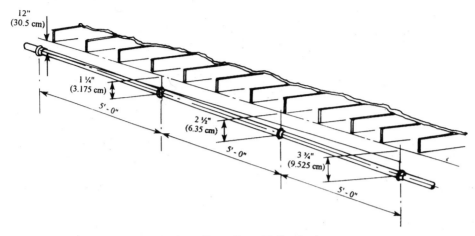

Figure 9.7e The proper pitch for drain lines. (*Leon E. Korejwo*)

backflow of sewer gas into the house. The trap for the toilet (water closet) is built into it.

5. Re-vents should be provided wherever there is danger of siphoning the water from a fixture trap or where specified by local codes.

Fixtures

The fixtures provide the required means for using the water (Fig. 9.9). In this sense, a faucet on the outside of the house (for attaching a hose) is a fixture, as well as a laundry tub in the basement, or a dishwasher, or a toilet. Each has a purpose connected with the homeowner's use of water, and each must have certain features to serve its purpose. Fixtures can be costly plumbing items and should exactly suit the homeowner's needs.

Plumbing codes

Be aware that most building codes require plumbers to have a special license, so unless you have that license, you need to hire a plumbing contractor. Some areas have exceptions; however, the work must still be approved and checked by the local building inspector. Although most plumbing systems are based on national codes, local building codes may vary from those in an adjoining city or town.

When you are installing a kitchen, all plumbing work must be done in accordance with local plumbing codes. The primary purpose of a plumbing code, as well as of a building code, is to protect the health and safety of the homeowner.

There are three major plumbing codes used in the United States: the BOCA National Plumbing Code (primarily used along the East Coast), the Uniform Plumbing Code (most prevalent in the western United States), and the

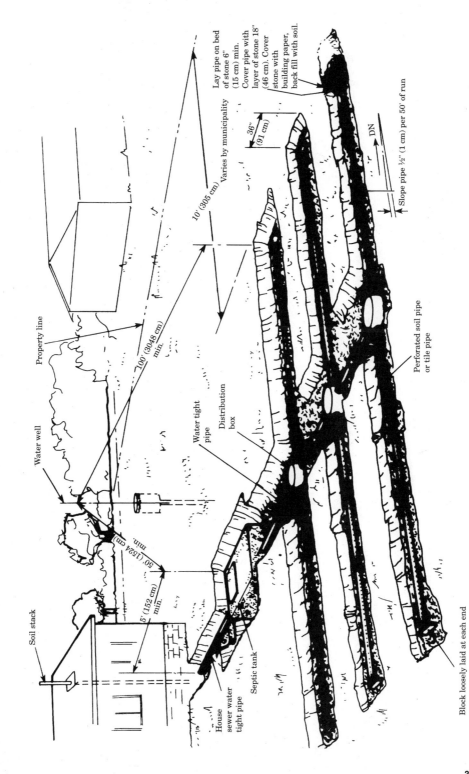

Soil stack

Water well

Property line

10′ (305 cm) Varies by municipality

Lay pipe on bed of stone 6″ (15 cm) min. Cover pipe with layer of stone 18″ (46 cm). Cover stone with building paper, back fill with soil.

36″ (91 cm)

DN

Slope pipe ½″ (1 cm) per 50′ of run

100′ (3048 cm) min.

Water tight pipe

Distribution box

Perforated soil pipe or tile pipe

50′ (1524 cm) min.

5′ (152 cm) min.

House sewer water tight pipe

Septic tank

Block loosely laid at each end

Figure 9.8 Septic tank tile field. (*Leon E. Korejwo*)

303

Figure 9.9 Various types of fixtures. (*Leon E. Korejwo*)

Standard Plumbing Code (primarily in the southern states). Other codes are used, but these are the three major ones. If plumbing work is done in accordance with the codes, the homeowner is protected against improper fixtures, materials, and installation (against contamination of the water supply, contamination of the air, and undesirable odor due to escape of sewer gas).

Plumbing materials

Regional preferences and area building codes generally dictate what type of pipe must be used for both new construction and remodeling projects. However, builders should be familiar with the major types of plumbing pipe used.

Plastic piping is lightweight and one of the easiest pipings to use. In new construction, it is becoming the water piping of choice. However, it is important to be aware that its use is restricted by some building and plumbing codes. As with all plumbing, be sure to check all codes before using these materials. Some types of plastic piping include polybutylene (PB), chlorinated PVC (CPVC), polyethylene, acrylonitrile-butadiene-styrene (ABS), and polyvinyl chloride (PVC).

PB and CPVC pipe are used in water-supply systems. PB is very flexible and extremely durable (Fig. 9.10a). This pipe is available in long rolls, eliminating the need for joints. It is flexible and resistant to damage from backfilling. If comparing costs, ease of installation, and effectiveness, polybutylene pipe is hard to beat. When installed properly, it is one of the best materials for carrying water. CPVC is a rigid pipe that tends to be brittle and is prone to cracking under stress (Fig. 9.10b). But once it is installed correctly, CPVC gives reasonably good service, as long as it is not subjected to abuse, and can last a long time. CPVC and PB are suitable for both hot- and cold-water applications. Whenever feasible, it is best to eliminate joints in underground piping, and with CPVC, you cannot eliminate the joints in long runs.

Polyethylene pipe has long been used as a water-service pipe (Fig. 9.10c). It also comes in large rolls, eliminating the need for underground joints. The plastic pipe is durable and resistant to backfilling accidents. Polyethylene is probably one of the most frequently used pipe for modern water-service installations.

ABS and PVC are both commonly used in drain (DWV) systems (Figs. 9.10d and e). They are very simple materials to work with and can be cut with any fine-toothed handsaw. ABS and PVC are, in some instances, used together in the same system.

Note: The home electrical system could be grounded through metal water pipes. When adding plastic pipes to a metal plumbing system, be certain the electrical ground circuit is not broken. Use ground clamps and jumper wires to

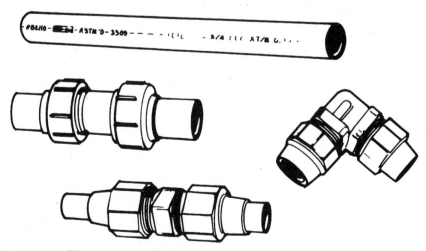

Figure 9.10a PB piping. (*Leon E. Korejwo*)

bypass the plastic transition and complete the electrical ground circuit. Clamps must be firmly attached to bare metal on both sides of the plastic pipe.

Copper (technically called tubing, not pipe), though somewhat more expensive than others, is lightweight, versatile, highly resistant to corrosion, and gives many years of good service (Fig. 9.10f). It is most commonly

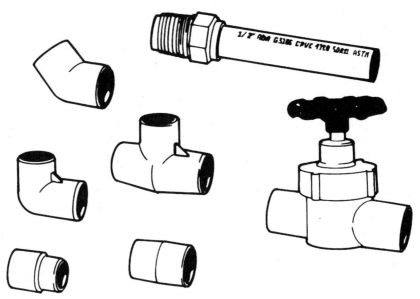

Figure 9.10b CPVC piping. (*Leon E. Korejwo*)

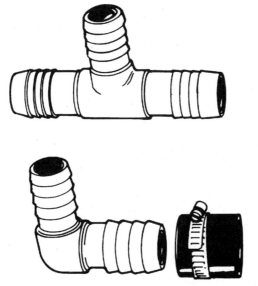

Figure 9.10c PE piping. (*Leon E. Korejwo*)

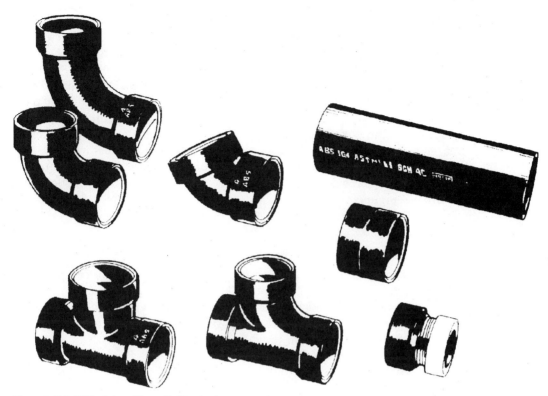

Figure 9.10d ABS piping. (*Leon E. Korejwo*)

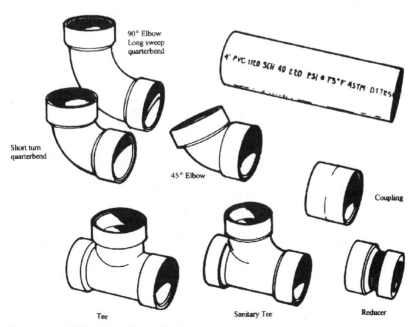

90° Elbow
Long sweep
quarterbend

Short turn
quarterbend

45° Elbow

Coupling

Tee

Sanitary Tee

Reducer

Figure 9.10e PVC piping. (*Leon E. Korejwo*)

used for water supply for remodeled kitchens. It is good water-service pipe, unless the water supply contains a high acid content. Be certain to verify that the pipe thickness is sufficient to meet all building codes. Many times, building codes specify the use of copper piping.

Galvanized steel piping (found in houses built before 1935) is notorious for its ability to rust out and clog up, and it is heavy (Fig. 9.10g). Whether used for drain, vent, or water service, all galvanized steel pipes should be replaced. It is outdated and serves no purpose in modern plumbing installations. It increases the risk of rust, leaking joints, and restricted water flow.

Cast-iron piping has a long life with few problems and is basically satisfactory for continued service (Fig. 9.10h). However, it is one of the heaviest and most difficult pipings to work with. It does only drain-waste-vent duty, though no-hub clamps make smaller jobs not overly difficult for an amateur.

Checking plumbing

Inspection of the plumbing system is not easy because much of the piping is concealed and inaccessible. Be sure to look under any cabinets that have plumbing (i.e., under the sink) to find out the types of materials used for water pipes, drains, and vents. It might also be possible to inspect some of the plumbing from the basement, cellar, or crawl space.

The American Institute of Kitchen Dealers suggests to its members that in checking a kitchen remodeling job, the checklist in Fig. 9.11 be used to evaluate the present plumbing system.

After completing the plumbing evaluation procedure, accurate assessments can be made on the condition of plumbing systems and how the problems can be corrected.

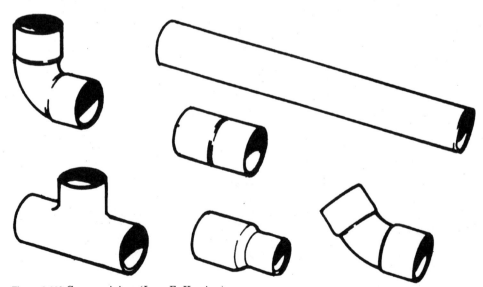

Figure 9.10f Copper piping. (*Leon E. Korejwo*)

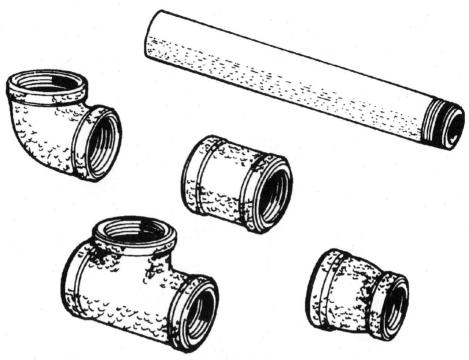

Figure 9.10g Galvanized steel piping. (*Leon E. Korejwo*)

Lead in plumbing

Lead has become a serious problem in plumbing, and one that must be addressed by kitchen builders doing remodeling projects. Its presence depends on when the house was built, what kind of plumbing pipes it has, and what types of water mains connect the building to the water-supply system. If you discover any lead bends, traps, or pipes, they should always be replaced when remodeling a kitchen. Soft lead deteriorates over time and problems are sure to arise.

Unsightly pipes

When remodeling a kitchen in an older home, unsightly pipes running up the walls can be a problem. There are two relatively simple solutions.

- Enclose the pipes in the wall (Fig. 9.12a). You must cut the pipe and install elbows to move the pipe into the wall itself, which means cutting a slot in the wall for the pipe, then patching the wall afterward. It is a lot of work, but it is the most satisfactory solution. If drains from upstairs bathrooms pass into the remodeled area, a 2-by-6-inch (5-by-15-cm) wall may prove adequate.

- Box in the pipes (Fig. 9.12b). This solution is easier. Simply build the wall out around the pipes using a simple wooden-box construction. Then paint or

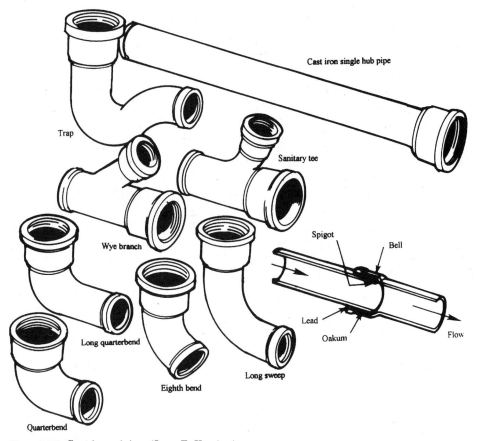

Figure 9.10h Cast iron piping. (*Leon E. Korejwo*)

paper the addition to match the rest of the wall. This method is more of a compromise solution, and it does leave a corner jutting into the room. Drain lines are usually larger than the 2-by-4-inch (5-by-10 cm) framing in most walls. Since most kitchens are on the first floor, drain lines run into the floor and basement.

Heating, ventilation, and air-conditioning (HVAC) systems

Heating, ventilation, and air-conditioning (HVAC) systems are designed to maintain comfort in the home. HVAC systems may be affected by your proposed kitchen project. Changes are governed either by the local plumbing regulations or a separate mechanical code. As with plumbing, a special trade license is usually required in this area of work. Therefore, unless you are a skilled HVAC technician, you should plan on hiring a competent subcontractor to work with these systems. It is always important to make your client aware of two important factors when installing a heating and air-conditioning system:

1. The cost of the purchase and installation
2. The cost of actually operating the various systems

As the installer, you are not responsible for the design of the system, but you need to review the design before the project begins for any potential problems, i.e., noise from HVAC systems or air from supply registers blowing on people while dining. Kitchen builders must check the type of heat and make sure the designer or architect took the existing system and ducts or radiators into consideration in the design and layout of the room. Probably the most difficult task in providing heating and cooling for a kitchen is finding a suitable location for the terminal device of the heating/cooling system. Generally, the outside walls of the kitchen are lined with cabinets, making it almost impossible to locate air registers, hydronic baseboards, or electric resistance baseboards on the wall.

The various heating systems used in a home are not discussed fully here. However, suggestions as to how they can be employed and installed in a kitchen—new or remodeled—are given here.

Both air-conditioning and heating ducts are fairly easy to reroute, as long as you can gain access from a basement, crawl space, garage wall, or unfinished attic. Radiant heat pipes or other slab-embedded systems may pose problems.

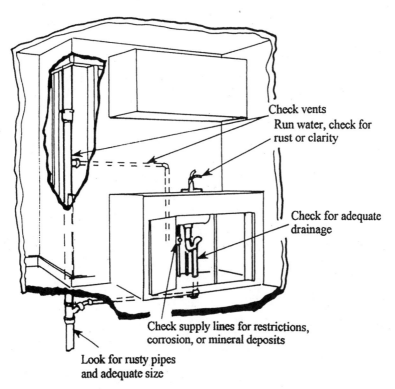

Check vents
Run water, check for
rust or clarity

Check for adequate
drainage

Check supply lines for restrictions,
corrosion, or mineral deposits

Look for rusty pipes
and adequate size

Figure 9.11 Kitchen checklist. (*Leon E. Korejwo*)

Pipe rerouted through soffit

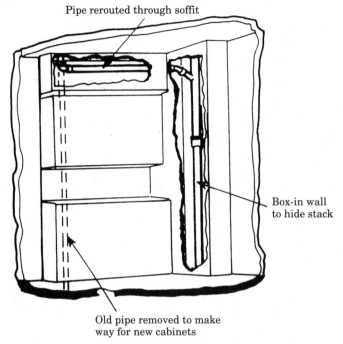

Box-in wall
to hide stack

Old pipe removed to make
way for new cabinets

Figure 9.12a Enclose unsightly pipes. (*Leon E. Korejwo*)

Moving the main drain stack
too wide for normal partition

2" × 6" (5 cm × 15 cm)
Studs used in place of
2" × 4" (5 cm × 10 cm)
provide wider wall space

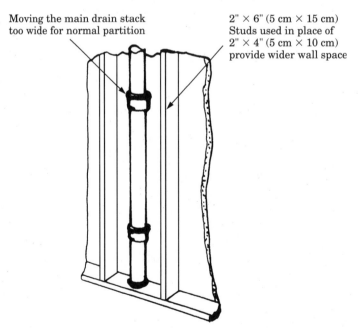

Figure 9.12b One method of concealing pipes. (*Leon E. Korejwo*)

Heating systems

Home heating systems can be classified by the fuel they use or by heating medium. The medium can be warm air, hot water, or steam. Fuels are natural gas, fuel oil, or electricity. Water and steam heat the house with radiators piped from a boiler. Warm air is generated by a furnace and is widely favored because it is easy to add a central air-conditioning unit and a central humidifier. Electric resistance heaters are easy to install, but electric rates make them expensive. Warm-air systems deliver heat to a room through wall, floor, baseboard, or ceiling registers or diffusers (Fig. 9.13a). Air returns to the furnace through return grilles. If the kitchen being remodeled does not have a register, one can be added in a kick space with a duct elbow to route heat into the room through a toekick grille.

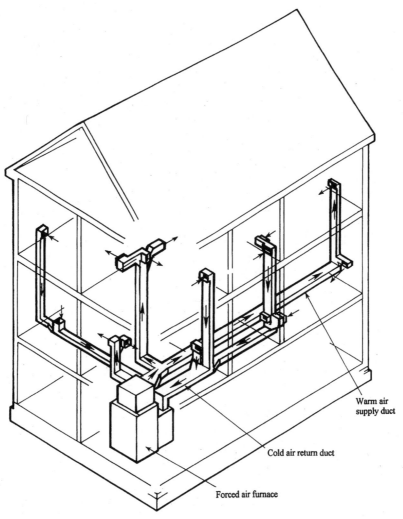

Warm air
supply duct

Cold air return duct

Forced air furnace

Figure 9.13a A typical forced-air heating system. (*Leon E. Korejwo*)

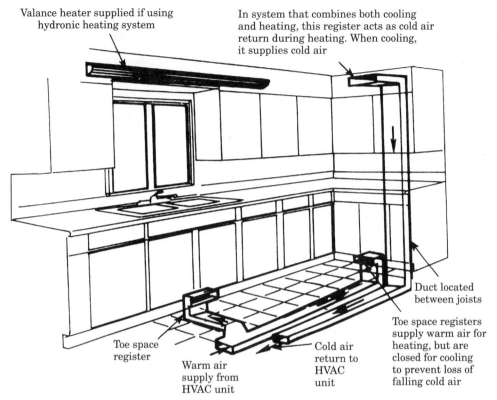

Valance heater supplied if using hydronic heating system

In system that combines both cooling and heating, this register acts as cold air return during heating. When cooling, it supplies cold air

Duct located between joists

Toe space registers supply warm air for heating, but are closed for cooling to prevent loss of falling cold air

Toe space register

Warm air supply from HVAC unit

Cold air return to HVAC unit

Figure 9.13b Adding registers. (*Leon E. Korejwo*)

Warm-air systems Registers for warm-air systems and small electric resistance blower-operated units can be located in the toe space beneath lower cabinets. These units are acceptable for heating, but registers in the toe space present problems of proper air distribution for cooling applications. If they are used, they should be considered for heating use only, and provision should be made so that they may be blocked or turned off in summer when the cooling system is in operation.

In new construction, provisions can be made for using toe-space registers for heating and auxiliary registers located on the side walls for cooling, if the installed system is one that provides both heating and cooling (Fig. 9.13b). For remodeling work, the relocation of ductwork can sometimes present serious difficulties, particularly if the house is built on a slab or if there is a finished ceiling in the basement below the kitchen. If the "new" kitchen is located in an addition to the house, it is possible to extend the ductwork to provide for heating and cooling the addition. Before this is done, the existing furnace should be checked to see that it has sufficient heating and/or cooling capacity for the addition. The blower unit must also be checked to see that it is capable of delivering the additional air needed in the new room. It may be necessary to replace the blower assembly or the blower motor to increase the air-handling capability of the system, even though the furnace burner has sufficient heating capacity. It is also possible to increase the air-handling capacity of the system by speeding up the blower. In general, this is not a good practice, since the increased blower speed normally increases the noise level from the furnace.

Be aware that when remodeling for a kitchen or room adjacent to a heating unit, moving walls or surfaces too close to the heater or ductwork is hazardous. Each heating unit has minimum safe clearances that are included with the manufacturer's installation information or, in some instances, shown on a sticker on the unit (Fig. 9.13c). If in question, consult the subcontractor or manufacturer before planning to move a partition. A heater must have adequate air for combustion. It must not be enclosed in such a way that cuts off its air supply or impedes its flue or exhaust.

Most forced warm-air systems can support another register or two if the furnace is centrally located. A zoned heating system designed to heat the addition may be the best answer. Quite often, the HVAC plan you receive with design plans may be very schematic, and the dedicated areas for running ductwork may not be indicated or planned for in the design. HVAC systems require a substantial amount of space, and cramping the ductwork areas only leads to awkward installations that can cause the HVAC system to be inefficient or noisy. If the addition of any bump-outs or chases for the ductwork will affect the cabinet layout, this information should be determined and reported to the

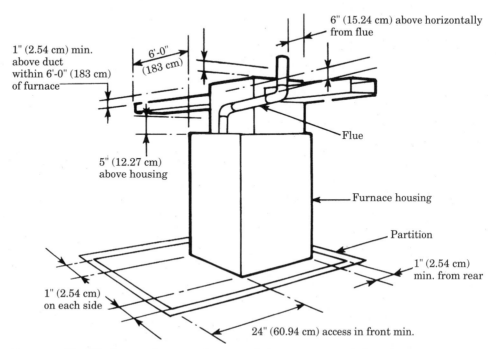

Figure 9.13c Possible furnace clearances (see manufacturers recommended specifications). (*Leon E. Korejwo*)

dealer or designer so that modifications can be made before cabinets are ordered. When remodeling around a hot-air system, registers can be added or relocated. When planning the kitchen, consult the manufacturer or a subcontractor before finalizing a design.

Hydronic (hot-water) systems   In a forced hot-water system, a boiler heats water that circulates through pipes (Fig. 9.14a). The pipes lead to fan coil units, convectors, or radiators. These units radiate heat to the room air. Convectors consist of a core of fins that are heated by hot water. The air passing over these fins is warmed. Fan coil units work on the same principle, but have small fans that push the warm air out into the room. Earlier hot-water systems operated by gravity and did not have a circulating pump. A circulating pump can be added on this type of hot-water system.

In small kitchens or in corridor or U-shaped kitchens, the walls are lined with cabinets, making it difficult to locate baseboard units. A valance unit can be used, or the radiation necessary to offset the heat loss of the kitchen can be located in an adjoining breakfast area or other space immediately

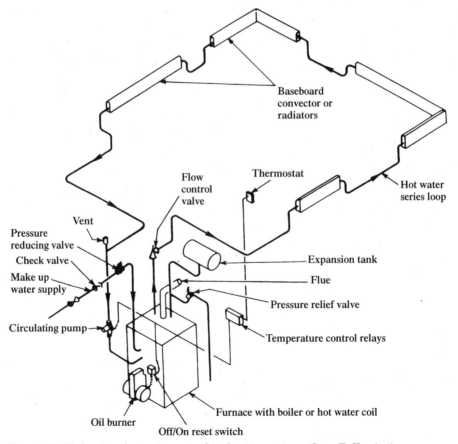

Figure 9.14a Hydronic or hot-water series loop heating system. (*Leon E. Korejwo*)

adjacent to and open to the kitchen (see Fig. 9.13b). The HVAC subcontractor can tell you if the system can be modified or if convector or fan coil units can be relocated. The pipes often run through walls, so you need to trail the path to determine if pipes inside a wall need to be removed or modified. When adding to a hydronic system, be sure the new radiator units are made of the same metal as the old, or it is a sure invitation to corrosion. In extreme cases where the equipment cannot be located in an adjacent space, part of the wall space must be left free for the installation of a convector with or without an integral fan unit to help distribute the warmed air.

Hydronic systems can be modified to suit if a kitchen is part of a new addition. The addition may be heated using the existing furnace; consult the subcontractor. Two methods to add on to an existing system are zone valves or an additional circulating pump and piping. Zone valves are electronically controlled units that are actuated when heat is required in the room. Heated water is pumped by a circulator through the zone valve to baseboard convectors (Fig. 9.14b). A circulating pump may be added to the existing system if that unit is not adequate to move the needed hot water (Fig. 9.14c). New piping, baseboard convectors, and a pump are installed with little disturbance to the existing heating system.

Generally, hydronic systems are not designed as combination heating/cooling systems. Chilled water pumped through the system can be used for cooling if valance units are installed throughout the house. This is not a typical installation, however. When a separate cooling system is installed, the registers may be located in the ceiling, and no difficulty will arise.

The piping and radiators of old-fashioned steam heat systems are best left alone. Because of the age of these systems, the radiators and fittings are quite worn and difficult to repair or replace should a problem occur during your installation.

Electric heat   If electric ceiling cables or panels are used in a kitchen, there is no particular problem except interference from the soffit work over the upper cabinets. Other electric resistance-type heaters are available, including panels of glass or metal that are mounted on the wall; units that have a small fan that circulates the heated air; and hot-water baseboards in which the water is heated by electrically heated elements similar to those used in a water heater. Also available are resistance units that may be inserted in the branch supply ducts of a central-air duct system.

For kitchen installations, a small unit or units that resemble warm-air system registers may be installed in the toe space of the base cabinets. These units are equipped with small blowers to provide air circulation over the resistance elements.

If wall space is not available for baseboard units and a toe-space installation is also not practical, electric cable can be applied to the ceiling. It is also possible to install electric-resistance panels on the ceiling to furnish radiant electric heat. They are used with plastered or plasterboard ceilings. The wires are fastened to the ceiling before installing the finished wall surfacing.

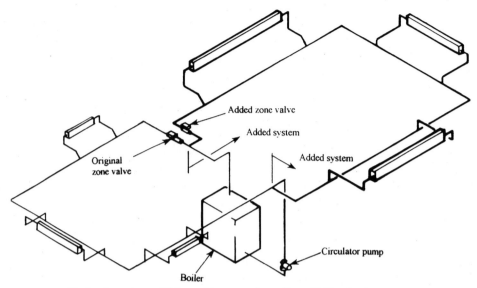

Figure 9.14b Hydronic system addition with zone valves. (*Leon E. Korejwo*)

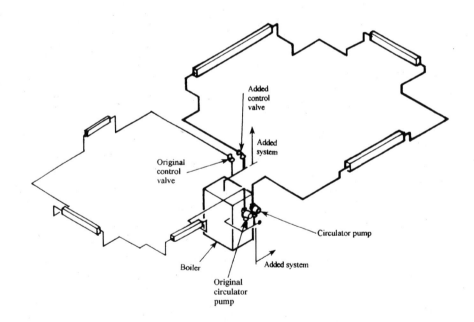

Figure 9.14c Hydronic system addition with new circulator. (*Leon E. Korejwo*)

These panels are rectangular in shape and resemble acoustical tile. With electric-resistance heating, a separate cooling system must be installed.

Subfloor radiant-heating systems   Many manufacturers have introduced electric cable floor warmers that act like heating coils, permanently placed beneath wood, tiles, and even carpet (Fig. 9.15). They are supplementary heating sources of warmth that evenly radiate heat upward, where it does the most good. Most of the new types on the market are mounted on top of the subfloor, have certified insulated cables, and are powered by their own 120-volt box with its own thermostat. In addition, rather than having to design around radiators or typical baseboard heat units, these heating systems are built in under the floor, giving designers more options. The homeowner has more even heating and more space. These systems rely on electricity to provide heat (from 8 to 15 watts per square foot, depending on the floor material) and are designed as a secondary heat source, not the main heat source for the home.

Major benefits to these systems is that, unlike forced hot-air systems, these radiant systems cannot spread dust, pollen, or germs throughout the house and, once installed, are maintenance-free. These subfloor heating systems can be placed under most any flooring materials—hardwood, ceramic tile, carpet—however, they are not recommended for under linoleum.

Radiators While radiators provide a good form of heating and are often found in older homes, they are a real problem when one sits right where new cabinets are desired. What can be done? There are three basic solutions to the problem of a radiator in the kitchen:

1. *Cover it.* Build a cover that matches or complements the rest of the new kitchen design. Make sure this cover allows proper circulation to heat the kitchen adequately.

2. *Move it.* Is there another location in the kitchen for the radiator? Perhaps moving it to another wall might improve the kitchen design and the heating, although this option can be costly. If a radiator is moved, it probably will still have to be covered.

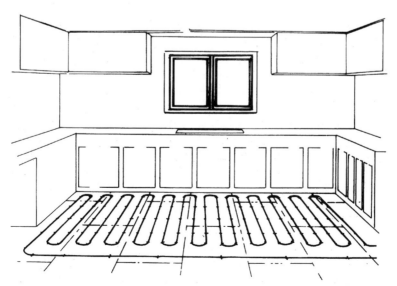

Figure 9.15 Radiant floor heating system. (*Leon E. Korejwo*)

3. *Replace it.* The old-style radiator can often be replaced with a new baseboard unit that gives more versatility in decorating the walls. These old radiators are generally cast iron and are of column, large-tube, or small-tube design. The output of these units is expressed in Btu per hour (Btuh), 1000 Btu per hour (MBtu), or in square feet equivalent direct radiators (EDR). Columns and large-tube radiators are no longer manufactured, but small-tube radiators are still available. When radiators are replaced by modern equipment such as convectors, finned tubes, or baseboard radiators, the replacement must be sized to supply a similar amount of heating. It is always best to consult the literature published by the manufacturer of the equipment for the exact amount of heat output of the unit.

Air-conditioning/cooling systems

As already mentioned, it is possible to combine heating and cooling in a central system. This is usually a good choice. It is also possible to have a separate central cooling system. In addition, window air-conditioning units or through-the-wall units may be used to cool the kitchen. This option eliminates the need for ductwork, but it is not as attractive, and it might not be as efficient. With individual units, occupants can control their own comfort. It a central unit is used, some people can be chilly while others are warm. Independent units are easy to install, unlike central units, which require extensive ductwork and considerable time to install.

Window units fit into the opening of a double-hung window, and special units are available for installation in casement windows. The units vary in cooling capacity from as small as 5000 Btuh to 35,000 Btuh. Room air is circulated through the unit, where it is cooled, dehumidified, and filtered. A condensate drain is not necessary since the moisture condensed from the air is evaporated into the outdoor air. The units are electrically operated, and many of the small ones can be plugged into existing electrical outlets. The larger units require 220 or 240 volts and use up to 200 amperes of electricity. They require a separate circuit installed specifically for the air-conditioning unit.

Through-the-wall units are simply window units that have been provided with a metal sleeve built into the wall, which makes the installation more permanent (Fig. 9.16). The chief advantage of the through-the-wall unit is that the window is not obstructed by the air conditioning unit, and the unit may be placed high on the wall so that the distribution of the cooled air is more efficient. The chief disadvantage of the through-the-wall unit is that when the unit must be replaced after some years of service, it may be difficult to find another unit that fits the sleeve. Some manufacturers have standardized the size of the sleeves, but there are still many nonstandard units.

Heat pumps   Heat pumps have become one of the most popular forms of home heating and cooling because of their excellent energy efficiency. They provide both heating and cooling in one system, and they can be installed almost anywhere (extremely helpful if space is limited). While the heat pump's source

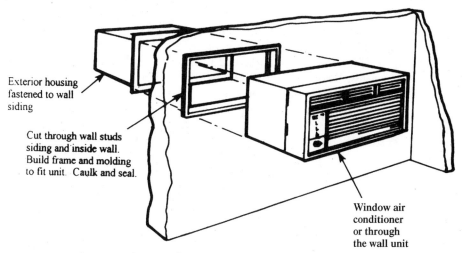

Exterior housing
fastened to wall
siding

Cut through wall studs
siding and inside wall.
Build frame and molding
to fit unit. Caulk and seal.

Window air
conditioner
or through
the wall unit

Figure 9.16 A through-the-wall air-conditioning unit. (*Leon E. Korejwo*)

of energy is electricity, it does not produce heat directly from the electric current. Heat pumps use electricity to move heat from one place to another. In the heating mode, the heat pump extracts heat from outside of the house and delivers it inside. In the cooling mode, it extracts heat from the house and takes it outside.

Flues

Sometimes the chimney flue seems to run right through the kitchen. If you're lucky, it is already in the wall. If not, there is a problem. A chimney flue cannot be readily moved and enclosed in the wall like a heating duct. A chimney flue, however, can be boxed in the same manner as pipes.

Another solution is to accept the limitations of the flue location and design around it. It can be boxed with drywall and hidden, or it can be dramatized and highlighted.

Electrical considerations

In the previous chapters, the basic principles of planning a kitchen layout, as well as of selecting appliances and cabinets, were covered. Another aspect of the kitchen of equal importance to its dual role as a work and living area is adequate electrical provision.

Because of all that can be done with electricity in a kitchen, the biggest part of a house's electrical load—up to 80 percent of the total house wattage—is drawn by kitchen appliances and lighting. Because electricity offers such convenient power (and appliance makers are so inventive), more and more power will be used to make the storage, preparation, and cooking of foods and cleanup after meals more convenient and time-saving for the homeowner. A new kitchen usually means more wiring for new appliances and relocated circuits.

A trade license is usually required to handle electricity, so in most instances, you need to hire a subcontractor. However, as the builder, you must become knowledgeable about electrical requirements and codes and be confident that you understand their impact on each project. It is important to have a general understanding of electricity so you can inspect the existing electrical service, if applicable, to recognize what may need to be modified, and intelligently discuss the project with an electrical contractor.

The National Electric Code

The National Electric Code is the most complete, detailed set of guidelines, specifying correct installation methods and types of materials acceptable for various jobs. In some communities, local regulations sometimes supersede the National Code, so it is important to know what the local regulations are. Ultimately, the last word on what can and can't be done electrically comes from the local building code. Also, be sure that materials you intend to use are approved by the power company. Most codes stipulate that the entire system be brought up to present-day standards anytime a change is made. If existing wiring is on the verge of being in violation of current codes, you, along with your client, must decide whether or not to upgrade the wiring. Some builders may choose to do nothing, although it is usually recommended to upgrade any wiring that is questionable. Prepare your client for the possible additional expenses of bringing in more power to the home's main service panel. Whenever major construction and wall removal is planned, remember to figure in adequate costs for electrical removal as well.

Electricity basics

Power from the utility comes in through a service entrance to a meter that keeps track of the consumption (total use measured in kilowatt hours). The power then continues on to a service panel (located inside the house near the electric meter or in the basement directly below the electric meter), which breaks it down into a series of circuits. These circuits deliver current throughout the house.

At the service panel, fuses or circuit breakers control individual circuits and protect against fire, which could develop if a circuit draws more current than it is designed to handle. In most newer installations, the safety devices are called circuit breakers. In an older home in which the original wiring is intact, the service panel may contain fuses rather than breakers. A single circuit breaker controls a 120V circuit. Any 240V circuit is controlled by two circuit breakers that might be connected with a plastic cap. Any 240V circuit serves only one outlet, such as a dryer or electric range. A 120V circuit might serve as many as a dozen outlets, as long as the total amperage of the lights and appliances does not exceed the amperage rating of the circuit breaker or fuse. If a circuit is overloaded or shorted, the breaker automatically breaks the flow of current. Ground circuits (discussed in greater detail later in this chapter) are an important part of the wiring.

Electrical current flows, under pressure, through the wiring in the house. The amount of current (measured in amperes, or amps) going through a wire at a given time is based on the number of electrons passing a certain point each second. The pressure that forces these electrons along their route is known as *voltage*.

Electrical current goes to the service panel on what is called "three-wire" service. Each of the two hot wires supplies electricity at approximately 120 volts. The third, or neutral, wire is maintained at 0 volts. These wires provide both 120-volt and 240-volt circuits in the house (Fig. 9.17). Some small appliances may operate on low-voltage current, usually around 10 or 12 volts, but possibly ranging from 6 to 24 volts. Their voltage is cut down by a small transformer wired in anywhere between the appliance and the service panel.

Portable appliances and devices are readily connected to an electrical supply circuit by means of an outlet called a receptacle. A duplex outlet is one that has two receptacles. A receptacle with two slots is a 2-wire receptacle. Newer houses have 3-wire receptacles, which have two slots and a round hole. The third wire on the three-wire receptacle is used to provide a ground lead to the equipment that receives power from the receptacle. The three-wire receptacle takes the regular two-prong plugs still found on some appliances. However, it is wise to ground a receptacle and is required by electrical codes. Another feature of modern receptacles is polarization, in which one slot is larger than the other to accept modern polarized plugs. The larger slot connects to the white neutral wire, the small slot to the hot wire, ensuring that the switch always interrupts the flow of current so no current can flow in when the switch is off. If polarity is reversed, an exposed socket can give a shock even when the switch is off and the light or appliance isn't running.

Wherever there is an outlet or any other place where wires are joined, it must be done in a junction box. These metal or plastic boxes come in many sizes and shapes, with knock-out holes so wiring can be brought in from any direction. They usually are nailed to studs, joists, or floor plates.

Expanding existing electric service

If you open the door of the service panel, you see all of the circuit breakers or fuses. If you see blank slots with no breakers, they are places where new breakers can be installed to expand the house electrical system with more circuits. In remodeling a kitchen, you almost surely want to add circuits. You might be able to combine some circuits that have light loads (such as for bedrooms) to create an open slot or two for expansion.

Older houses often were rated at only 60 amps for the entire house. Today, houses are likely to need 150 or 200 amps. The main breaker on the service panel should show the house rating. It is usually a double breaker in line with

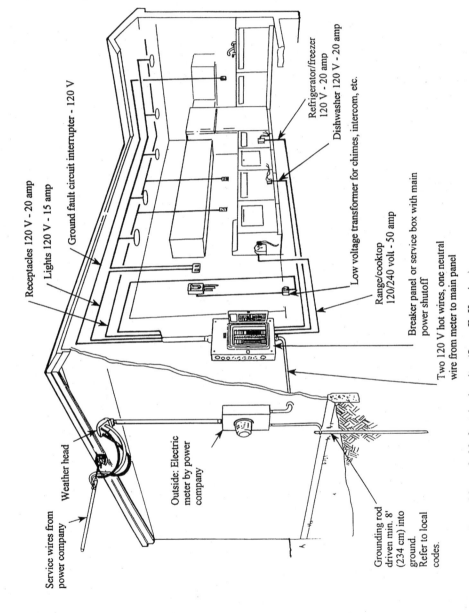

Receptacles 120 V - 20 amp

Lights 120 V - 15 amp

Ground fault circuit interrupter - 120 V

Refrigerator/freezer
120 V - 20 amp

Dishwasher 120 V - 20 amp

Low voltage transformer for chimes, intercom, etc.

Range/cooktop
120/240 volt - 50 amp

Breaker panel or service box with main
power shutoff

Two 120 V hot wires, one neutral
wire from meter to main panel

Service wires from
power company

Weather head

Outside: Electric
meter by power
company

Grounding rod
driven min. 8'
(234 cm) into
ground.
Refer to local
codes.

Figure.9.17 Typical household electric circuits. (*Leon E. Korejwo*)

all the others that is larger and a different color. This rating might be the number "60" or "100" molded or pressed into the tops of the two breakers. In any kitchen remodeling, consider upgrading to 150 or 200 amps to provide circuitry for future needs. The service ampere rating is obtained by adding together the current required for general lighting, major appliances, motors, appliance circuits, and other special circuits. Circuit breakers (which should be used in place of a fuse box) should be placed in an easy-to-reach location in or near the kitchen, so the homemaker does not have to hunt for it in a dark basement or garage.

Electrical circuits

Most residential wiring today is done using nonmetallic sheathed cable (NM cable). NM cable contains three or more individual wires, each with their own color-coded insulation. The number of the conductors grouped together in the cable is dependent on the type of circuit you are running. For example, most 110-volt circuits require three wires (including the ground wire), while some types, such as those for three-way lights, require a four-wire cable. Likewise, most 220-volt circuits require a four-wire cable. The size of wire used in the conductors determines the available amperes in the circuit (the amount of electrical flow available). The sizes are designated by gauge number. As the numbers increase in numerical value, the amperes decrease. Be sure to check the incoming wire size. Every builder should know how to determine capacity or at least know when to call the electrician. Refer to Table 9.2 for the most common household uses and to determine the size of the wire needed.

Kitchen wiring

Modern kitchens require an adequate supply of electricity. Of course, adequate wiring in a kitchen includes general-purpose circuits for lighting, branch circuits for small appliances, and individual major-appliance branch circuits. The recommendations may vary from city to city, depending on standards and codes.

There are several basic wiring requirements common to all kitchen layouts. Depending on its size and layout, your client's kitchen will probably contain all or most of these, perhaps more. Using the cabinet layouts and floor plan, be sure to plot out an electrical plan to ensure that your clients receive all the wiring needed in their kitchen while helping to eliminate waste. Give some

TABLE 9.2 Wire size as determined by circuit amperage

Wire size	Circuit size
No. 6 gauge	55 amperes
No. 8 gauge	40 amperes
No. 10 gauge	30 amperes
No. 12 gauge	20 amperes
No. 14 gauge	15 amperes

thought to other kinds of wire that might need to be installed in the walls. For example, prewire phone jacks, wires for cable television, security systems, door chimes, and thermostats. They should be placed at the safest location possible. Running future-use wires when walls are open is a lot easier than working with them after everything is sealed up. If a designer is drawing up the plans, he or she should also provide an electrical plan (Fig. 9.18a), as well as a lighting plan for the electrician to work from (Fig. 9.18b); these plans include a table of electrical symbols (Fig. 9.18c). It is important to remember that no matter who provides the plan, it is best to have it verified by a local electrical inspector.

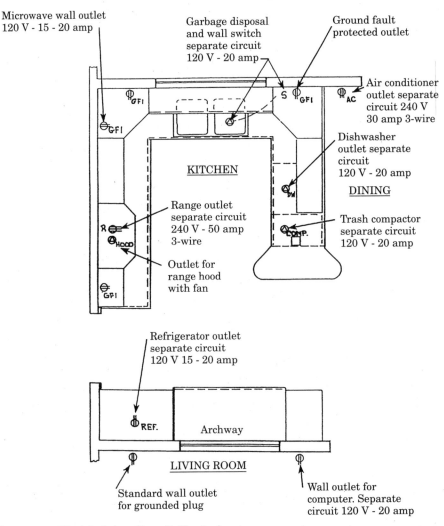

Figure 9.18a Electrical plan. (*Leon E. Korejwo*)

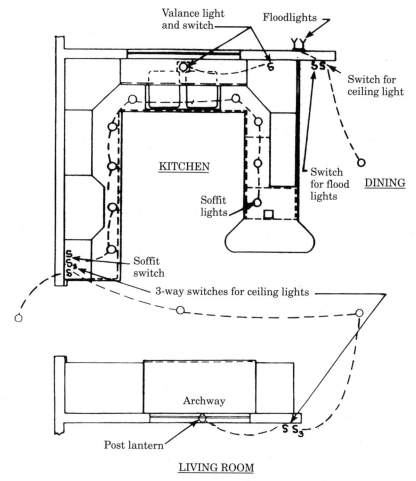

Figure 9.18b Lighting plan. (*Leon E. Korejwo*)

Wiring color coding

To ensure universal wiring and safety, all electrical wiring today is color-coded. The individual wires in the cable each have a colored jacket to simplify the proper installation. The neutral wire is always white. The ground wire is either bare or color-coded green. Hot wires are colored black, red, or any color other than green, white, or bare.

Grounding and ground-fault circuit interrupters (GFCIs)

All new wiring in a home is required to be grounded (a safety measure built into every home electrical system), whether the existing wiring is grounded or not. In some instances, officials will make you upgrade the existing wiring if it is ungrounded and appears unsafe. Grounding is accomplished through the use of a third wire in the circuit (the bare copper or green-jacketed wire) and pro-

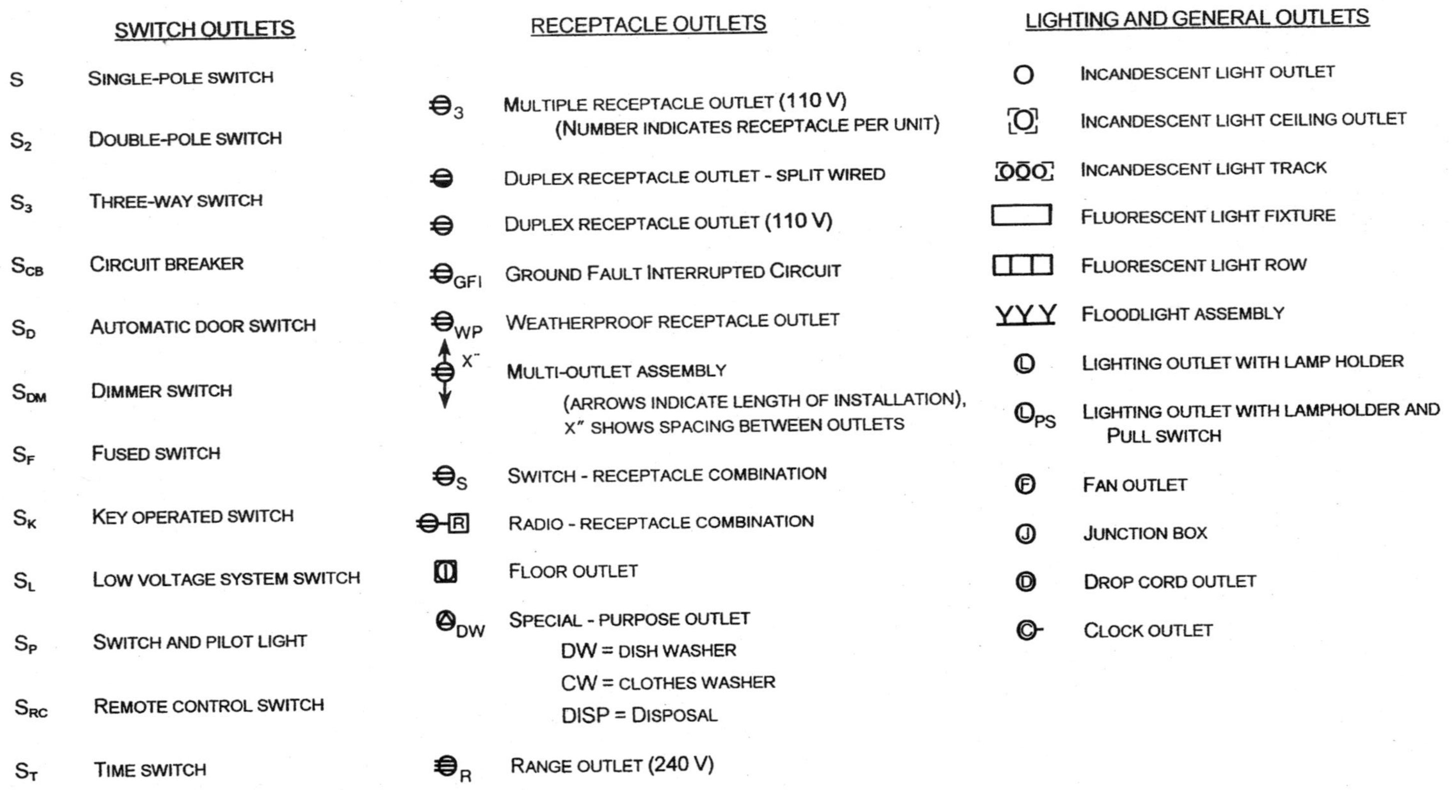

SWITCH OUTLETS

S	SINGLE-POLE SWITCH
S_2	DOUBLE-POLE SWITCH
S_3	THREE-WAY SWITCH
S_{CB}	CIRCUIT BREAKER
S_D	AUTOMATIC DOOR SWITCH
S_{DM}	DIMMER SWITCH
S_F	FUSED SWITCH
S_K	KEY OPERATED SWITCH
S_L	LOW VOLTAGE SYSTEM SWITCH
S_P	SWITCH AND PILOT LIGHT
S_{RC}	REMOTE CONTROL SWITCH
S_T	TIME SWITCH

Figure 9.18c Electrical symbols. *(Leon E. Korejwo)*

RECEPTACLE OUTLETS

$\oplus_3$	MULTIPLE RECEPTACLE OUTLET (110 V) (NUMBER INDICATES RECEPTACLE PER UNIT)
⊕	DUPLEX RECEPTACLE OUTLET - SPLIT WIRED
⊕	DUPLEX RECEPTACLE OUTLET (110 V)
$\oplus_{GFI}$	GROUND FAULT INTERRUPTED CIRCUIT
$\oplus_{WP}$	WEATHERPROOF RECEPTACLE OUTLET
◄⊕► x	MULTI-OUTLET ASSEMBLY (ARROWS INDICATE LENGTH OF INSTALLATION). x'' SHOWS SPACING BETWEEN OUTLETS
$\oplus_S$	SWITCH - RECEPTACLE COMBINATION
⊕-Ⓡ	RADIO - RECEPTACLE COMBINATION
▣	FLOOR OUTLET
⊕$_{DW}$	SPECIAL - PURPOSE OUTLET DW = DISH WASHER CW = CLOTHES WASHER DISP = DISPOSAL
⊕$_R$	RANGE OUTLET (240 V)

LIGHTING AND GENERAL OUTLETS

O	INCANDESCENT LIGHT OUTLET
[O]	INCANDESCENT LIGHT CEILING OUTLET
OOO	INCANDESCENT LIGHT TRACK
▭	FLUORESCENT LIGHT FIXTURE
▥	FLUORESCENT LIGHT ROW
YYY	FLOODLIGHT ASSEMBLY
Ⓛ	LIGHTING OUTLET WITH LAMP HOLDER
Ⓛ$_{PS}$	LIGHTING OUTLET WITH LAMPHOLDER AND PULL SWITCH
Ⓕ	FAN OUTLET
Ⓙ	JUNCTION BOX
Ⓓ	DROP CORD OUTLET
Ⓒ-	CLOCK OUTLET

vides an alternate route for any leaking current, protecting the circuit and family members from shock. The main grounding wire connects all the metal parts within the electrical system to the service panel and from there to a ground source, usually a metal cold-water pipe, or a grounding rod buried in the earth (driven at least 8 feet—244 cm—into the ground) or in the building's foundation. Older homes may have the metal plumbing used as a ground. Code compliance is vital when remodeling this type of home. Previously installed plastic pipe that may have been inserted between sections of metal plumbing may have destroyed the grounding to sections of the house not within your contracted area. Be aware that although your project does not involve major rewiring, you may still want to consult a professional electrician during the estimating phase. When connecting appliances, fixtures, receptacles, or any other metal electrical component, always look for the green grounding screw and be certain the ground wire in the NM cable is connected to it.

According to the National Electrical Code, all kitchen outlets located above countertops must have GFCI protection. In the past, only receptacles near sinks or all outlets within 6 feet (183 cm) of a water source required GFCIs. Some local codes may be even more stringent. A GFCI breaker, which is located at the main breaker box, is so sensitive that when it detects an imbalance, it trips the breaker and shuts down the circuit when even a tiny leakage occurs. A GFCI outlet, which has its breaker and reset included right in its own wall unit, accomplishes the same effect. The shutoff action is so fast, there is not enough time to be injured. As little as 200 milliamperes (about enough to light a 25-watt bulb) can kill if you happen to be touching plumbing components or standing on wet earth. On either a GFCI breaker or outlet is a test button and a reset button. It should be tested regularly by pushing the test button, then reset by pushing the reset button.

Basic circuits

The National Electrical Code requires that every kitchen be provided with two 20-ampere, 120-volt circuits for the operation of kitchen appliances and ground-fault circuit interrupters (GFCIs). A remodeled kitchen almost certainly requires more circuits than that and a lot more outlets. These circuits should be divided so that approximately half the kitchen is on each circuit and can also extend into areas immediately adjacent to the kitchen (i.e., laundry areas, eating areas, dining rooms, living room, etc.). These two circuits should be for receptacles only and should not include any lighting. While two 20-ampere circuits are generally sufficient, some thought should be given to providing one or even two additional 20-ampere appliance circuits. Keep in mind that each year the number of portable or small kitchen appliances increases, making increased demands on the electrical capacity of the kitchen.

The accepted modern standard for wiring in new houses is to put an outlet at least every 12 feet (366 cm) in every wall and within 6 feet (183 cm) of every door. These outlets usually are on a 20-amp circuit with 12-gauge wire, although the slightly smaller 14-gauge is adequate for lighting, radio, and television.

When adding one light or moving an existing light, simply adding wiring from an existing switch may be easily accomplished (Fig. 9.19a) when an additional receptacle is desired and the circuit you plan to put it on will not be overloaded. You may run new wiring from an existing receptacle as shown in Fig. 9.19b. Each stationary appliance, such as a cooktop or an oven, requires a separate, dedicated (serving that appliance only) circuit. The circuit size must be sufficient to serve the appliance's amperage rating as listed on the appliance nameplate.

To wire a kitchen for convenience, each of the major work centers, as well as the various supplementary ones, makes its own electrical demands, as follows:

1. *Storage centers* should have separate 120-volt circuits for each freezer and refrigerator. These units typically draw less than 15 or 20 amps. The separate circuit is a must, because if other appliances are operated from the

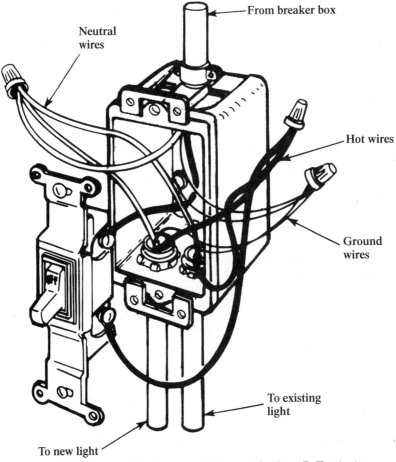

Figure 9.19a Wiring a new light from an existing switch. (*Leon E. Korejwo*)

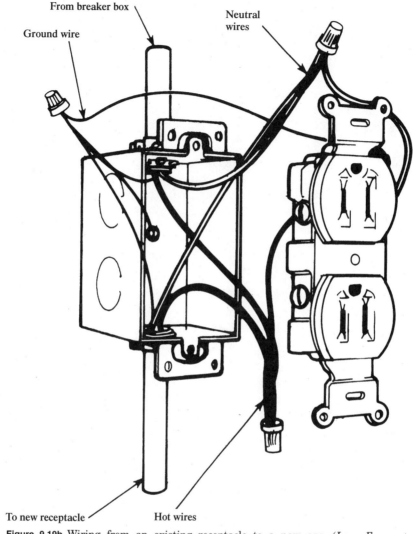

From breaker box

Ground wire

Neutral wires

To new receptacle   Hot wires

Figure 9.19b Wiring from an existing receptacle to a new one. (*Leon E. Korejwo*)

same circuit as the freezer, they might inadvertently trip the breaker without the knowledge of the homeowner. Also, it is a good idea when installing the freezer outlet to use a locking-type receptacle to prevent accidental disconnection. When choosing panel-box locations, choose them carefully. The panel boxes should be as close as possible to the wiring paths.

2. *Cook centers* alone can use enough watts to heat a house. A double oven and a high-speed cooktop can add up to 16,000 watts. This situation calls for a minimum of a separate, three-wire, No. 6 circuit fused to handle 50 amps. While most electric ranges are designed to operate on 240 volts, a house wired for 208-volt service can obtain better cooking results if the selected

range (and oven, if separate) is rated for 208 volts rather than for 240 volts. If a gas range is used rather than the electric type, it needs a convenient 120-volt wall outlet for its clocks, lights, and rotisserie. A microwave oven typically draws less than 15 or 20 amps and requires only a 120-volt circuit.

3. *The cleanup center* creates a second major load. Heating water may take up to 4500 watts and operating a dishwasher, disposer, and compactor takes at least another 2600. Dishwashers, garbage disposers, and trash compactors usually run on 120-volt circuits; provide a separate, 20-amp circuit for them. A water heater takes 240 volts, 20 to 30 amps, and requires a separate circuit.

4. *The laundry center* forms the third major load. A dryer needs about 4500 watts, a washer 700 watts, and a hand iron draws approximately another 1500 watts. The dryer needs a three-wire, No. 10 circuit fused for 30 amps, 120/240 volts. An electric dryer requires its own 240-volt receptacle. The washer typically draws less than 15 or 20 amps and requires only a 120-volt circuit. The other appliances can be taken off one side of a three-wire, No. 12 circuit. At least one receptacle other than the one for the washer is also required.

5. *The mix center* works off the convenience outlet loop. Motor-driven power appliances like mixers, blenders, knife sharpeners, can openers, etc., draw relatively light loads—100 to 200 watts—but small cooking appliances draw up to 1500 watts. Two cooking appliances in use at the same time may overload one 120-volt circuit.

6. *A planning center* is usually a well-lighted desk area, close to the telephone, intercom master station, computer, and climate center (with heating, cooling, filtering, and humidity controls). The area is served by part of one convenience outlet circuit and by the low-voltage wiring for the intercom and climate center.

7. *Air conditioner* window units vary according to their Btu capacity—from less than 15 amps at 120 volts up to 30 amps at 240 volts. Wire size is No. 12 for 20 amps or less and No. 10 for 30 amps or less.

8. *Lighting circuit.* Provide at least one 15-amp circuit for the lighting, including overhead and undercounter lights. (For more on lighting, see Chap. 11). A kitchen needs more artificial light than any other room in the house. It is usually in use for several hours every day. Wiring should provide for special area lighting, either from adequate ceiling fixtures or a luminous ceiling. About 1000 watts of lighting is not too much. It can be supplied by one side of a three-wire, No. 12 loop.

Areas in the kitchen other than the work area should not be neglected. Provide ample outlets around eating spaces for toasters and other small table-top appliances. Over-the-counter power sources should be about 2 to 4 feet apart unless a continuous strip system is used. Over-the-counter wall outlets should be installed 44 inches (112 cm) from the floor.

Table 9.3 summarizes the wattages needed for typical kitchen uses.

Table 9-3   Typical wattage for kitchen uses

Major appliance	Volts	Wattage	Circuit
Range and oven	14,000	Three-wire No. 6,	50-amp, 120-240
Dishwasher	13,300		
Disposer	800	Three-wire No. 12,	20-amp, 120-240
Trash compactor	500		
Refrigerator	400	Two-wire No. 12,	20-amp 120
Freezer	400	Two-wire No. 12,	20-amp 120
Washer	700	Two-wire No. 12,	20-amp 120
Hand iron	1500	Two-wire No. 12,	20-amp 120
Dryer (standard speed)	4500	Three-wire No. 10,	30-amp 240
Water heater	2500–4500	Consult your utility	240

Small appliances	Volt	Circuit	
Can opener	110		
Mixer	120		
Blender	300		
Toaster	750	Two-wire No. 12, 20-	
Waffle iron	1100	amp 120	
Coffeepot	900		
Intercom-radio	100		
Vent hood	150		
Microwave unit	1500		
Rotisserie	1500		
Deep-fat fryer	1300	Two-wire No. 12, 20-	
Electric frying pan	1100	amp 120	
Broiler	1500		

Smoke detectors

Under current code regulations, a smoke detector is required in a kitchen. Requirements can mandate that the smoke detector be hard-wired and connected so that if one detector goes off, they all sound an alarm. Battery-powered detectors may be sufficient; however, check the requirements of local authorities.

Noise abatement

Noise can be a problem in both a new and a remodeled kitchen. During the meal-preparation hours, it is not unusual for several appliances to be used at the same time. With all of this activity, the noise level in the kitchen can reach a point where it causes great discomfort, as well as disturbs other areas throughout the house. Noise can be kept at a reasonable level if the following antinoise steps are taken.

1. One of the first things to consider is the installation of sound-insulating walls. Application of an extra layer of material on a wall helps, but a staggered stud wall that allows weaving of insulation material through the studs gives a great deal of sound protection.

2. Level the refrigerator and freezer to eliminate annoying vibrations. Either appliance is properly balanced if the door closes automatically from a half-open position.

3. Dishwashers are very noisy because of the tremendous water activity. Choose a dishwasher that has more insulation and other sound-deadening features built into it or wrap the sides with insulating material to prevent transmission of sound to cabinets and countertops.

4. Mount the dishwasher, food waste disposer, compactor, and other similar appliances on springs or pads to prevent vibrations from being transmitted through the floor, walls, and countertops.

5. Use rubber isolation gaskets at the mount of food waste disposers to prevent the sink bowl from amplifying the grinding noise. The better-quality disposers have sound-deadening jackets of insulation.

6. Heavy-gauge stainless-steel sinks vibrate less than thinner ones; porcelain on cast iron is still quieter. Wooden cabinets reflect less noise than metal ones.

7. Place rubber bushings behind cabinet doors to stop banging. Check the drawer slides, and if they are noisy, replace them with quiet ones.

8. The hum of a fluorescent light often annoys people. Use rubber mounts combined with ballast to eliminate hum.

9. Make sure room air conditioners and vent fans are properly mounted to avoid excess vibration. Ductwork should be designed without reduction in size and with a minimum of turns.

10. Install air chambers or pneumatic antihammer devices in the water lines to stop any hammering noises.

11. Long runs of hot-water supply creak or snap as they expand or contract. Differences up to 100 degrees F can exist in piping and can cause expansion of up to $\frac{1}{8}$ inch in 10 feet. Noise occurs when expanding pipe is forced against its hangers or surrounding structural members. Allow space for growth at the end of long runs where pipes change directions.

12. Holes cut through common walls for plumbing or heating may leak noise. Seal all holes with a resilient material to isolate noise and seal against air leaks, either vertical or horizontal.

13. Remember that acoustical ceiling panels or tile, carpeting, and fabrics used for curtains and tablecloths absorb most of the sound that reaches them.

14. Move the vent fan from the range in the kitchen to an outside wall in the attic, utility room, or basement.

15. Noise transmits to other rooms through lighting fixtures, electrical outlets, and heating ducts. Seal these with insulating material. Use insulating panels over the ceiling wallboard and under floor underlayment.

16. Avoid placing any switches or outlets back-to-back. Position junction boxes on opposite sides of the same wall so there is 36 inches (91 cm) of lateral distance between them.

17. Attach ceiling fixtures so that fasteners are set into furring strips attached to the ceiling joists with spring clips. The clips will absorb any noise that may develop in the fixture.

Installation of cabinets, countertops, and major appliances

Once the room is thoroughly prepared, you are ready to begin installation of the cabinets, countertops, and appliances. All of the plumbing and electrical wiring should be complete, all rough carpentry should be out of the way, and all floor, wall, and ceiling preparation work should be accomplished. The kitchen should be completely clean. Remove any debris and unneeded materials, and lay out your tools where they are accessible, but not in the way. Protect all finishes that have been completed prior to this point.

Whether the installation of the kitchen equipment is troublesome or easy depends on good preparation and proper installation techniques. Out-of-square walls, insufficient clearances, and incorrect installation procedures can create costly problems and delays.

Kitchen cabinets

All cabinets, hardware, moldings, and other materials must be on hand; verify that the shipment is complete. Most manufacturers perform rigid quality inspections; however, damage may occur through shipping and handling. Carefully compare the cabinets you received with the kitchen plan, ensuring the proper type of cabinet and that quantities are correct for the installation. Store the cabinets in an out-of-the-way place where they will be safe and protected, such as in the client's basement or garage. Keep the cabinets in the room overnight to allow them to adjust to the temperature of the room.

Checking for square and plumb

All cabinets must be installed perfectly level and plumb from a standpoint of function as well as appearance. Because each floor and wall has uneven spots that will affect the installation, it is necessary to locate these uneven areas and shim or cut the cabinetry to make the installation plumb, true, and square. Never assume that a room is perfectly square and plumb; in most cases, no two walls intersect at a corner with a perfect 90-degree angle.

The first step in laying out the kitchen for cabinet installation is to determine whether the floor under the cabinets is level. To check the floor, lay a long, straight 2-by-4 with a 4-foot (122 cm) or longer carpenter's level on top of it (Fig. 10.1). Examine the room in any area where cabinets will be installed, paying particular attention to the area between the wall and about 24 inches (61 cm) out. If the kitchen floor is uneven, which is common, determine where the highest point is. Measure up from this high point to draw the reference lines. Mark a continuous level line on the wall along the floor. This line, called the base line, indicates where the bottom of the base cabinets will rest and gives you an idea of where shimming will be needed and in what amount. If the subfloor or finish floor is to be replaced and will extend under the area where the base cabinets are installed, the rough-in work should be completed before checking for level. Protect the floor with cardboard or tarps during the rest of the installation.

The next step is to determine if the ceiling is level. This step is very important if you are installing cabinets that run all the way to the ceiling or if you are planning to build soffits above the wall cabinets. Perform the same process as you did for the floor, but rather than determine the high point, determine the low point of the ceiling. From the low point, mark a continuous level line on the wall along the ceiling.

Probably the most common installation problem is the out-of-square wall. It is most troublesome at the end of a wall or base cabinet run and can prevent drawers and sliding shelves from being opened their full length, cabinet doors from being opened a full 90 degrees, and countertops from being properly fitted. If an existing wall is out-of-square with the cabinets, a filler strip (see more on filler strips later this chapter) should be used at that wall, which brings a finished look to the shimmed cabinet. It moves the cabinet door or drawer away from the out-of-square wall enough to allow the full use of the cabinets. Since many oven doors are the same width as the oven cabinet, an oven placed against an out-of-square wall at the end of a cabinet run will not have enough room for the oven door to be opened fully.

To check the squares of the corners of a room, use the triangulation method often referred to as the 3-4-5 rule. Measure 36 inches (3 foot) (91 cm) from a corner along one wall, make a mark; measure 48 inches (4 foot) (122 cm) out from the corner along the second wall and make a mark (Fig. 10.2). If the diagonal distance between the two marks is exactly 60 inches (152 cm), the corner walls are square. If it measures less than 60 inches (5 foot) (152 cm), the walls intersect at less than 90 degrees, and the cabinets must be shimmed at the corner. If it is more than 60 inches (152 cm), the walls intersect at an angle greater than 90 degrees, and the cabinets are shimmed at the far end from the corner. Mark the actual cabinet outlines of all wall cabinets on the wall to check wall dimensions against the layout. Be accurate in the layout and include any fillers you will be using to make up odd inches.

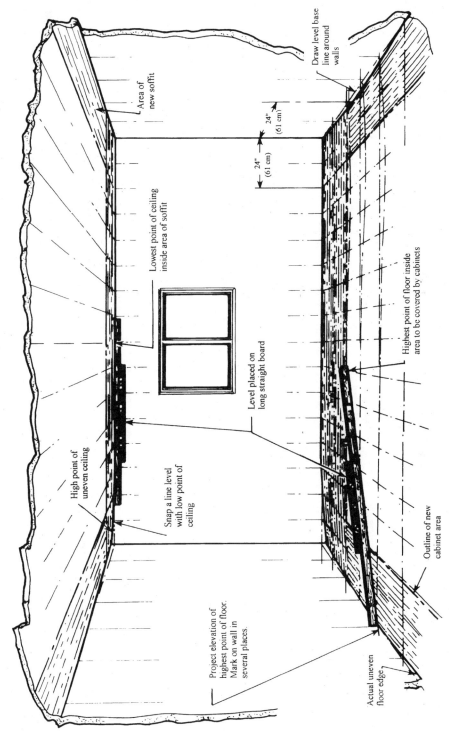

Figure 10.1 Find high point of floor and low point of ceiling. (*Leon E. Korejwo*)

Area of new soffit

Draw level base line around walls

24" (61 cm)

24" (61 cm)

Lowest point of ceiling inside area of soffit

Highest point of floor inside area to be covered by cabinets

Level placed on long straight board

High point of uneven ceiling

Snap a line level with low point of ceiling

Project elevation of highest point of floor. Mark on wall in several places.

Outline of new cabinet area

Actual uneven floor edge

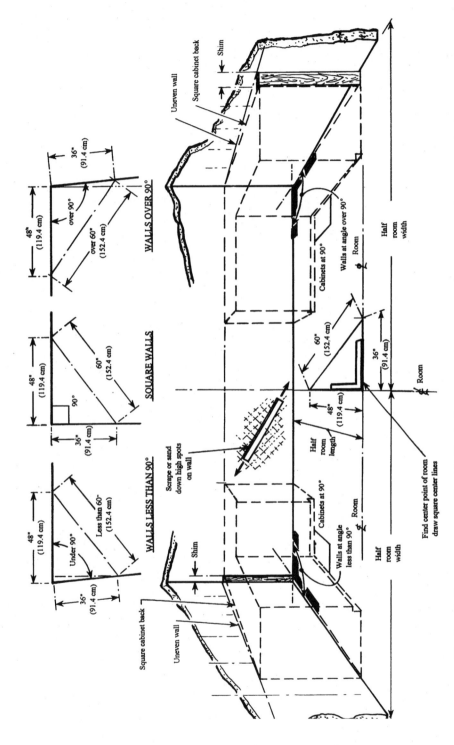

Figure 10.2 Check walls for square corners. (*Leon E. Korejwo*)

Next check the walls for uneven spots. Wall unevenness can cause cabinets to be misaligned, resulting in racking or twisting of doors and drawer fronts after they have been screwed tight against the wall. Some high spots can be removed by sanding. Otherwise, it will be necessary to shim the low spots to provide a level and plumb installation. Stretching string across a wall from corner to corner is a quick reference to the unevenness of a wall.

The tops of wall cabinets are generally installed 84 inches (213 cm) above floor level. Locate this height from the base line, which is the highest point of the floor. Check the width of the room first at the high point. Next check the level of the floor at this point, 24 inches (61 cm) from the wall (which is the width of the wall cabinet). If the floor is higher than 24 inches (61 cm) away from the wall, mark this height on the wall. Then, proceed up the wall and mark off the 84-inch (213 cm) height and, using the level and straightedge, continue this line around the room. This line will show the location of the tops of the wall cabinets at 84 inches (213 cm) above floor level. Mark another horizontal line on the wall at 34½ inches (88 cm) above the high point to locate the tops of the base cabinets. Mark the walls for each cabinet. Begin at any corner and fix the position of each cabinet by marking its exact location.

On the walls where cabinets are to be installed, remove the baseboard and chair rail for a flush fit. If the kitchen is utilizing tall units, such as oven or pantry cabinets, they require a full 84 inches (213 cm) in height. If the room has an existing soffit and the height from the baseline to the soffit is less than 84 inches (213 cm), you need to work to the soffit and trim off the bottom of the tall units to make the adjustment.

Again, take several measurements and use the highest mark for the reference point. When the typical 1½-inch (4-cm) countertop is added, the top of the countertop should be 36 inches (91 cm) above the finish floor. Use a level to mark a reference line on walls. Base cabinets will be installed with top edges flush with the 34½-inch (88-cm) high line.

Lay out the entire cabinet arrangement on the walls and floor. Doing so helps you visualize the configuration of the cabinets and quickly points out dimension problems or conflicts. Marking full plumb lines for each cabinet unit is not necessary. Start with a vertical plumb line using an accurate level and double-check with a plumb bob. Use this line as a reference point. It may not be necessary to repeat the leveling for all your lines. Because the face frames extend slightly beyond the sides of most base cabinets, there will be about a ¼-inch (0.64-cm) space between the lines drawn for each base cabinet.

It is important to take the time to draw the cabinet layout on the walls and floor, measuring carefully the entire kitchen before you just jump in and start installing the first cabinet or piece of equipment. Knowing where to start installing the cabinets, and in what order they will be installed, is vital to a successful installation.

Wall cabinets are installed with top edges flush against the 84 inch (213 cm) high line. Measure down 30 inches (76 cm) from the wall cabinet reference line and draw another level line where bottom of the cabinets will be (Fig. 10.3). Temporarily, ledgers (or cleats) can be installed against this line. Double-check

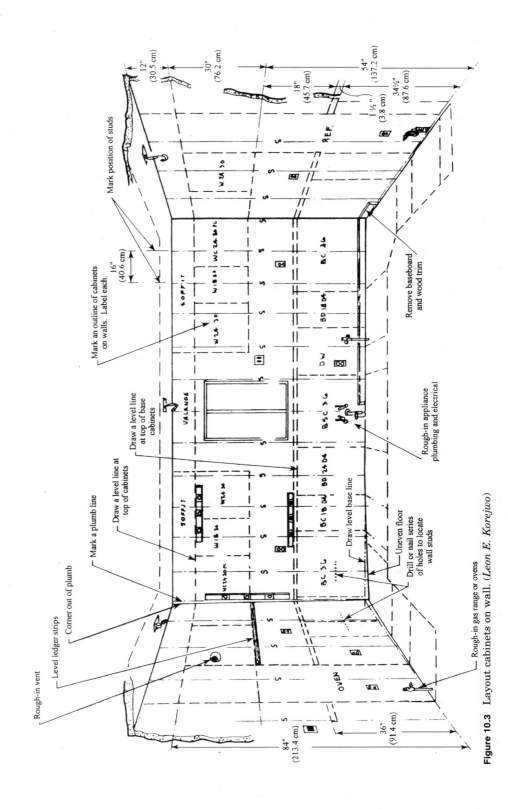

Mark position of studs

Mark an outline of cabinets on walls. Label each.

12" (30.5 cm)

30" (76.2 cm)

18" (45.7 cm)

54" (137.2 cm)

1½" (3.8 cm)

34½" (87.6 cm)

REF.

16" (40.6 cm)

Draw a level line at top of base cabinets

Draw a level line at top of cabinets

Mark a plumb line

Corner out of plumb

Level ledger strips

Rough-in vent

Draw level base line

Uneven floor

Drill or nail series of holes to locate wall studs

Remove baseboard and wood trim

Rough-in appliance plumbing and electrical

Rough-in gas range or ovens

SOFFIT

VALANCE

OVEN

84" (213.4 cm)

36" (91.4 cm)

Figure 10.3 Layout cabinets on wall. (*Leon E. Korejwo*)

this point by measuring up from the base line. The measurement should be 54 inches (137 cm).

If you are working tight to the ceiling or a soffit, make sure to have established a level line at the low point. There is generally 16 to 18 inches (41 cm to 46 cm) between the countertop and the underside of the wall cabinets, but this distance must be checked to ensure that the wall cabinets align with the top of the tall cabinets. With the bottom line of the wall cabinets established, mark vertical lines for each wall cabinet unit in the same manner as the base cabinets. Allow for the required clearances of appliances and equipment in the layout of the room.

You need to check for vertical-alignment relationships between the base and wall cabinets, especially where an appliance such as a refrigerator sits adjacent to both a base and wall cabinet or where a base and a wall cabinet abut a tall cabinet. Because it is extremely important to support wall cabinets with a solid mounting, you also need to determine the layout of the wall studs to which you will be attaching the cabinets. Cabinets must be attached to studs for full support. Remember that studs are usually laid out at 16 inches (41 cm) or 24 inches (61 cm) on center. If you know the location of any studs from the rough work you did earlier, measure out on 16-inch (41 cm) centers to locate the rest of them. If you don't know the location of any of the studs, use a stud finder.

Two types of stud finders are available. The simple magnetic type is attracted to the nails in the studs. The electronic designs, while more expensive, are more accurate, and in some cases capable of detecting studs more deeply embedded behind the wall. Certain models allow several modes, finding wood, a higher concentration of material, and electrical wiring. It is not uncommon to find studs with irregular center-to-center locations and also not plumb. Also, tapping on the wall with a hammer to locate a solid sound generally tells you where the wall framing members are located. This location can be verified by driving a small finishing nail into the wall (make sure this is an area that will be covered by wall cabinets). Mark the stud locations with a vertical line that won't be concealed by the cabinet units being installed. With the studs located, installation can begin.

Soffits

To close the gap between the tops of the wall cabinets and the ceiling, soffits are frequently employed (Fig. 10.4). They are framed with 2-by-3s (5-by-7.6 cm). A ceiling plate is spiked to the joists above, and the wall cleat is spiked to the studs. The short 2-by-3s between the main members are toenailed in place. Afterward, the face of the soffit is covered with wallboard or any convenient sheet material. When the cabinets are hung, gaps remaining between the bottom of the soffit and the tops of the cabinets are hidden by a cove molding that is the last item to be affixed.

If soffits are to be built, they should be constructed and installed before installing the cabinets. Soffits in a room with 8-foot (244 cm) ceilings are normally 12 inches (30 cm) in height and 14 inches (36 cm) in width. If the wall cabinets are to be installed against an existing ceiling soffit, it should be

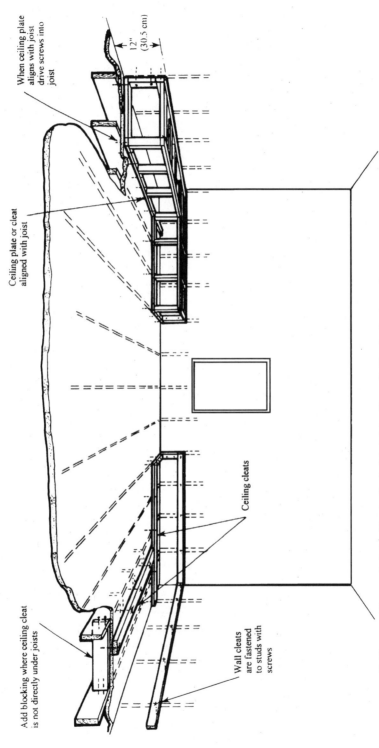

When ceiling plate
aligns with joist
drive screws into
joist

12"
(30.5 cm)

Ceiling plate or cleat
aligned with joist

Add blocking where ceiling cleat
is not directly under joists

Ceiling cleats

Wall cleats
are fastened
to studs with
screws

Figure 10.4 Soffits or ceiling-mounted cabinet attached to joists. (*Leon E. Korejwo*)

examined to determine if it is level. Use a straightedge and level much the same as you did when looking for the lowest spot (closest to the floor) on the ceiling. Corner cabinets should be shimmed down to this "lowest" height. By comparing measurements from the base level line, you can locate any unevenness in soffit height.

The installation

Proper installation means better performance of equipment, a less costly installation, and few service complaints.

Ceiling-mounted cabinets

Not all wall cabinets are mounted to the wall. Peninsula wall cabinets must be hung from the ceiling or from overhead soffit construction. If there is existing ceiling framing, determine if the layout of the framing aligns with the location of the cabinets. To adequately support the cabinets, it may be necessary to open up the ceiling and install additional blocking. If constructing a new soffit, lay out the soffit framing to provide proper means of connection. Peninsula wall cabinets must be fastened to the overhead structure with at least four screws, as close to the outside corners as possible.

Most manufacturers recommend installing wall cabinets before base cabinets. The reason most prefer this order is because working around the wider base cabinets to hang the wall units may be awkward. Also, to support the wall cabinets on jacks during installation the installer needs to have the floor space free below the wall cabinets. There is also the risk of damaging the base cabinets if a wall cabinet or a tool is dropped during the installation of the wall cabinets.

Some installers do, however, prefer to begin with the base cabinets. Many feel it is best to set these cabinets first, because base cabinets must be arranged to accommodate appliances such as dishwashers, ranges, and refrigerators. With this approach, once the base cabinets are set, the vertical relationships between the base and wall cabinets can be established and the wall cabinets installed. Countertops cannot generally be measured for or ordered until the base cabinets are set. Therefore, while the countertops are being fabricated, the wall cabinets can be installed. Installers who prefer to install base cabinets first can still use jacks to install wall cabinets. In this instance, shorter jacks are used. Be sure to protect the base cabinets. Plywood sheets serve this purpose and also provide a surface to support the jacks.

If a full-height backsplash is specified, the best fit is achieved by installing the base cabinets first. Once the base cabinets and backsplash have been set, the wall cabinets can be installed to sit tight against the backsplash.

Whether you choose to begin installing wall cabinets or base cabinets first is up to you. One way is not necessarily better than the other, although each has its merits. Consider all of the options and choose what makes the most sense for each individual project.

Wall cabinet installation

To make the installation of wall cabinets easier, install a 1-by-2 (2.5-by-5-cm) or 1-by-3 (2.5-by-7.6 cm) strip of wood as a temporary support cleat or ledger strips at the 54-inch (137-cm) cabinet line mark. Check the trueness of the ledger after the screws are tightened to the wall. This ledger strip allows you to align and support the wall cabinets as they are being installed. Attach ledgers with 2½-inch (6-cm) wallboard screws driven into every other wall stud. Mark stud locations on ledgers. Cabinets rest temporarily on ledgers during the installation, which are then removed. If, however, you are working against a wall that has already been finished or one that will not later be covered with a backsplash,you should not use a ledger strip, since the holes will show. Use the 54-inch (137-cm) line as a visual reference point, a location to raise the cabinets to, and use a floor T-brace.

Remove doors from cabinets to ease access to hanging strips, lighten the load, and ease handling. If possible, remove the cabinet doors by their hinge pins and interior shelves from the cabinet unit before lifting the cabinets into position. However, it is important to bundle the shelves and mark them so that you know which cabinet unit they came from.

Start the cabinet installation from an inside corner and work outward. If you have a diagonal corner wall cabinet, it is an excellent cabinet to install it first (Fig. 10.5). The back of a corner wall cabinet is usually beveled at a 45-degree angle so that it can be fitted into nonsquare corners. This angled back can make corner wall cabinets very awkward to set in place. Check with the manufacturer for pull spacing, as alignment may require a space behind. Ledger strips help keep the unit from shifting. Once the cabinet is supported, measure from the corner to the first stud mark and transfer this measurement to the inside of the cabinet to be installed in the corner. Mark the next stud, etc. Drill through the hanging strips (or mounting rails), which are horizontal wood strips built into the backs of the cabinet at both the top and bottom. Locate holes about ¾ inch (1.9 cm) below the top shelf and ¾ inch (1.9 cm) above the bottom shelf from inside the cabinet. Holes can also be marked on the back of cabinets and holes drilled before cabinets are raised into place. When drilling, be sure not to split the finished face of the inside of the cabinet.

A variety of jacks and props are available to assist in holding the wall cabinets in place for a one-person job. However, some installers choose to have a helper brace the cabinet while checking for plumb and level. Wooden shims can be driven between the cabinet and the wall, as needed, to bring the cabinet perfectly into position. C-clamps can be used to connect cabinets together to obtain proper alignment.

Lift the first cabinet into place atop the ledger strip and brace it. Drill through previously located marks (if you have not already done so). Loosely attach the cabinet to the studs with installation screws long enough to extend minimum 1½ inches (4 cm) into the studs. Depending on the thickness of the hanging rail, cabinet back, and drywall or plaster wall surface, the screws should be between 2½ (6 cm) and 3 inches (8 cm) long. Some manufacturers supply installation fasteners; others do not. (If walls are plaster and thicker

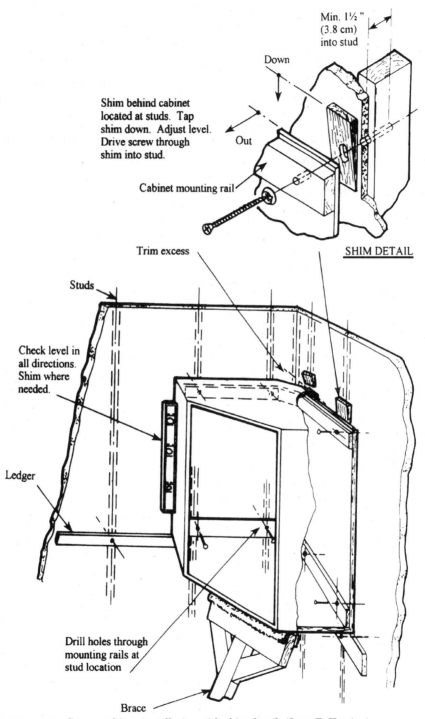

Min. 1½ " (3.8 cm) into stud

Down

Shim behind cabinet located at studs. Tap shim down. Adjust level. Drive screw through shim into stud.

Out

Cabinet mounting rail

SHIM DETAIL

Trim excess

Studs

Check level in all directions. Shim where needed.

Ledger

Drill holes through mounting rails at stud location

Brace

Figure 10.5 Corner cabinet installation with shim detail. (*Leon E. Korejwo*)

than normal, run a test screw and remove it to see its depth of engagement). In most instances, the fasteners used are #8 and #10 screws, with the recommended head types varying by cabinet manufacturer. Drywall or wood screws are two popular choices, and a power driver can speed the work.

Wall cabinets are generally attached to wood studs with #8 or #10 wood screws or drywall screws of sufficient length. If you were not able to locate a wall stud, or if a stud does not exist at the point where the cabinet is being mounted, install what you can and then use the next cabinet to add support. Toggle bolts can be used in some cases in lieu of attaching to a stud, but in no case should a wall cabinet be installed without being secured to at least one wall stud. Masonry walls may require use of expansion anchors. Adjust the screws only loosely at first so that the final adjustments can be made after the cabinets have been checked with a level. A minimum of two screws should be placed in the top and bottom mounting rails. Each cabinet under 24 inches (61 cm) wide should receive four screws, a minimum of two screws per stud, one top rail, one bottom rail. Cabinets larger than 24 inches (61 cm) should receive two screws per stud. Cabinets over 42 inches (107 cm) wide should have six screws, depending on the stud layout. Since studs are usually 16 inches (40 cm) apart center-to-center, a 24-inch (61 cm) cabinet should be located over at least two studs. Occasionally studs are 24 inches (61 cm) apart; then toggle bolts must be used through the wall. Always keep in mind that careful attention to detail ensures a first-rate installation.

Remember the increased weight of kitchen storage can pull fasteners through wallboards. Use good judgment in adding toggle bolts or fasteners. Use fasteners with finishing washers to prevent the screw from sinking into the wood or splitting it and to provide a professional finish. To adjust and align cabinets, use shims made from tapered wood such as shake shingles, which can be driven between the wall and cabinet back. Adjust by tapping wedges and loosening or tightening screws until everything is plumb and level.

Place the shims at the stud locations. Drill them, running mounting screws through them after final adjustment to prevent them from working loose. Trim off the excess with a handsaw. When tightening, look for gaps behind the cabinets. Do not tighten a screw that is at an unshimmed gap. Tightening screws can cause the cabinet to bow. Shim as needed (refer to Fig. 10.5). The first cabinet sets the accuracy for the entire row—time spent here saves time later. Before screwing the cabinet to the wall, recheck the sides of the cabinet to ensure that they are level and plumb.

A blind corner cabinet must be adjusted horizontally to line up with the row of cabinets it abuts (Fig. 10.6). The manufacturer recommends a dimension. Mark the required distance on the wall, and locate the studs. Lay the cabinet in place on the cleat, adjust the position, referred to as *pull*, away from the wall, and proceed as before.

Note: Cabinets must always be attached to walls with screws. Never use nails.

With the first cabinet and any necessary fillers in place, the second wall cabinet in the run is placed in position next to the first (Fig. 10.7). Set it on the cleat or prop it into place; then carefully align it with the first cabinet. When

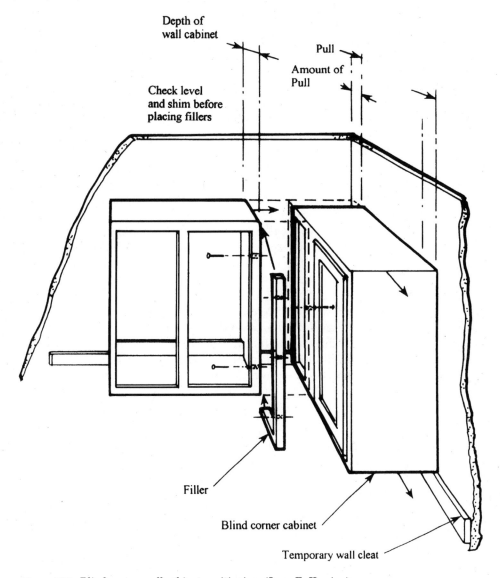

Depth of
wall cabinet

Pull

Amount of
Pull

Check level
and shim before
placing fillers

Filler

Blind corner cabinet

Temporary wall cleat

Figure 10.6   Blind corner wall cabinet positioning. (*Leon E. Korejwo*)

the bottoms and faces of the stiles are flush, use C-clamps (or handscrew clamps) to clamp the two cabinets together. (Use pads with the clamps so that the cabinet finish will not be damaged.) Drill through the first stile with a diameter just wide enough to allow the screw to slide in without binding its threads. Countersink this hole slightly to recess the screw head. Connect the cabinet with two screws in from one side at the top and bottom and one in from the other side in the center. Two screws are usually adequate for wall cabinets up to 36 inches (91 cm) high. When the cabinets are secured to each other, check the second cabinet for plumb and level, and shim it as necessary. Drill through the upper and lower nailing strips into the studs and screw the cabinets to the wall. Once again, do not fully tighten the screws. The seams between the cabinet units are not generally covered by any trim or molding, as it is imperative that the alignment of the face frames is exact and that the fasteners pull the units snugly together.

Most manufacturers recommend the use of electrician rails or some type of lag-bolt system for mounting to masonry walls, with the quantity required being the same as for regular screw installations. Alternatively, lag-bolt a hanging rail to the masonry wall first, and then mount the cabinets to the rail. The hanging rail is usually a 1-by-4 (2.5-by-10-cm). Some installers notch out the wall finish so that the rail can set directly against the masonry and thus reduce the gap between the back of the cabinets and the wall.

To attach adjacent wall cabinet units, fasten #8 by $2\frac{1}{4}$-inch (5.7-cm) or $2\frac{1}{2}$-inch (6.4-cm) wood screws or drywall screws through the vertical stile of one cabinet into the vertical stile of the adjacent cabinet. The length is dependent on the thickness of the stiles, which varies from manufacturer to manufacturer. Two screws per pair of wall cabinets being connected are generally sufficient for cabinets up to 36 inches (91 cm). Taller cabinets should have three screws.

This process continues until the full run of wall cabinets has been installed. Check often during the installation to make sure that the face of the cabinets are aligned, plumb, and level. It might be necessary to shim at the wall and between cabinets to correct for uneven walls or floors. When everything looks

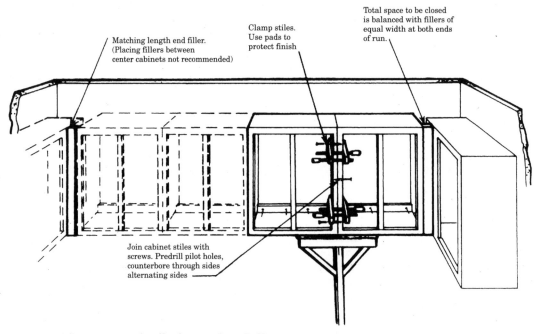

Figure 10.7 Continue run of wall cabinets. (*Leon E. Korejwo*)

correct, finish tightening the screws. Finish installing all the wall cabinets before beginning the base runs. A series of two or three wall cabinets can also be attached together before lifting them into place, allowing the alignment work to be done without having to support the weight of the cabinets. The use of jacks is recommended for this installation method.

Base cabinet installation

Base cabinets should be installed next, in the same manner as wall cabinets, starting at the same inside corner and working outward in each direction from that point. Before you begin installing them, remove any baseboard, moldings, or wall base that might interfere. Since support is not a problem for the base cabinet, it is a good idea to bring in all the cabinets for one run and set them together. If a base blind corner cabinet is used, pull it out of the corner to the proper dimension as shown on the kitchen layout. Position the cabinet so that the top is flush with the reference line. Measure the cabinet carefully for level and plumb—from side to side and front to back. If necessary, adjust by driving wood shims under cabinet base (Fig. 10.8). Scribing strips may be included along the sides to allow full alignment with the wall. Both shims and irregularities in the floor can be hidden by baseboard trim, vinyl wall base, or new flooring. Be careful not to damage flooring.

Fasteners for base cabinets are similar to those used for wall cabinets. However, because base cabinets are resting on the floor surface, fasteners are used basically to secure the unit in place once it has been leveled. If the base cabinet backs up to only one wall surface, two screws placed through the center of the mounting rail into wall studs are generally sufficient to stabilize the unit. Adjacent units are screwed together through the vertical stiles or through the adjoining cabinet side panel, depending on whether they are framed or frameless cabinets. Brads or other suitable types of nails are recommended only for use with wood trim, panels, and other types of moldings. Nails are not considered sufficient fasteners for either base or wall cabinets.

Attach cabinets loosely to the wall with screws. If necessary, attach filler strip to adjoining cabinets. Clamp adjoining cabinets to the corner cabinet. Make sure the cabinet is plumb, and then join the cabinets with screws. In some cases the filler may not be required, i.e., where a drawer unit is the adjacent cabinet or where the full opening of doors on the adjacent cabinet is not necessary. Use a jigsaw to cut any cabinet openings needed for plumbing, wiring, or heating ducts. If you need to cut access holes in a cabinet's back or bottom for plumbing supply and drain pipes or for electric wire serving the sink complex, do so before installing the cabinet.

If using a base lazy Susan cabinet, apply a 1-by-2 (2.5-by-5cm) or 1-by-3 (2.5-by-7.6cm) strip to the wall at the proper height. To support the countertop, apply strips to both walls at right angles (Fig. 10.9). This frame should be nailed or screwed to the studs and will provide support for the countertop once it is installed into the corner. It is usually better to fasten at least one cabinet in both directions to the base lazy Susan cabinet and be sure they are shimmed properly before attaching the wall cleat.

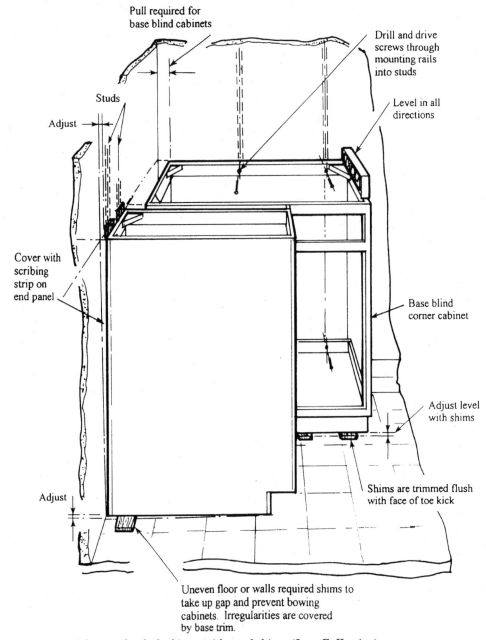

Pull required for
base blind cabinets

Drill and drive
screws through
mounting rails
into studs

Level in all
directions

Studs

Adjust

Cover with
scribing
strip on
end panel

Base blind
corner cabinet

Adjust level
with shims

Shims are trimmed flush
with face of toe kick

Adjust

Uneven floor or walls required shims to
take up gap and prevent bowing
cabinets. Irregularities are covered
by base trim.

Figure 10.8 Adjusting level of cabinets with wood shims. (*Leon E. Korejwo*)

Drill holes through the mounting rail into each stud, then drive #8 flat-head wood screws that are $2\frac{1}{2}$ inches (6 cm) long. Two screws are usually sufficient to secure the base cabinet in place. Screws for base cabinet installation are the same as those used for wall cabinets. If the spot where you need to screw the base cabinet is not flush with the wall, place a shim between the cabinet and the wall so that as the screw is tightened, the cabinet is not pulled out of square.

Adjacent base cabinets are installed in the same manner as the wall cabinets. Continue installing base cabinets one next to the other, and at the end of the cabinet run, where necessary, attach filler strips. Continue to shim up the base cabinets so they are all level and follow the base cabinet line on the wall. Set the second cabinet, level, align, and then screw it to the first cabinet through the adjoining stiles. The second cabinet then can be screwed to the wall. Once again, if necessary, adjust by driving wood shims underneath the cabinets. Place wood shims behind cabinets near the stud locations wherever there is a gap. As you tighten the wall screws, be sure face frames remain square and in exactly the same place.

Frameless cabinet installation

Frameless cabinets originated from a European technology that centered on portability and a metric standardization. The standard was based on unitized 32-mm modular spacing of components. Domestic frameless cabinets combine features of the European unitized manufacturing with American practice.

The difference between framed and frameless cabinets is that framed cabinets have a face frame attached to the cabinet sides, and the doors are hinged to the frame. With frameless cabinets, doors and drawers cover the entire front

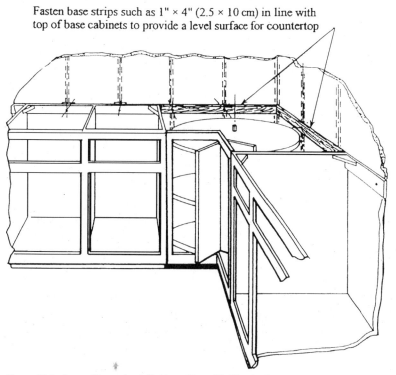

Fasten base strips such as 1" × 4" (2.5 × 10 cm) in line with top of base cabinets to provide a level surface for countertop

Figure 10.9 Lazy Susan installation. (*Leon E. Korejwo*)

of the cabinet, and doors are attached directly to the sides of the cabinet with concealed hinges.

Frameless cabinets are essentially an open-fronted box with large doors and drawers that cover the entire front. Multi-adjustable concealed hinges and mounting hardware present different installation procedures for some of these cabinets and vary among manufacturers. Refer to the manufacturer's instructions for specific details.

In general, the wall cabinets are mounted on a wall track. The track is leveled and screwed to the studs, then the cabinets are hung on the track. Adjustments in the hanging clips allow the cabinets to be raised or lowered to level them. Most frameless wall cabinets attach to the wall similarly to framed wall units. But some frameless manufacturers' wall cabinet units themselves are not permanently attached to the wall.

A hanging rail (or track) of steel that is approximately $1\frac{1}{4}$ inches (3 cm) wide holds the wall cabinets. It has an offset channel that is designed to accept the adjustable hooks on the back of each wall cabinet unit. The hanging rail is drilled and screwed to wall studs. Holes that are $\frac{1}{4}$ inch (0.6 cm) in diameter are predrilled in the rails for mounting. Use #14, $2\frac{1}{2}$-inch (6-cm) panhead screws to attach the rails to the wall studs.

Some cabinets have the hanging rail behind the wall cabinet or run above the wall cabinets. Crown molding or trim can conceal the hanging rail. If the hanging rail runs behind the wall cabinets, the backs of the cabinets should be notched so that the cabinets sit tight against the back wall. Do not notch the end panel of a wall cabinet at the end of a run of cabinets. Stop the rail at the inside end panel.

When the alignment of cabinets is complete, they can be bolted together with the partially drilled holes for this purpose. Finish drilling the holes for screws or hardware included with the cabinets. This specialized hardware has connectors that are threaded to pull the cabinets tight. The holes are then covered with plastic caps (Fig. 10.10).

Attach the doors to the cabinets and adjust them. Hinges provide for slight adjustment and can move the doors up, down, in, and out, and side-to-side for perfect alignment, but not enough to correct an out-of-square condition.

Frameless base cabinet installation

Some styles of base cabinets have exposed screw legs, which simplify leveling. These legs allow the cabinets to be leveled and adjusted during installation without the use of shims. The leveling legs are slipped into plastic sockets on the underside of the base cabinets. They are adjusted with a screwdriver through holes located in the bottom of the cabinet. You can also twist the legs by hand when more than a minor adjustment is required. After the cabinets have been leveled and screwed to the wall, the toekick cover panel is simply snapped in place to provide a very clean, finished look.

Both upper and lower cabinets are screwed together as conventional cabinets are, but the screw heads are recessed and then concealed with decorative

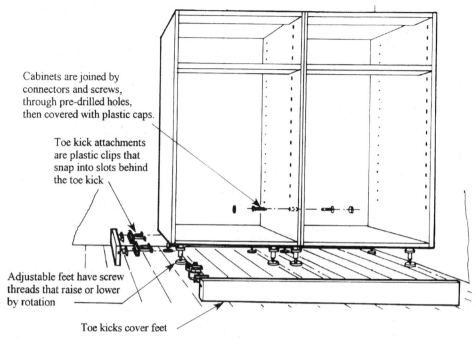

Cabinets are joined by connectors and screws, through pre-drilled holes, then covered with plastic caps.

Toe kick attachments are plastic clips that snap into slots behind the toe kick

Adjustable feet have screw threads that raise or lower by rotation

Toe kicks cover feet

Figure 10.10 Leveling frameless base cabinet. (*Leon E. Korejwo*)

snap-on tops that match the surrounding laminate finish. Final adjustment of the doors is done with the hinges, which can move up, down, in, out, and side to side for perfect alignment.

Mount the cabinets to the wall or a continuous blocking strip behind the cabinets. European frameless base cabinets are not as deep, and standard countertops extend beyond the face more than usual. Blocking behind the cabinets moves them out to 24 inches (61 cm) from the wall and also provides a means of attaching the cabinets to the wall. You need special end panels on exposed cabinet ends to cover the gap between the wall and the base cabinets.

Note: If you choose to screw the base cabinets to the wall, make sure you use washers or some other means of keeping the screws from pulling through the back of the cabinet.

Toekicks are supplied in long sections. You need to rip the toekick to the proper width once the cabinets are in place. Scribe the toekick right to the flooring, and seal it to the floor using a vinyl strip, similar to weather stripping, that is usually supplied with the cabinets.

End caps are generally provided for the connection of inside and outside corners. If corners meet at angles other than 90 degrees, abutting corners need to be mitered and glued. When the toekick has been assembled, the clips that attach it to the legs are inserted in a groove in the back of the toekick to correspond to the locations of the legs. Some side-to-side adjustment is possible once these clips have been attached to the toekicks.

The leveling leg system may not be suitable for some island or peninsula cabinets (Fig. 10.11). Order cabinets without the leveling legs. A subbase can be built on the job. Blocking can be installed underneath the cabinets, and the cabinets can be attached to the blocks.

Tall cabinet installation

Tall cabinets, which usually follow the wall and base cabinets, can add difficulty to the installation that must be planned for in advance. If the tall unit is at the end of a run of cabinets, which is most often the case, the installation

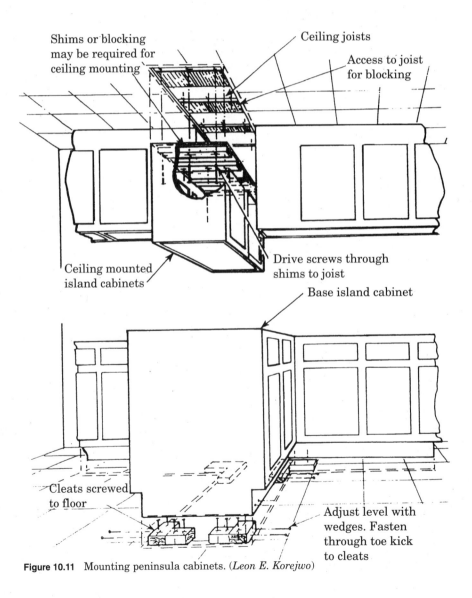

Figure 10.11 Mounting peninsula cabinets. (*Leon E. Korejwo*)

methods previously explained are still acceptable. However, if the tall cabinet unit is in the middle of a run of cabinets, it may be necessary to install the base cabinets, along with the tall cabinet unit, before installing the wall cabinets. This order of installation ensures that the wall cabinets and the tall cabinet unit are perfectly aligned.

Tall cabinets are mounted in the same manner as base and wall cabinets (Fig. 10.12). The screws are driven through the cabinet mounting rails into the wall studs. The position of the mounting rails varies by manufacturer, so follow the manufacturer's instructions closely. If the tall cabinet will house an oven unit, the rough-in information for the oven should have been supplied with the cabinet order. Using this information, the cabinet manufacturer provides the appropriate opening for the oven, as well as a support kit to provide full support for the oven unit. It is important to be sure that the weight of the oven is supported within the cabinet directly to the floor of the cabinet, rather than on the front frame.

Appliance garages and other counter-height cabinets

Appliance garages with roll-up tambour doors are a popular option for the kitchen. These cabinet units are positioned below the wall cabinets and rest directly on the countertop. Therefore, they cannot be installed until the countertop is set in place. Some appliance garages are separate units mounted under the wall cabinets while others are a part of a one-piece wall cabinet unit. The installation technique varies depending on the type of unit. Generally, separate under-cabinet units are mounted by screwing up through the mounting rails into the frame of the wall cabinet above. Just as standard wall cabinets, the one-piece units are mounted to the wall studs through the mounting rails.

Appliance garages often have a roll-up tambour door. Care must be taken when installing the unit so that the fasteners do not interfere with the operation of the door. For specific installation recommendations, follow the manufacturer's instructions.

Most appliance garages are manufactured so the bottoms can be trimmed to fit tight to the countertop surface. Most installers prefer to scribe the unit tight to the countertop without any molding; however, molding can be applied at the intersection of the cabinet and the countertop. Before ordering an appliance garage, determine the thickness of the countertop material and mounting height of the wall cabinets. Both of these factors can cause the distance between the countertop surface and the underside of the wall cabinets to vary by as much as 2 inches (5 cm). Make sure that the unit will be tall enough to fill the space or will be trimmable if larger than the space. The flexible design of the tambour door adjusts to the available height without having to be trimmed or modified.

Another factor to consider when installing an appliance garage is the countertop backsplash. A good method is to set the appliance garage in place when the countertop measurements are being taken, which allows the countertop to be fabricated to accommodate the unit. Also, to achieve a custom look, many installers scribe the sides of the cabinet around the backsplash.

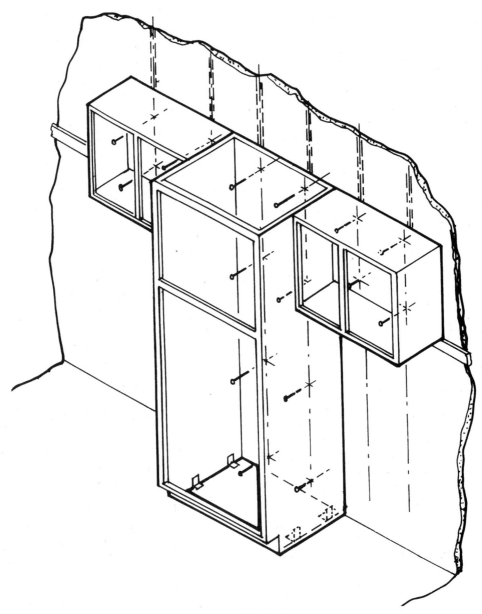

Figure 10.12   Tall cabinet installation. (*Leon E. Korejwo*)

Refrigerator and dishwasher panels

If the dishwasher, refrigerator, or trash compactor is to be placed at the end of a run of cabinets, a panel needs to be installed to finish off the run and protect the appliance. Dishwasher panels, which can also be used next to a trash compactor, are 34½ inches (88 cm) high and 24 inches (61 cm) wide. Most have a 1-by-2 (2.5-by-5cm) attached to the front to create a finished edge, and some have a toekick cutout. Refrigerator panels, which are sometimes used alongside oven and pantry cabinets, are 84 inches (213 cm) high and 24 inches (61 cm) deep. In most cases, they also have a finished front edge.

The panels are installed as the cabinets are being placed and before the appliances are installed. The refrigerator panel is secured to the upper cabinets, to the floor, and sometimes to the wall. The dishwasher panel is secured to the floor and to the back wall. If it falls between studs, toggle bolts or molly bolts can be used to hold it in place.

Cabinet accessories

A wide range of accessories are available to customize kitchen cabinets, as discussed in Chapter 5. The construction and installation procedures for each of these accessories varies according to the manufacturer; therefore, follow manufacturer-supplied instructions closely. Take note of the clearances that each of these accessories require. Make certain adjacent cabinets, appliances, and equipment will not interfere with the accessories.

Fillers

Planning a kitchen on paper is one thing; on the job site, you may run into some realities. Fillers are placed at the end of a run of cabinets that abut an irregular wall surface. The filler can then be attached to the cabinet stile and scribed to the wall. They are also used to make up odd inches in a run of cabinets and to adjust the portion of wall that remains exposed around doors and windows. Some manufacturers offer cabinets with extended or side stiles that eliminate the need for fillers in many cases. Be certain to plan for and lay out all fillers shown on the plans. The need for fillers varies with the size of the cabinet's face frame and the depth of the handles being used (Fig. 10.13).

For runs that start in a corner and end at a window or door, place all the fills at the starting corner. If the cabinet run goes from corner to corner, fillers should be evenly divided and placed at each end of the run, rather than placing them all at one end or in the middle.

Additional trim and finishing

To complete the cabinet installation, use moldings to cover gaps between the cabinet and the wall or floor, as well as to provide an attractive transition. Finish trim offers a custom look to any type of cabinet.

Prefinished moldings are usually available from the same manufacturer that supplied the cabinets. Some cabinets come with decorative panels that finish the visible end of a cabinet run and some are designed with scribing strips along the sides. Both include extra material to be shaved down to a perfect fit between the end cabinet and an irregular wall.

Scribe molding are flat trim pieces that have relatively little profile and can be applied along the edges of cabinets where they meet the walls. They help conceal the gaps that occur because of shimming and irregular wall conditions.

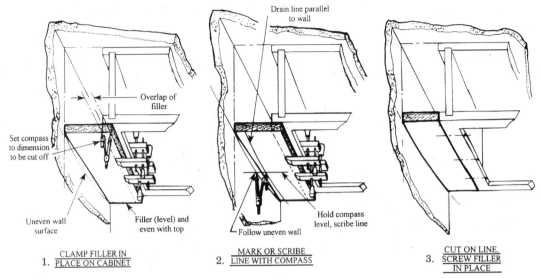

Figure 10.13 Scribing filler. (*Leon E. Korejwo*)

Trim or scribe molding is recommended to hide small gaps between the top of the cabinets and the soffit or ceiling.

Batten molding is a flat trim available to mask joints between adjacent cabinets where desired or required. The edges of this trim are usually chamfered.

Outside corner molding, used to seal joints between two panels at right angles to one another, is an L-shaped molding tacked onto the corner of the cabinet like an edge band or edge strip.

Inside corner molding is available to mask open joints between cabinet units or cabinets and wall surfaces. This molding is often milled with a concave cove profile.

Trim can be used with considerable impact at the top of wall cabinets, either at the ceiling or the soffit, depending on the type of construction and the method of installation. Many cabinet manufacturers provide one- and two-piece crown moldings expressly for this purpose. The trim is factory-finished to match the cabinet finish. Crown molding is generally sold by the lineal foot, and can be quite expensive. Special care must be taken in the cutting and fitting together of crown molding to produce tight-fitting joints. The joints for this type of molding are usually mitered and coped. Some manufacturers produce specialty trim such as window and door trim, which can integrate all the wood trim used in the room. This trim is prefinished to match the cabinet finish. Other specialty trim pieces include items such as galley rails, which are placed on top of wall cabinets without soffits to provide a display area as well as a finished look. Because trim pieces have to be nailed onto the cabinets in most cases, the manufacturer-applied finish will need to be touched up when the trim has been installed. You may find a few areas on the face of the cabinets that require touch-up also. Manufacturers provide finish touch-up kits in all the various finishes they produce for this purpose. Include touch-up kits in the cabinet order so that you have it on hand during installation.

Handles and pulls

Many manufacturers provide hardware for drawers and doors. Because the homeowner may not like the hardware selections available from the manufacturer, cabinets usually are ordered with the least expensive hardware, which is later discarded for hardware purchased from another source. Few manufacturers predrill the holes required for the hardware installation because the orientation and position of the handles and pulls varies according to the personal taste of the end user and the designer. Quite often the hardware supplied with the cabinets is not used at all. If the holes were predrilled for the hardware provided, the substituted hardware might not work in the same holes.

Jigs are available to aid in the consistent alignment of hardware holes. Some installers who have a large quantity of holes to drill make a template to ensure that the holes are consistent. Typically, one template would be required for doors and another for drawer fronts. Some handles and pulls require one hole each, while others require two or more. Review the hardware carefully before you begin any drilling. Check the spread between holes on all hardware requiring multiple holes. Do not assume that the spread is the same for all hardware on the entire job, although generally that would be the case.

Use a punch or awl to mark the hole before drilling to keep your drill from drifting. You do not want to damage the face of the cabinet. Always drill from the finished face of the door or drawer to avoid any splintering of the wood or chipping of the finish. The installation of the handles and pulls should be the last step in the cabinet installation process. If the handles have a protective coating, leave it in place until you are ready to do the final cleanup.

Installation of countertops

You need to know how to install the various countertop types as described in Chapter 5 and how to provide proper substrate (base materials) for them. Generally, countertops are fabricated by countertop fabricators from measurements you supply. However, some manufacturers come to the project site and take the field measurements themselves. Because countertops are not usually cut to fit in the field, measurements must be accurate for proper fit. If a manufacturer is fabricating the countertops and does not take the field measurements, the most accurate method is for you to draw a rough sketch to scale, showing the exact dimensions. This sketch can then be given to the manufacturer to fabricate the countertops to your exact measurements. It is generally recommended that measurements be taken only after all the base cabinets have been set.

Depending on the countertop material, the level of difficulty required to construct the countertops, and the manufacturer, lead time for countertops can range from a few days up to two weeks. It is important for you to establish a good relationship with a countertop fabricator who has a reputation for reliability and prompt delivery. It is common to use one fabricator for laminate and solid-surface countertops and a different fabricator for other countertop materials, such as marble or granite.

Before beginning the countertop installation, be certain that all cabinets have been properly prepared to receive the countertop. All cabinets must be level, and the kitchen must be measured carefully, especially from corner to corner and up to window and door openings.

Unless the countertops are being fabricated in the field, as with ceramic-tile countertops, the countertop section or sections are installed onto the base cabinets. Be certain that the substrate or support system is in place for the type of countertop being installed.

In most cases, base cabinets have triangular corner struts (or gussets) at the four upper corners of the cabinet (Fig. 10.14). The countertop frame is screwed to these struts from below. Be careful to measure the distance and select the proper length screws. Also, be extremely careful to drill the proper depth hole into the countertop. Do not drill through the countertop.

If the countertop is in multiple sections, the sections should be set in place to determine proper fit and alignment of edges and patterns. To install the countertop sections, start with the section that fits into a corner, if applicable, and work outward in the same way that the cabinets were installed. The sections of countertop should be joined with connecting bolts that fit into a special slot under the countertop surface (Fig. 10.15). Remember to follow any instructions provided by the countertop fabricator.

Backsplashes

Typically, 4-inch-high (10-cm) standard backsplashes are not attached to the wall. Backsplashes, which are not an integral part of the countertop, should be installed after setting the horizontal portion of the countertop. To provide

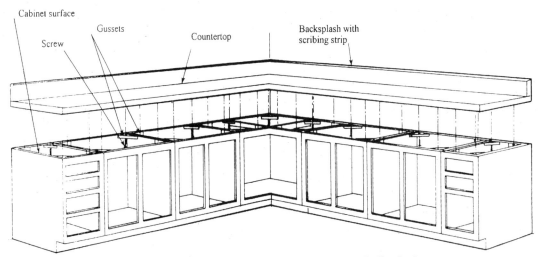

Figure 10.14 Attaching the cabinet to the countertop with gussets. (*Leon E. Korejwo*)

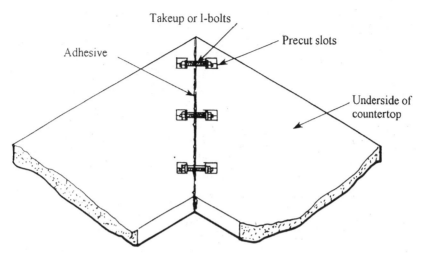

Figure 10.15 Take-up bolts. (*Leon E. Korejwo*)

proper alignment, the standard 4-inch (10-cm) backsplash is often provided with thin, oval-shaped disks, called *biscuits*. These biscuits slide into slots in both the countertop and the backsplash. They are also sometimes used between sections of countertop being joined together to aid in alignment. The backsplash is joined to the countertop surface with the same adhesive used to join sections of the countertop.

Full-height backsplashes, which often extend from the countertop surface to the underside of the wall cabinets, are attached to the wall surface. Solid-surface and laminate materials are adhered directly to the wall surface with an adhesive. Other backsplash materials, such as ceramic tile, are field-assembled and installed. The techniques and types of adhesives for preparing the edges to be joined vary according to the type of countertop material. Closely follow the recommendations provided by the material manufacturers.

Installing plastic laminate countertops

The most crucial step with laminate counters is the measuring; order the counters to exact size to minimize cutting on the job site. If possible, as with all countertops, it is best to wait until the base cabinets are installed before taking the measurements. If not possible, order the counters long and cut them after the cabinets are installed and you are able to verify the measurements. As explained earlier, in most instances, most of the work has already been done by the manufacturer. However, if your dealer has a backlog, there may be a delay of several days in the installation, so you may have no choice but to make the order beforehand. In either case, measure the kitchen very carefully, especially from corner to corner and up to window and door openings. If you are planning to have all the cutouts done for you, provide the exact location and sizes of the sink, cooktop, range, etc. Once the countertops arrive, you need only to assemble the pieces and attach them to the base cabinets.

Plastic laminate countertop sections have a wood framework already attached to the underside of the countertop. This framework supports the countertop and provides a way to attach the countertop to the base cabinets without screwing into the laminate surface. There are two types of plastic laminated countertops—post-formed and self-rimming.

Post-formed Post-formed countertops are premolded, one-piece tops, from curved backsplash to bullnosed front (Fig. 10.16). They are bought as blanks in stock sizes, are made to order, and cannot be reshaped because of the forming. Since post-formed countertops come only in standard sizes, you normally need to buy one slightly larger than needed and cut it to length. (To cut the countertop with a handsaw, cut from the top side of the counter. If using a power saw, or any saw that cuts on the up stroke, make the cut from the back side.) If the countertop has an exposed end, you need an endsplash kit that contains a preshaped strip of matching laminate. For a precise fit, the backsplash must be trimmed to fit any unevenness in the back wall (a process called scribing). Post-formed countertops have a narrow strip of laminate on the backsplash for scribing.

Measure the span of the base cabinets, from corner to outside edge of the cabinet, to determine the size of the counter needed. The standard overhang on a laminate top varies between ¾ inch (1.9 cm) and 1 inch (2.5 cm) in front and on open ends. Add these dimensions to the dimensions of the cabinet.

If you are planning to include an endsplash at one or both ends, check the endsplash kit. Since most endsplashes are assembled directly above the end of the cabinet, generally subtract ¾ inch (1.9 cm) from the length of the countertop on that side. Endsplashes are assemblies used at the cut or exposed ends of the countertop (Fig. 10.17). They consist of added strips of material called battens, which are screwed to the underside of the countertop. The edge is covered by an endcap, a preshaped section of laminate joined to the exposed end by silicone sealant or adhesive recommended by the manufacturer. Holding

Figure 10.16   Post-formed countertop. (*Leon E. Korejwo*)

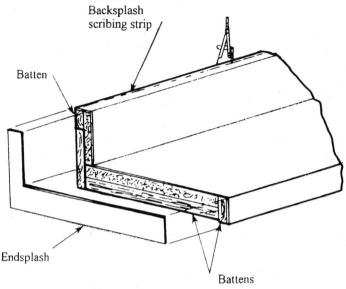

Backsplash
scribing strip

Batten

Endsplash

Battens

Figure 10.17   Endsplash kit. (*Leon E. Korejwo*)

the endsplash in place with C-clamps, drill pilot holes, if needed, and drive in the screws. File the edges of the new strip flush with the top and front edges of the countertop or use an electric router and laminate-trimming bit.

If you have installed new base cabinets, as explained earlier in this chapter, they should be level. Older cabinets may have settled somewhat. Lay a level at several points along the front, rear, and sides to see if you need to shim under the new countertop or realign the cabinets.

Position the countertop on the base cabinets. Make sure the front edge of the countertop is parallel to the cabinet face and check for level. It is difficult to adjust drawer alignment once countertops have been fastened in place. Therefore, before installing tops, check drawer alignment in all base cabinets to be sure drawer fronts meet the face frames evenly all around. All drawers and doors must open and close freely. Test-fit one of the counters to be certain the overhanging front lip does not interfere with the drawers or undercounter appliances.

If needed, adjust the countertop with wood shims. Remove the countertop. Use a belt sander to grind the backsplash to the scribe line.

If the manufacturer has not already cut out the area for the sink, and you are installing a self-rimming sink, mark the cutout for the sink. Position the sink upside down on the countertop and trace an outline. Draw a cutting line $\frac{5}{8}$ inch inside the sink outline. To make cutouts for the sink, cooktop, or other appliance, cutouts can be made with the counters out and sitting on sawhorses or after the counters have been installed, provided there is enough clearance within the cabinets for the saw blade to move.

Carefully measure and mark the cutout. Position the metal frame on the countertop, using the sink rim as a template or the manufacturer's instructions to establish the exact size, and trace an outline around the edge of vertical flange. Remove the frame. Drill pilot holes just inside the cutting line. Make cutouts with a jigsaw. Support the cutout area from below so that the falling cutout cannot drop down before the cut is finished, ruining the laminate or bending the saw.

Precut laminate and post-formed tops have miter joints at their corners. Lay the pieces upside down on a soft surface to prevent scratching them. If the cabinets are U-shaped or L-shaped, you need mitered countertop sections. The mitered sections should have small slots along the bottom edges. They are connected with takeup or draw bolts. Coat the edges with silicone sealant, align them carefully, and tighten the bolts. Fasten the backsplashes together with wood screws. As mentioned earlier, countertops, like cabinets, rarely fit uniformly against the back or side walls because the walls rarely are straight. Use a scribing strip that can be trimmed to the exact contours of the wall. Apply adhesive caulking to the edges of the miters, and press them together. From underneath the cabinet, install and tighten the miter takeup bolts. As always, check the alignment before tightening each bolt. To set the assembled counter in position, you may need assistance. Push the backsplash tightly up against the wall. Temporarily shim underneath if the cabinet tops are not perfectly level. Any high spot (a bulge or unevenness) on the wall's surface will create a gap between the wall and backsplash on either side of the bulge.

Measure to be sure there is at least $34\frac{1}{2}$ inches (88 cm) in height between the underside of the counter and the floor. This height is the minimum rough-in height for undercounter appliances. If the counter is less than $34\frac{1}{2}$ inches (88 cm) above the floor or if its front edge interferes with the operation of drawers, you need to raise the counter with riser blocks—2-inch-square spacers located about every 8 inches (20 cm) around the perimeter. When the counters are correctly in place, secure them by drilling pilot holes up through the corner braces into the top's underside. Be careful that these holes do not penetrate more than two-thirds of the counter's thickness. Finally, drive #10 wood screws through the braces into the countertop. To prevent moisture from seeping behind the counter, once the top is secured, run a bead of tile caulking along the joint between the backsplash and wall.

Self-rimmed   The term *self-rimmed* simply means that the laminate is applied over an old countertop or new core material (Fig. 10.18). Though post-formed countertops are simpler to install, a self-rimmed countertop can be tailored to fit any space and can be customized with a decorative edge treatment. They are made on the job and are easily shaped into curves.

Laminates are sold in 6-, 8-, 10- or 12-foot lengths that are about $\frac{1}{16}$-inch (.16) thick (standard), and range in width from 30 to 48 inches (76 to 122 cm).

Cut the core material to size from $\frac{3}{4}$-inch (1.9-cm) plywood or high-density particleboard. The perimeter is built up with strips of particleboard screwed to the bottom of the core. Laminate pieces are bonded to the countertop with contact cement. Edges are trimmed and shaped with a router.

Measure along the tops of the base cabinet to determine the size of the countertop. Allow for overhangs by adding 1 inch (2.54 cm) to the length for each exposed

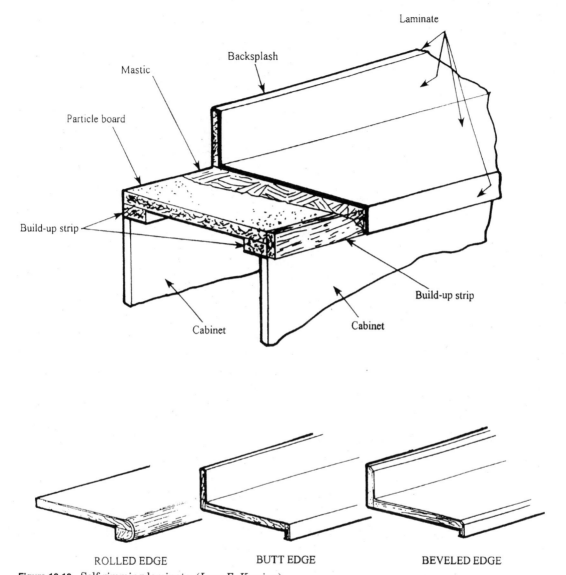

Laminate

Backsplash

Mastic

Particle board

Build-up strip

Build-up strip

Cabinet

Cabinet

ROLLED EDGE   BUTT EDGE   BEVELED EDGE

Figure 10.18 Self-rimming laminate. (*Leon E. Korejwo*)

end and 1 inch (2.54 cm) to the width. If an end butts against an appliance, subtract $\frac{1}{16}$ (.16 cm) inch from the length to prevent scratching the appliance.

The substrate surface must be clean, dry, and smooth. Cut 4-inch (10-cm) strips of particleboard for the backsplash and for joint support where sections of countertop core are butted together, and 3-inch (8-cm) strips for edge buildups. Join the countertop core pieces on the bottom side using $1\frac{1}{4}$-inch (3-cm) wallboard screws.

To determine the size of the laminate countertop, measure each surface to be laminated, adding at least $\frac{1}{4}$ (6 mm) to $\frac{1}{2}$ inch (1 cm) to all dimensions to allow for trimming after bonding. Mark and cut the laminate by scoring and breaking it. Place masking tape over the cut line and cut through the tape and laminate; the masking tape prevents chipping. Cut with a fine-toothed saw, face down with a circular or saber saw, face up with a table saw or handsaw.

Build down the edges of the core with 1-by-3 battens. Laminate the countertop, sides, and front strips first, and then the top surface.

Apply contact adhesive to both the laminate back and the substrate to be joined, and allow to dry according to manufacturer's instructions. Check alignment carefully before joining the two; once joined, the laminate cannot be moved. Press the laminate into place and bond with a J-roller.

Trim all of the laminate with an electric router flush with core's edge. Where the router can't reach, trim with a file. File all edges smooth with downward strokes to avoid chipping. Make cutouts for the sink and cooktop. Use templates provided to mark the top, make straight cuts, and then radius corner cuts with a router or keyhole saw. Install the sink and appliances according to manufacturer's instructions. Apply back- and endsplashes as needed. They should be cut from the same core material as the main countertop and butt-joined to the countertop with sealant and wood screws.

Installing solid-surface countertops

The installation of subsurface countertops requires fine tools, care, and expertise. Solid surfaces can be sawed, drilled, and shaped with customary woodworking tools and carbide-tipped blades. To form corners and long countertops, sheets are welded together with color-matched joint adhesive. Some of these materials are sold only to distributors or other qualified firms with trained fabricators on staff. One slip with a saw or router can ruin a very expensive piece of material. It is important to know that many distributors will not sell solid-surface materials unless they do both the fabrication and the installation. They recommend that only factory-trained fabricators install the tops. However, as the kitchen installer, you should be familiar with some key factors.

This type of countertop surface is made from ½-inch (1-cm) solid-surface material. The countertop edges are built up with ¾-inch (1.9-cm) solid-surface strips attached with special joint adhesive (Fig. 10.19). Prior to installation, blocks are usually bonded to the surface of the tops with silicone on either side of the joint. A notching-joint adhesive is applied to the joint, and clamps are then slipped over the blocks and tightened. After the joint has set, the blocks are pried off and the joint area is sanded smooth. Cutouts can be cut with a scroll or saber saw, especially where curves are needed, or a portable circular saw for straight cuts. The edge is then shaped with a router, and the surface is smoothed with sandpaper. The ½-inch (1-cm) thickness must be continuously supported by the cabinet frame or closely spaced plywood blocks.

Unsupported overhangs should be kept to a minimum. A maximum overhang of 6 inches (15 cm) for a ½-inch (1-cm) sheet or 12 inches (30 cm) for a ¾-inch (1.9-cm) sheet is recommended. For more than that, add corbels or other additional support.

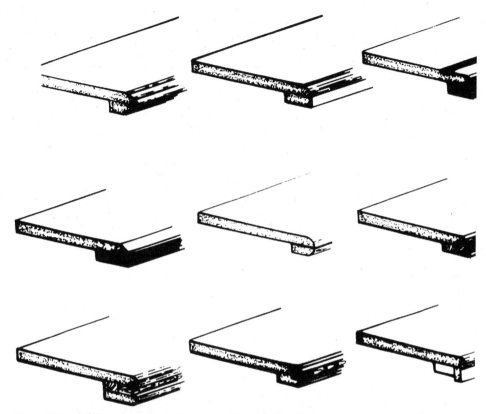

Figure 10.19 Solid surface countertop edges. (*Leon E. Korejwo*)

To install the top, attach three 1-by-3-inch (2.54-by-8-cm) wood strips to the cabinets for the length of the run. Install cross-supports to frame each cutout and to support each seam. Run a bead of neoprene adhesive along the top of the frame and on any supports. A full-height backsplash of solid surface should be installed before the top is installed; it is placed behind the top. The backsplash is attached to the wall with panel adhesive and sealed with silicone caulk. A standard backsplash should sit on top of the countertop, affixed with silicone sealant, and must be sealed to the top, not to the wall.

When installing a solid-surface top between walls, allow ⅛ inch (.32cm) for expansion every 10 feet. The countertop can be edged with wood trim, strips ofsolid-surface material, or a combination thereof.

Tiling a countertop

Of all of the countertop options, ceramic tile is the most long-lasting. Problems with the tiles themselves are rare. Most tile countertop problems stem from improper installation of the underlying decking material. A ceramic-tile countertop installation is only as good as what is installed below the tile surface. Although stated many times before, the cabinets must be level and plumb and the substrate set level as well. If this is not done, the backsplash grout will not be straight.

Tiling a countertop does not necessarily require the services of a professional tile setter. The most common choice for countertops is glazed ceramic tile. Wall tiles, which are lighter and thinner than floor tiles, are the normal choice for countertops and backsplashes. A variety of sizes and thicknesses are available. Ceramic tiles are available individually, or connected with mesh backing to form mosaic sheets. Do not use pregrouted sheets of tiles. This type may contain a mildewcide that is not FDA-approved for food-preparation areas. When your client invests in ceramic tile, they expect it to last, which means it must be properly installed. Proper planning promotes a quality installation and reduces time and labor spent.

Ceramic-tile countertops are installed directly on a deck or substrate by one of two installation methods: mastic (organic adhesive) or conventional mortar bed (mud).

Mastic With the mastic method, generally referred to as a thin-set installation, tile is directly applied to the substrate material with troweled-on mastic (Fig. 10.20). The surface is raised only by the thickness of the tile. A mastic installation may use any of the following base surfaces: existing tile, gypsum board, fiberglass, wood, paneling, brick, masonry, concrete, plywood, or vinyl—virtually any sturdy surface, as long as the surface is structurally sound, level, and moisture-free. However, most countertop installations are done over plywood. A mastic installation does not hide any dips or bows in the substrate material; any imperfections show up in the finished tile installation.

For an installation using mastic on plywood, before laying the tile, remove any old countertops; then install an underlayment of ¾-inch (1.9-cm) exterior-grade plywood, cut flush with the cabinet top, screwing it to the cabinet frame from below. If the cabinets were installed correctly, the plywood top should be level. If it is not, shims can be placed on top of the cabinets to level the decking. Test-fit a piece of cap tile on the front edge of the plywood to be certain drawers and appliances will clear the tile. If they don't, block under the plywood as necessary to raise it. Make all necessary cutouts in the decking for sinks and appliances, and test-fit them for accuracy. Double-check that the decking is level and secure and that the cut ends of the decking around any openings are adequately supported to carry the weight of the sink or appliance. Because kitchen surfaces are exposed to moisture, use moisture-resistant adhesive, followed by a layer of cement backerboard on top, and glazed tiles. A successful tile job requires a solid, flat base, and careful planning. Surfaces may need to be primed or sealed before tile is applied.

Conventional mortar bed (mud) For many years, the standard installation method was to first lay a thick, level mortar base, allow it to dry, and then adhere the tile to that base. Traditional decking is often used to provide flexibility under the tile. In this method, the tile is installed on a bed of mortar ¾ inch to 1¼ inches (1.9 cm to 3.2 cm) thick. This procedure is still the method preferred by professional tile setters, however, perfectly acceptable results can

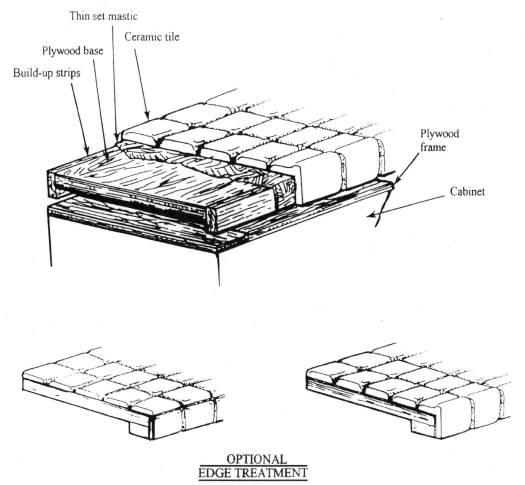

Thin set mastic

Ceramic tile

Plywood base

Build-up strips

Plywood frame

Cabinet

OPTIONAL
EDGE TREATMENT

Figure 10.20   Thin Set Method–ceramic tile. (*Leon E. Korejwo*)

be obtained working with adhesive over a plywood base (Fig. 10.21). With this method of installation, the countertop height is raised to the thickness of the tile and the mortar bed.

Surface preparation   The wood base installed over the top of the base cabinets that the tile rests on is called the deck. Its proper installation is essential for a good job. Solid decking and careful planning and layout are the keys, along with selecting the proper materials for each particular job.

For a mortar-base installation, the decking is done with grade-one or grade-two, kiln-dried, 1-by-4-inch (2.54-by-10-cm) or 1-by-6-inch (2.54-by-15-cm) Douglas fir. The decking boards can be applied either parallel or perpendicular to the face of the cabinets, with a spacing of about $\frac{1}{4}$ inch (0.64 cm) between the boards. Fasten the boards securely to the tops of the cabinets, using screws (preferred) or ring shank nails. Plywood should be avoided as a decking under

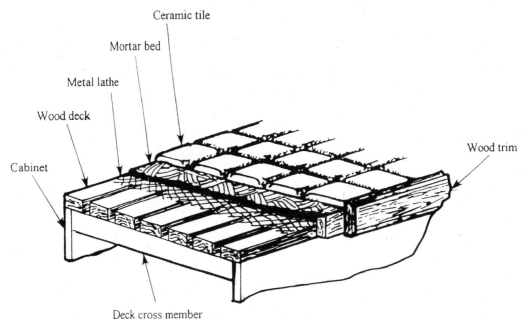

Ceramic tile

Mortar bed

Metal lathe

Wood deck

Cabinet

Wood trim

Deck cross member

Figure 10.21　Mortor Method–ceramic tile. (*Leon E. Korejwo*)

mortar-base tile. However, its use is necessary in some cases, such as on wide overhangs and eating counters. The decking should be delivered to the project site several days before the installation to allow the wood to reach the relative humidity of the room. The decking should overhang the cabinets and be flush with the face of the drawers and doors. Fixture cutouts are made during the tile decking installation. Actually place the fixture in the cutout to ensure a proper fit. When possible, any cutout should be a minimum of 2 inches (5.08 cm) away from a wallboard or plastered backsplash.

Before you start laying the tile, you must decide how you want to trim the countertop edge and the sink. Ceramic tile comes with several edge trim options—in areas of meal preparation, a curved edge of bullnose trim; for the sink area, install sink cap tiles that have a raised edge to prevent water from dripping on the floor. In some installations, wood furring strips are used to frame the tile surface. If using wood trim, seal the wood and attach it to the cabinet face with finishing nails. When in place, the wood strip's top edge should be positioned at the same height as the finished tile.

Trim pieces are available with a $\frac{3}{4}$-inch (2-cm) radius for conventional mortar installations and a $\frac{1}{4}$-inch (0.64 cm) radius for organic-adhesive installations. These trims are generally more expensive than field tile.

Elimination of stress is crucial when the countertop overhangs and eating bars are planned. The tile must have a solid base. If any movement occurs when pressure is placed on the top, the tile or grout will crack. The underside of the decking should be finished to match the cabinets or correspond with other products used in the project. The wall backing up the ceramic-tile backsplash must be solid.

To support the top, install cross-braces of either 1-by-2s (2.5-by-5-cm) on edge or 2-by-4s (2.5-by-10-cm) laid flat. Do not place the braces more than 3 feet apart. Drill two pilot holes, evenly spaced, into each brace. Then, working from below, screw each brace to the top. To hold an apron or other drip-edge trim, nail a 2-by-2 (5-by-5-cm) furring strip to the front edge of the top. This strip will also be covered with tile.

Dry tile layout   Carefully lay out the tiles on the top to determine the best location for the cuts. Dry-laying tiles shows how many tiles are needed and which tiles must be cut to fit and to make sure the finish layout is pleasing to the eye.

The row that falls between the edge trim and the last full row of field tile is cut to fit. Work from the front to the back so that cut tiles are the last row, with the cut edge against the back wall. Find the center point of the area and snap perpendicular chalk lines through it to divide the space to be tiled into equal quarters. Check the intersection of the angles to make sure they are perfectly square.

Lay out one vertical row and one horizontal row in a quarter, allowing for even grout lines. If the end tiles are less than one-half a tile wide, reposition the vertical center line. Pieces smaller than one-half the width of a tile are difficult to cut. When possible, try to position cut tiles in an inconspicuous area. Measuring from end to end, locate and mark the center of the countertop. Lay the edge tiles out on the countertop, starting from the mark.

To allow for grout lines, some tiles have small ceramic lugs molded onto their edges to keep spacing equal; if not, use plastic spacers. Use a carpenter's square to check that the courses are straight. Carefully position the rest of the field tiles on the countertop. Observing the layout, make any necessary adjustments to eliminate narrow cuts or difficult fits.

If the countertop has a backsplash or turns a corner, be sure to figure the cove or corner tiles into the layout. Mark reference points of the layout on the plywood base; then remove the tile.

Placing the trim and setting the tile   Set all trim tiles before spreading adhesive for the field tiles. Keep the dry-laid tiles in place while you adhere the edge trim along the front of the lip. Use the countertop tiles to maintain the spacing you have decided on. With a notched trowel, apply adhesive (thin-set adhesive mixed with latex additive is water-resistant and easy to use) along the front edge of the furring strip. Then cover the backs of the tiles with adhesive and press them into place, aligning them with the reference marks. Next apply adhesive to any back cove tiles and set them against the wall. With the edge complete, install any trim tiles for sink openings or corners. Be sure to caulk between the sink and the base before setting the trim.

Next spread adhesive over a section of the countertop with a notched trowel. Begin laying the field tiles, working from front to back. The installation will begin at the center point and progress outward, one quarter at a time. Cut tiles to fit as necessary. As you lay the tiles, check the alignment frequently with a carpenter's square. Mark and cut all partial tiles prior to applying any adhesive. Doing so saves time because you can then tile in one uninterrupted operation.

Press firmly so that only beads of adhesive from the trowel notches (forming ridges for better adhesion) are evident on the surface. Spread no more adhesive than you can cover before it starts to harden. If the adhesive starts to skin-over before you have a chance to lay the tile, scrape it off and apply new adhesive. Periodically check to see that all joints are straight and even.

Place the first tile at the intersection of the lines. Press it into position with a slight twisting motion to ensure a good bond. To set the tiles and level their faces into the adhesive, use a block of cloth-covered plywood over them and gently tap the scrap with a hammer. Lay all the full tiles and leave spaces for the cut tiles.

In the case of a sink set parallel to the wall, work from the center of the sink toward the ends of the top. An L-shaped countertop with a corner sink must be handled differently than other settings. Since there is no way to avoid very odd cuts of tile around the sink, work from the ends of each leg of the top toward the sink opening.

Because of variations in wall runs, each tile should be butted, measured, and cut individually, rather than taking a single measurement and cutting a number of tiles at once. To make straight cuts, place the tile face up in tile cutter to score a continuous line on the tile; then press down on the cutter's handle to snap the tile. Smooth the cut edges of tile with a tile sander. To cut irregular or curved lines, score a crosshatch outline of the cut with a tile scoring tool. Use tile nippers to gradually break away small portions of the tile until the cutout is complete.

Bullnose tiles, which have rounded edges, are used to cover edges of countertop and backsplash. Use them to finish off the sidewall installation, creating a smooth, round edge. Backsplash tiles can be installed over a separate plywood core or directly to the wall behind the countertop.

Tiling the backsplash If the backsplash area is to be tiled, it must be patched and solid. To set the backsplash, begin one grout joint space above the cove tiles or countertop tiles. Cover the area with adhesive. Also, for a better grip, cover the back of each tile, and set the tiles in place. If desired, finish off the backsplash top edge with trim pieces, or continue tiling up to the undersides of the wall cabinets or window sill. Run a wide strip of masking tape along the underside of the front trim tiles to keep the grout from dripping out before it has a chance to set up. Also mask any surrounding wood surfaces to protect them from grout stains. Turn off the circuit breaker or remove the fuse in the line that feeds any light switch or fixture before setting tiles around it.

Applying the grout Before grouting, make sure adhesive is set and tiles are held firmly in place. Grout lines can be as thin as $\frac{1}{16}$ inch (.16 cm) up to $\frac{3}{8}$ inches (.95 cm) or more. These lines look better with small tiles, while $\frac{1}{4}$-inch (0.635-cm) lines look better with larger tiles (i.e., 6 inches square).

Carefully remove any spacers and clean the tile surface and grout joints until free of adhesive. Joint depth should be at least two-thirds the thickness of the tile. Check to see that all tiles are level. Wait at least 24 hours for the tile to set. Work the grout diagonally across the grout lines to fill the tiles completely. Mix the grout and apply it with grouting trowel, working at an angle to the grout lines. Once the grout has set, use a damp sponge to remove any excess grout. Caulk the seams between the tile and the sink. After the grout has dried, approximately 24 hours, thoroughly wipe (in a diagonal direction) the tile with a soft, lightly dampened sponge. Lightly buff the tile to a shine.

Grouting procedures are the same for mastic or mortared tile. Grout is porous and can absorb moisture and stain. Sealing it prevents this absorption; sealing should be periodically reapplied. After installation, seal the tile and grout with a quality silicone sealer to prevent water damage.

Installing wood countertops

Wood countertops are easy to install. Classic wood butcher-block tops, laminated from strips of hardwood laid on edge, are sold in 24, 30, and 36-inch (61, 76, and 91 cm) widths and incremental lengths. For a butcher-block insert, add new framing supports to match existing supports. Position them so that the top of the new insert sits flush with the existing counter. If required, cut the top to size with a power saw. Sand or rout the edges before adding framing support. Fasten the countertop to the cabinets with screws from below. Run a bead of silicone caulk along the seam between the countertop and walls. The wood can be sealed with mineral oil.

Installing natural-stone countertops

Traditionally, marble used for countertops is supplied in large slabs. Suppliers differ on the size and thickness of countertops that they stock. Many slabs are available that are $1\frac{1}{4}$ inches (3.18 cm) thick. Other suppliers, however, stock $\frac{3}{4}$-inch (2-cm) -thick countertops. Some carry $1\frac{1}{2}$-inch (3.81-cm) slabs. The appearance of a $1\frac{1}{2}$-inch (3.81-cm) counter can be achieved by joining the $\frac{3}{4}$-inch (2-cm) -thick counter to a $\frac{3}{4}$-inch (2-cm) edge treatment in much the same manner as solid surface countertops are constructed. The pieces of marble can be glued together so that the seam is unnoticeable. For proper installation of natural stones, specialized tools and skills are required.

Installation of appliances

The installation of appliances, countertops, and other pieces of equipment happens near the end of the kitchen installation. However, it is extremely important to plan for these installations early on in the project. All the necessary clearances, cutouts, and rough-ins must have been determined and provided well before installation. If subcontractors are needed to return to the job site to assist with final hookups and connections of plumbing fixtures and appliances, be sure to have planned ahead and scheduled for their visits. These details are all critical to the timely completion of the project. Mistakes at this point in the project (such as having to order additional cabinets) are very costly for you, as well as disturbing to your client. Although situations do occur, with proper planning, the majority of problems can be avoided.

Appliance delivery inspections

At the time it is accepted from the carrier, the installer should carefully inspect the unit for visual, as well as concealed, transportation damage. File a damage claim immediately with the carrier's claim agent if visual damage is evident; transportation damage is the responsibility of the carrier. It is important that the claim agent sees both the appliance and the crate it arrived in.

The exterior finish of most kitchen appliances, with proper usage, lasts indefinitely. However, some are more fragile to impact and bending strains than others (i.e., porcelain). Therefore, all appliances should be uncrated and moved with extreme care. The appliance is usually packaged in a carton-type crate and fastened to a wood base by four shipping bolts (the range leveling feet). Cut the bottom metal band, then lift the carton off the appliance. With the appliance on its back, remove the wood base, and install the leveling feet. Careful handling of the appliances in the user's home is also very important since most kitchen floor coverings scratch easily.

Most appliances come with manufacturers' installation instructions; therefore, in all cases, be sure to read and abide by them. The following information is based on general conditions. Some fixtures may require other installation procedures.

Sink installations

The installation of a kitchen sink must be made in accordance with local plumbing codes. In remodeling work, the location of existing plumbing is important in deciding where to place the sink in the new kitchen. Unless the remodeling plans call for a new vent stack, the sink should be located within the limits of the existing stack. One way of increasing the distance between the sink trap and vent stack is to increase the drain line pipe size.

As described in Chapter 6, a variety of kitchen sinks are available. Each has its own method of installation. Therefore, before beginning, read the manufacturer's instructions. The procedure for installing a sink depends partly on what the counter is made of and partly on the type of sink being installed. With wood and laminate countertops, install the counter first, then make a cutout for the sink. With a tile counter, make the cutout and, in some cases, install the sink before tiling.

To locate the sink position on the countertop, first mark the center of the sink at the sink edges. Marking the edges where the centerline occurs helps you line the sink up with its desired position on the countertop.

When installing a new kitchen sink, cut an opening in the countertop. With a self-rimming type, use the template provided by the supplier. That is, lay the template on the countertop so that the back template line is roughly 1 inch (2.54 cm) from the backsplash and at least 2 inches (5 cm) from the front. When centered, tape it down with plenty of masking tape. Then, just inside the template line, drill a $\frac{3}{8}$-inch (.95 cm) or larger hole through the template and countertop. Before beginning to cut, look under the cabinet top to see where the holes are. If they penetrated a countertop support, that support should be moved just far enough from the sink that it does not interfere with the sink or its hardware. Using this hole as a starting point, cut around the entire template with a power jigsaw, being careful to stay on the line.

The method of fastening of the sink to the countertop depends on the type of rim. For example, self-rimmed sinks are fastened in the countertop with metal clips that fit into channels on the underside of the sink (Fig. 10.22). The number of clips varies between sink models, but in any case, the key is to locate one approximately in the middle of each side. Distribute other fasteners uniformly around the sink. The more clips you use the better, because the forces holding the sink in place will be more uniform.

To hang the sink, install a ¼-inch (0.64-cm) bead of plumber's putty around the top edge of the countertop opening. The putty should be placed so that it uniformly seals the area where the sink is pressed to the counter. Lower the sink into the opening.

Tighten each clip screw all around the sink snugly, but not all the way. Check the fit of the sink all around. Some clips may need to be slightly tighter than others to bring the sink and counter together without gaps. Continue tightening the clip screws, gradually, until they are just tight enough to hold the sink firmly to the counter. Do not turn the slip screws too tight or you may damage the sink. Clean up any excess putty that squeezes out beneath the rim.

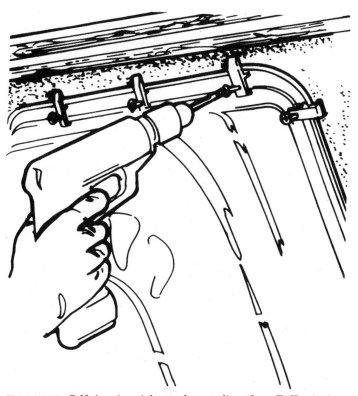

Figure 10.22 Self-rimming sink attachment clips. (*Leon E. Korejwo*)

Steel-rimmed sinks, as the name implies, have a steel rim that fits around the sink. The rim performs the same function as the channel on self-rimmed sinks, and the clip-screw hardware of both types of sinks is about the same. Before positioning the sink and rim through the counter, install a $\frac{1}{4}$-inch (0.64 cm) bead of plumber's putty around the edges of the hole. Put the sink in position and tighten the clip screws all around. Place another putty bead between the sink edge and the metal rim before you draw the sink up tight. Tighten the clip screws all around, gradually, until the sink and the counter fit tightly together. As with a self-rimming sink, do not tighten the sink too much to prevent damaging it. Clean away any excess putty.

With ceramic tile, there is the option to lay tile up to the opening, then install the sink, or to make a cutout in the counter core and, with a router, notch around its perimeter. Hang the sink, then lay tile, using quarter-round trim tile to cover the sink rim.

If you can handle the installation, it saves the time and expense of scheduling the plumber to come back to the job again. If you must use a plumber, carefully schedule the work to avoid delays in completing the job. If you have an integral sink that is a part of the countertop, or if the sink was under-mounted by the countertop fabricator in the shop, all that is left to do is hook up the plumbing. Check to make sure that all holes or cutouts required for installation of the faucet and any other accessory items have been made before you schedule the plumber.

Faucet and drain hookups   Before lowering the sink into the opening, it is a good idea to make the faucet and drain-assembly hookups. Faucets sometimes come with the sink; other times only holes are punched in the sink so that faucets can be purchased separately. There are many different styles of faucets available—standard two-faucet types, single-lever types, color-coordinated faucets, and so on. Other attachments such as sprays and hot-water dispensers may also be installed in the sink unit.

Most modern kitchen faucets are the deck-mounted type, seated on the rear of the sink and secured from below. All are interchangeable as long as the new faucet's inlet shanks are spaced to fit the holes on the sink.

Food-waste disposer installations   A food-waste disposer mounts in the drain hole of a kitchen sink and drains into a P-trap or into a continuous waste on double-bowl installations. These appliances have become popular in new construction. Because the disposer affects the configuration of the sink waste piping, it should be part of the sink installation. The disposer eliminates the need for a basket strainer. When not mounting a garbage disposer to the sink, a basket strainer is used as the drain assembly. The finished part of the disposer's drain is installed like that of a basket strainer. The waste outlet from the disposer must always be located above the trap. The electrical service for the food waste disposer should have been provided during the electrical rough-in earlier in the installation. At this point, all that is required is to have the electrician make the hard-wire connection.

Ventilation ductwork

A properly selected vent hood and fan can only do its work if it is correctly installed. The main rules are simple, but important:

1. Follow the manufacturer's instruction exactly, including the recommended duct size and the corresponding elbows and fitting.

2. If possible, locate the fan across the room from incoming air.

3. Make ducting as short and direct as possible to the outside, with a minimum number of elbows.

4. Do not reduce ducting in size from the fan discharge.

5. Install the proper terminal accessories.

In many instances, the ductwork for ventilation hoods and other forms of ventilation runs through a cabinet unit. For a range hood, the ductwork often passes through the wall cabinet directly above the hood and then runs through the soffit area to an outside wall. The ductwork would have been run at the time the soffit was constructed and stubbed down directly above the hood.

Determine the point at which the ductwork will pass through the bottom of the wall cabinet. Cut a hole in the cabinet. This hole should be planned for in advance, and the cut should be made before the wall cabinet is installed. Manufacturers provide detailed rough-in measurements so the required cutout can be easily located. As recommended by the manufacturer, align the duct outlet from the hood at the cutout, and secure the hood to the cabinet or wall above. If you have successfully located the stubbed-down duct in the soffit, the amount of cabinet space lost to ductwork is minimized.

On peninsula or island installations, the hood should overhang the cooktop 3 to 6 inches (8 to 15 cm) on all four sides. Wall-mounted hoods should extend at least to the front of the cooktop and 2 to 6 inches (5 to 15 cm) out on each side. In these wide-open spaces, because of air turbulence, which spreads cooking vapors, the more overhang the better. Factors affecting the depth of a hood are its length, width, and distance from the cooking surface, as well as placement of the exhaust inlet. Cooking vapors should be picked up from the top of the hood where they have concentrated, and the larger the hood, the more depth is necessary to hold these vapors. A 5- or 6-inch (13- or 15-cm) depth has very little holding power, allowing cross-drafts to spread the vapors. A 9-inch (23-cm) depth holds vapors in a wall-mounted hood that is fairly well closed in on the sides with cabinets. A 14-inch (36-cm) depth is needed where the hood is placed as high as a person's head. On large peninsula and island installations, an 18- or 20-inch (46- or 51-cm) depth is necessary.

The safe distance from the cooktop to the hood is 27 to 30 inches (68 to 76 cm); it generally makes the most attractive installation. The maximum distance is 34 inches (86 cm) and should be used only for wall-mounted hoods.

Rather than snaking the ductwork through the base cabinet below the appliance, downdraft ventilation systems can be run in the space between wall studs or floor joists. Again, any cutouts in the base cabinet necessary to allow the downdraft ductwork to pass through to the below-floor ductwork should be planned in advance and cut prior to installation of the cabinet (Fig. 10.23). In addition, determine if the ductwork for the downdraft ventilation system will interfere with any drawers in the base cabinet. If this is the case, the drawers need to be shortened or eliminated completely, with false drawer fronts in their place. Remember to plan for access to the ductwork.

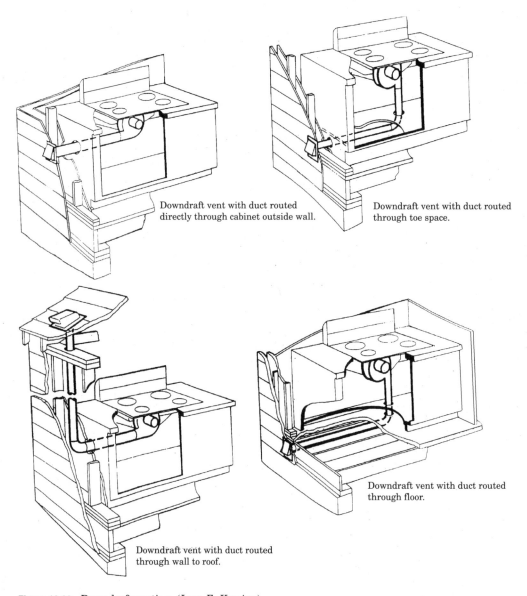

Downdraft vent with duct routed directly through cabinet outside wall.

Downdraft vent with duct routed through toe space.

Downdraft vent with duct routed through wall to roof.

Downdraft vent with duct routed through floor.

Figure 10.23 Downdraft venting. (*Leon E. Korejwo*)

As previously mentioned, if possible, locate the fan across the room from the incoming air. Do not reduce ducting size from the fan discharge. Minimize turns and curves in ductwork. If turns are necessary, to prevent a backflow of air, they should be at least 45 degrees. Be sure that all seams are airtight. If duct runs exceed those recommended by the manufacturer, the manufacturer's warranty may be voided.

Major appliance installation

The hookup, connection, and installation of major appliances is one of the final tasks in the installation process. Because they are difficult to work around and can easily be damaged, it is best to keep appliances out of the kitchen area as long as possible. Connection requirements, of course, vary according to the type of appliance and its features. Manufacturer-supplied instructions usually describe the hookup process at length and should be followed closely.

Dishwasher installations Installing the dishwasher is one of the last jobs in the kitchen installation. Hooking up the water supply and waste lines under the kitchen sink is fairly straightforward. The hard part is routing the lines through the dishwasher chassis and cabinet wall. You may have to slide the appliance in and out several times to get everything lined up.

A typical space requirement for a built-in dishwasher is 36 inches (91 cm) high, at least 24 inches (61 cm) wide, and 24 inches (61 cm) deep. The dishwasher fits into the space of a standard 24-inch (61 cm) base cabinet. Test-fit the dishwasher in the space intended for it and make any adjustments needed. Pull the machine out again and make an opening in the side of the cabinet between the dishwasher and the undersink compartment. The opening should be large enough to accommodate the supply and drain lines.

A dishwasher requires a hot-water connection, a drain connection, and an electrical hookup. Because a dishwasher is usually located adjacent to the sink, that is the most likely place to tap into a hot water line. If the dishwasher must be around the corner from the sink, allow an intervening space of at least 18 (46 cm), preferably 24 (61 cm), inches from the corner. Also, because the refrigerator generates cold and a dishwasher produces heat and steam, these two appliances will last longer and work better if they are separated by a 3-inch (8 cm) insulated filler strip.

A dishwasher requires its own 20-amp grounded circuit. Most dishwashers take a 120-volt, 60-hertz ac individual circuit, fused for 20 amperes. Three-wire electrical service to the dishwasher is recommended for connection to the terminal block and for grounding.

The water supply needed is 140 to 160 degrees F at 15 to 120 pounds per square inch (psi) pressure. The water pipe should be $\frac{1}{2}$-inch (1.27 cm) outside-diameter copper, with a $\frac{3}{8}$-inch (.95 cm) female pipe thread connection at the valve. A $\frac{1}{2}$-inch-by-$\frac{3}{8}$-inch (1.27-by-.95-cm) male compression elbow is provided as an accessory. An 8-foot (244 cm) flexible drain hose with a $\frac{1}{2}$-inch (1.27-cm) inside diameter is furnished. It is not recommended that the drain line be extended beyond the length of the hose provided, but should this be necessary, attach the hose to a line with a larger inside diameter.

The most desirable drain system for a built-in dishwasher is through a drain air gap mounted at the sink or at countertop level. An air gap prevents sink water from backing up into the dishwasher. A rubber connection pipe is usually run from the air gap back to the food waste disposer, allowing any back-up to flow into the disposal. For this reason, the food waste disposer should always be emptied before the dishwasher is used. Many plumbing codes now require such an arrangement and prohibit a direct connection.

Make drain-line connections with hose clamps. The dishwasher can drain either into the sink drain or into a food waste disposal. For the sink drainage installation, a dishwasher waste fitting must be installed. To do this, loosen the slip nuts tying the tailpiece to the trap and the sink's basket strainer. Remove the tailpiece and insert the waste fitting into the trap. Secure it by tightening the slip nut onto the trap. Cut the tailpiece so it is slightly longer than the distance between the basket strainer and the top of the waste fitting. Reinsert the tailpiece to complete the connection. To drain a dishwasher into a food waste disposer, simply remove the metal knockout inside the dishwasher drain nipple on the disposer's side. The knockout, when freed, will fall into the grinding chamber. Be sure to take it out.

Once it is hooked up, level the dishwasher by adjusting the height of its legs. The dishwasher unit is secured to the underside of the countertop. Dishwashers installed under countertops that have a wood substrate can be screwed directly to the wood. However, solid-surface and stone countertops may require special anchors to allow connection of the dishwasher. While this is a considerable amount of work, the dishwasher must be attached to the countertop to prevent it from tipping forward. Finally, restore water pressure and check for leaks. Install the kickplate cover that hides the plumbing and electrical connections.

Refrigerators Most refrigerators can be plugged into a standard duplex receptacle. The receptacle should be located behind the refrigerator, out of reach of daily use. A few commercial-type refrigerators require their own circuit. If the kitchen installation project includes this type of refrigerator, check the electrical requirements closely. Appliances with extra power requirements need to be planned for in advance. The additional circuit for this refrigerator could wind up being a costly addition to the project. It is important at the bidding stage of the installation to know what types of equipment and appliances are being installed.

Many manufacturers do not take into account the door or hinge clearance required for opening the refrigerator. A refrigerator can require several inches of additional space to allow the door(s) to swing back far enough to allow bins and drawers to be pulled out. If the refrigerator is located adjacent to a wall running perpendicular to the front of the refrigerator, the problem is even worse.

Be sure the kitchen configuration allows the refrigerator to fit into the room and its intended opening. Because of its large size, it often requires a considerable amount of maneuvering space.

Today, the majority of refrigerators require a source of water to service the ice maker. At a panel on the front of the door, some dispense chilled water and ice. Ideally, a water service box with a shutoff valve is installed in the wall behind the refrigerator where a flexible copper water line (connected by $\frac{1}{4}$-inch (.6cm) copper tubing) can be attached to serve the refrigerator. It is important to provide a sufficient length of copper water line so that the refrigerator can be pulled out for cleaning or service without having to disconnect the water line. The excess water line can be coiled behind the refrigerator. If the refrigerator can be easily reached from the sink complex, tap in there through the sides of the base cabinets. Unless the refrigerator is next to the sink base cabinet, the flexible copper water line must pass through base cabinets until you reach the refrigerator. Again, be certain that you provide a sufficient length of water line so that the refrigerator can be pulled out.

If custom front panels have been specified and ordered for the refrigerator, these should be installed before the appliance is pushed back into its opening.

Ranges   As described in Chapter 3, there are several different types of ranges; each is installed differently. Slip-in ranges rest on the floor and can be adjusted with leveling legs to square up tightly with the countertop and cabinet floors. Free-standing ranges sit on the floor, but even when placed within a run of base cabinets, they lack the custom-installed look of the slip-in range. Drop-in ranges have flanges on the side to support the range from the countertop. In some cases, there may be cabinet framing or a drawer unit below the range. You must be sure that the countertop substrate and framing system can support the weight of the drop-in range unit.

The roughed-in hookups for both gas and electrical connections must be accurately placed so that the unit can be connected and pushed back into its opening. If the rough-ins are not correct, the connections will not fit into the recessed area at the back of the appliance. Some ranges require hand-wiring, while others can be plugged into special range outlets. Depending on the type you are installing, hookup may be as simple as plugging in the cord. (Electric ranges plug into a combination 120/240-volt receptacle. Gas ranges require a gas line and a 120-volt receptacle for an electric cord to bring power to the lights, electric ignition devices, and clock.) Otherwise, the services of an electrician are required. Gas ranges need to be connected to the gas source roughed-in for the appliance. If this can be done by the installer rather than a plumber, it saves the project time and expense.

If the range has built-in ventilation, such as a downdraft exhaust system, the ductwork for this system needs to be installed as described earlier in this chapter. Gas ranges also require venting to ensure that gas fumes are safely removed from the home.

The installation location for the range must meet distance requirements between the appliance and combustible surfaces. Depending on the particular range being installed, these distances range from 0 to 12 inches (30 cm) side to side (with some commercial ranges requiring up to 18 inches or 46 cm) and 30 to 36 inches (76 cm to 91 cm) vertically.

The local power company should always be consulted before a gas or electric range installation is made. The electric range should be installed in accordance with the National Electric Code (NEC), state and local ordinances, and the rulings of the local power company. In the average residence, unless an electric range has previously been used, the existing service and meter may be found to be of insufficient capacity. Under such circumstances, a new service and meter with adequate capacity for both range and lighting circuits must be installed.

Cooktops   There are many types of cooktops (see Chapter 3), and each has specific installation instructions. However, some basic guidelines apply to most cooktop units. Cooktops are generally hard-wired to the electrical service, requiring the services of an electrician. Coordinate all other hookups and connections that require the electrician so they can be accomplished in one trip.

Most cooktops are clamped to the countertop and have an integral rim to seal them to the countertop surface. It is generally recommended that a bead of caulk be run under the edge of the rim so that spills cannot seep under the edge and into the interior of the cabinet below. The cutout in the countertop for the cooktop must be exact. If it is too small, it will be difficult to enlarge the opening in the field without damaging the countertop surface. If the cutout is too large, the cooktop will not cover the opening. You need to provide the countertop fabricator with the cutout dimensions supplied by the cooktop manufacturer. Most manufacturers recommend that the corners of the cutout be rounded. Refer to the manufacturer's instructions, and instruct the countertop fabricator to do this. If the cooktop has built-in ventilation, such as a downdraft exhaust system, the ductwork for this system needs to be installed as described earlier in the chapter.

Gas cooktops need to be connected to the gas source roughed-in for the appliance. If not restricted by local code regulations, this connection can be handled by the installer. Otherwise, this task should be scheduled for the plumber.

A thin cooktop generally allows for drawers beneath it, but a thicker cooktop interferes with the operation of the drawers. If drawers are not feasible, have the base cabinet below the cooktop be supplied with false drawer fronts instead. Remember that the underside of a cooktop must always be accessible. A standard base cabinet with doors provides the required access.

If the cooktop is wider than the base cabinet beneath it, anticipate it and have the sides of the cabinet notched to accommodate the cooktop in the same manner as for a sink. It is easily done prior to the cabinet installation but is much more difficult if it must be done at this stage of the project.

Ovens   Wall ovens can be installed in cabinets especially designed for this purpose or placed directly in wall openings that have been framed to accommodate the oven (Fig. 10.24). If the correct rough opening size was given to the cabinet manufacturer, installing the wall oven should be relatively easy. Because the sizes of wall ovens vary somewhat by manufacturer and model,

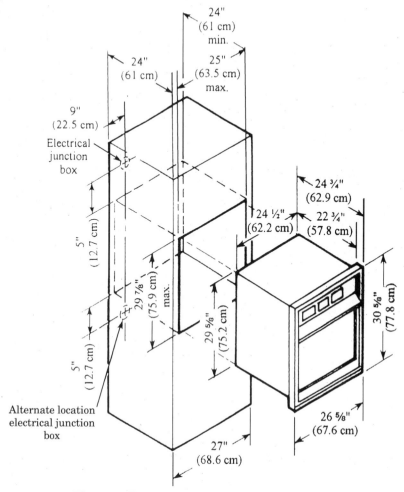

Figure 10.24 Electric wall oven. (*Leon E. Korejwo*)

the opening usually needs to be custom-fit to the oven. Wall ovens generally come in 24-, 27-, and 30-inch (61-, 69-, and 76-cm) width models.

Some oven manufacturers provide a unit dimension that does not take into account unit overlap outside of the cutout. While the oven fits in the opening provided for it, the overlapping area may interfere with cabinet doors. This overlapping is especially problematic with full overlay cabinet doors. Insist on obtaining all overlap dimensions in addition to the cutout opening dimension.

Wall ovens can also be installed below cooktops in a base cabinet specially fabricated for this purpose. If this configuration has been specified, make sure there is no interference between the underside of the cooktop and the top of the oven. Also, make certain that the controls for a wall oven installed below the counter are on the top of the unit rather than at the bottom. Not all wall

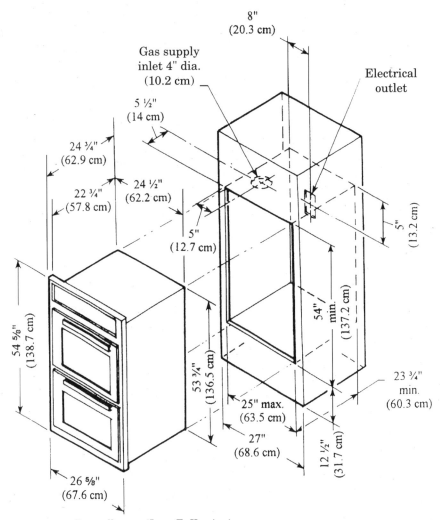

Figure 10.25 Gas wall oven. (*Leon E. Korejwo*)

ovens can be accommodated below a cooktop. Check to make sure that the unit specified fits in this configuration.

Most wall ovens are hard-wired and require the services of an electrician for connection. If the oven is gas, you need to connect it to the gas service roughed-in for the appliance (Fig. 10.25).

Microwave ovens The basic microwave oven requires little more than setting it on a counter and plugging it in. The combination microwave oven and ventilation exhaust system is designed to be installed over a range or cooktop. This unit offers the convenience of an eye-level microwave oven in an area that is typically of limited use because of the range hood. It provides task lighting and ventilation for the cooktop below. The above-range microwave oven is mounted to the back wall using a wall plate provided with the microwave oven. While the unit is generally mounted directly below and adjacent to wall cabinets, it is not intended to be supported by these cabinets. A ductwork connection has to be provided for the ventilation system if it is to be ducted to the exterior. If a shelf is being provided for the microwave oven, make sure there is an outlet at this location. This factor is commonly overlooked.

Utilities

The electrical, gas, and water services may have been totally disconnected or shut off at some point in the installation. Turning on these utilities may be as simple as flipping a switch or opening a valve; on the other hand, the process could be more extensive.

Most importantly, you must understand your responsibilities and know what regulations and code provisions you must follow. They may be stipulated by code-enforcement officials or utility companies and may cover anything from required inspections to licensing requirements for tradespeople working with the various utilities.

Prior to the issuance of an occupancy permit, a final inspection is usually required for electrical service. The electrical inspector will review electrical outlets, fixtures, and equipment. If your kitchen installation involved an upgrade of the existing electrical service or addition or modification of the electrical panel, before restoring electrical service to the home, the local electrical utility company may require an inspection of the power panel. If the installation project included running a new water line from a municipal water meter before restoring the water service, code regulations may require an inspection of the meter and the new water line. Before the kitchen is released to the client, if the water system is a private one, as in the case of a well, the water should be tested. Each system should be thoroughly checked and tested once it is turned back on to see that it is operating properly. Do not turn on all the utilities for the first time and leave the job site for the day. One leaky pipe joint can destroy weeks of work.

Before restoring electrical service to the home, make sure all the circuits are correctly identified in the circuit-breaker box. This identification will be very helpful should you find it necessary to shut off the power to a particular appliance, fixture, or piece of equipment. Check all gas-fired burners and ovens to ensure that no gas fumes are leaking from them. If you suspect that there is a leak, contact the local gas utility company.

As a general rule, all modifications and new connections to gas lines should be inspected by the plumbing inspector before the gas service is turned on. This inspection includes all appliance connections, as well as water heaters, gas fireplaces, and barbecue grills.

Finishing the Project

There has been a revolution in kitchen furnishings and materials, with a wider-than-ever selection, available in an amazing variety of materials, textures, patterns, and colors, and a new emphasis on easy-care materials (Fig. 11.1). Some installers may or may not be involved when the homeowner decides on the type of floor, wall, ceiling, or lighting finishings for the kitchen. However, it is very important for you to know what is available, as well as learn some basic facts about floor, wall, ceiling, and lighting finishings to inform your client, if and when the need arises. Whether you are subcontracting the finishing work or have excluded it from the contract altogether, the ultimate responsibility for the client's satisfaction lies with you. Therefore, being knowledgeable in the following areas is important to you. This chapter covers the most important information about the types of floor, wall, ceiling, and lighting finishings available; however, it does not cover how to completely install each particular type of finishing.

Floor finishings

Innovations in design and manufacture have made many flooring materials easier to install. Ceramic tile, once thought to require the skills of a professional, can be installed today by anyone with patience and average building talent. Hardwood flooring can be purchased with durable factory-applied finishes. Vinyl tiles come with adhesive backing or can be set in easy-to-apply adhesives.

Floors, like walls, are not easy to change. Whether applying a floor material to a new home kitchen or reconditioning a floor in an existing portion of the structure, the finished flooring must be chosen carefully and wisely. The term *finished flooring* refers, of course, to the material used as the final wearing surface that is applied to a floor.

What kind of floors should be used in the kitchen? The kitchen floor should be beautiful and should complement the decor, since it is one of the most

Figure 11.1 Today's kitchen has unlimited finishing possibilities with the wide variety of materials, patterns, textures, and colors available. (*Bruce Wood Floors*)

important design features tying the kitchen together. When choosing floor finishings, consider the following factors:

- *Durability.* Determine the amount of traffic the flooring must bear. For high-traffic areas, select the most durable materials.
- *Cost.* All standard flooring materials come in various grades, with the cost directly related to the quality.
- *Comfort.* The flooring should have some resiliency. Some flooring materials are softer underfoot than others. If the homeowners object to a cold, hard surface, they may not want to use ceramic tile or masonry. Wood or resilient flooring provides a more comfortable surface. Working on a solid floor (i.e., concrete, ceramic tile) can prove to be very tiring to stand on. On the other hand, a floor that is too soft can also be tiring.
- *Noise.* Realize that soft flooring materials, such as vinyl and carpeting, deaden sound. Wood, ceramic tile, masonry, and other hard surfaces tend to reflect sound, rather than absorb it.
- *Safety.* Safety is an important factor to consider when choosing flooring. It should be smooth without being slippery, and it should be level. Keep in mind that although glazed tile is common in kitchens, it can become very slippery when wet.

□ *Cleanability.* New floor materials, protective finishes, and cleaning techniques make maintenance less of a consideration now than in the past. Be certain the flooring materials are resistant to heat, dirt, and grease.

Resilient vinyl

The development of resins and synthetics has created a family of floor coverings called *resilient flooring*. Soft underfoot, durable, and easy to maintain, vinyl has long been a favorite in the kitchen. Resilient vinyl is one of the easiest floor surfaces to install. Almost all of the vinyl floors produced today are the no-wax type with a durable shine that needs minimal maintenance, is stain-resistant, and is impervious to water. There are a seemingly endless variety of color textures, patterns, and styles available, in a wide range of prices. Several quality grades are produced by any vinyl flooring manufacturer. The three common types of vinyl are

□ *Inlaid flooring.* Inlaid flooring is solid vinyl made of tiny vinyl granules fused together so the color and pattern go all the way through to the backing. Wear does not erase the vinyl pattern. These vinyls are the most costly, but they can last 30 years or more.

□ *Rotovinyl.* Rotovinyl is produced by covering a photographic image with a layer of clear urethane. The wear layer's thickness affects price and determines how long the floor will look good. However, in areas of heavy traffic, even the most expensive rotovinyls can wear through over time.

□ *Vinyl composition.* Vinyl composition coverings combine vinyl resins with filler materials. This least-expensive vinyl floor type can show signs of wear in about five years.

The one major disadvantage to resilient vinyl flooring is that it is relatively soft, making it vulnerable to dents and tears. Often, though, such damage can be repaired.

Resilient vinyl floors are manufactured in two basic types—sheet materials and tiles. Whether the homeowner chooses to have tiles or sheet vinyl installed depends, to some extent, on the use the new floor will receive.

Resilient sheet flooring

Most resilients are made with a solid vinyl coating called polyurethane. Sheet vinyls come in widths from 6 to 12 feet (190 to 366 cm) and generally cost more than comparative grades of vinyl tiles. The principal advantage of sheet vinyl in the kitchen is seamlessness. The result is a beautiful wall-to-wall sweep of color and design (Fig. 11.2). Be aware that large rooms, usually over 12 feet (366 cm), are going to require seams with most types of sheet vinyl. However, some manufacturers do produce wider rolls of vinyl.

Sheet flooring can be fully adhered (laid in adhesive over the entire floor), peripherally fastened (attached with adhesive), or laid loosely like a rug. The introduction of better materials and the manufacturer's ability to improve

Figure 11.2   Resilient sheet flooring provides a beautiful seamlessness for any kitchen. (*Mannington Mills, Inc.*)

photographic realism has improved vinyl's ability to mimic natural materials. The hundreds of patterns available include authentic-looking imitations of all types of flooring—brick, slate, wood, marble, terrazzo, flagstone, and ceramic tile. Resilient flooring can be laid on concrete slabs on grade (directly on the ground) or below grade (underground, as in a basement), on wood subfloors made of plywood panels or individual boards, or directly over an old floor (to prepare the room for installing resilient flooring refer to Table 8.1 in Chapter 8).

Vinyl tiles

Vinyl (the most popular material for resilient tiles) tiles are still used, but are not nearly as popular as sheet goods. Vinyl tiles are available in 9- or 12-inch (23- or 30-cm) squares, although other sizes are available, up to as large as 36 by 36 inches (91 by 91 cm). Individual tiles either come with a self-stick backing or are laid in adhesive. Vinyl tiles are fairly easy to install, but they do take longer to install than sheet vinyl. If improperly installed, moisture can seep between tiles and cause damage to the underlayment and subfloor. Vinyl tiles can come loose, and the cracks between the tiles can catch dirt, making them difficult to clean.

With the tremendous variety of materials, designs, and colors available, it is possible to create just about any floor scheme that strikes one's fancy. Tiles can be mixed to form custom patterns or provide color accents. As with vinyl sheet flooring, raised surfaces give many vinyl tile products the realistic look and feel of brick, stone, slate, wood, terrazzo, marble, or ceramic tile. Faux-wood vinyl planks come in 3-inch (8-cm) -wide strips that can be installed much like the real thing—in random or herringbone patterns (Fig. 11.3).

Ceramic tile

Ceramic tile is easy to maintain, tough to harm (it is fireproof and waterproof), and is available in a wide variety of patterns, styles, and colors that can complement any room (Fig. 11.4). When properly installed, ceramic tile has timeless elegance and is one of the most durable floor coverings available (to prepare the room for installing ceramic tile flooring refer to Table 8.2 in Chapter 8).

Figure 11.3 Faux-wood vinyl planks provide the beautiful look of real wood. (*Perstorp Flooring, Inc.*)

Figure 11.4 Ceramic tile provides a lifetime of durability and style for the kitchen. (*Florida Tile*)

Three types of ceramic tiles are common in kitchen floor use today: glazed tiles, ceramic mosaics, and quarry tiles.

Glazed ceramic tile can be shiny or bright-glazed, matte-glazed, or textured, and has a frosting-like layer on top that was sprayed on before firing. It is a good choice in the kitchen because liquids can't soak in. The tiles are made in various sizes and shapes and a variety of designs and colors. Some are so perfectly glazed that they form a monochromatic surface. Others have a softer, natural shade variation within each unit and from tile to tile. There are also extra-duty glazed floor tiles suitable for heavy-traffic areas. Sizes generally range from 1 square inch (3 square cm) to 12 square inches (30 square cm), with the most common size being $4\frac{1}{4}$ square inches (11 cm) by $\frac{5}{16}$ inch (0.79 cm) thick.

Ceramic mosaics are made from clay mixed with pigment. They are also available in a large assortment of colorful shapes. Mosaics are usually sold mounted in 1-by-1-foot squares or 1-by-2-foot octagons, $\frac{1}{4}$-inch (0.6 cm) thick, with or without a glaze.

Quarry tiles are unglazed ceramic tiles that are colored through and through in natural clay colors and pastel shades. Quarry tiles are available in shapes ranging from 6-inch to 8-inch (15- to 20-cm) square tiles and 4-by-8-inch (10-by-20-cm) rectangles, and are normally $\frac{1}{2}$ inch (1.27 cm) thick. Some quarry tiles, however, must be protected with a masonry sealer after installation to prevent staining. Quarry tile is suitable for kitchen use but requires more care than other types to keep it looking great. Today's market offers decorative tile products that simulate natural materials, surfaces, and textures. Many manufacturers bring the outdoors inside with realistic, rustic stone textures in colors indicative of nature's minerals. These beautiful tiles merge well with a wide array of decorative applications and color schemes.

There are, however, a few drawbacks to ceramic tile floors in kitchens. First of all is its cost. Ceramic tile is very expensive when compared to vinyl floorings. Also, people complain that their legs get tired more quickly when standing on a hard-surface floor, and fragile items dropped on a tile floor will break. The tile itself is susceptible to cracking if the floor shifts or if hard objects are dropped on it. The floor can be noisy, and, if glazed, very slippery underfoot. If they are not properly grouted, tiles can leak moisture, grout spaces can be tough to clean, and tiles can come loose. In addition, some tiles will stain unless properly sealed.

Wood

Wood flooring in kitchens is becoming a popular choice once again. New manufacturing processes have produced wood flooring that is much more durable and moisture-resistant than in the past. Prefinished hardwood flooring creates a warm, natural, and inviting look in the kitchen (Fig. 11.5), and, importantly, provides comfort underfoot. Wood floors that are properly sealed resist stains, scuffs, and scratches.

Oak, durable and readily available, is the most popular wood. Pine is a favorite in country settings because dents and nicks give it an old-time look. Other common types are walnut, cherry, ash, maple, birch, hemlock, redwood, and fir. Hardwood flooring is available as strips, planks, or parquet tiles.

Strip floors (the most common) are made of narrow tongue-and-groove boards laid in random lengths, typically $2\frac{1}{4}$ inches (6 cm) wide, and the widths don't vary. They are among the most economical wood floors to buy. Strip flooring actually contributes to the structural strength of a house.

Plank flooring comes in many widths, all wider than strips (3 to 8 inches, or 8 to 20 cm). Planks are often installed as a combination of several different random widths and random lengths (use same-width planks or vary the plank width to add interest). Like strip flooring, plank floors contribute to the structural strength of a house.

Figure 11.5 Today's prefinished hardwood floorings provide moisture resistance and durability. (*Premier Wood Floors*)

Parquet (block) floors are made up of small wood pieces glued together in various patterns, including herringbone and blocks (Fig. 11.6). It is considered a floor covering only and does not contribute to the structural strength of a house. This type of flooring may be solid or laminated of several layers. Parquet floors are generally easy to install.

Wood flooring comes unfinished or prefinished. *Prefinished* products let the installer keep working without waiting for the stain and finish coats to dry. Prefinished hardwood flooring locks together with tongue-and-groove edges and is usually installed over a troweled-on adhesive. When fastening boards down with nails, a thin layer of building paper or thin foam (to act as a cushion) between the underlayment and finished flooring boards may be used. Prefinished wood flooring costs more and must be installed with extreme care.

Unfinished wood flooring must be sanded to smooth minor surface irregularities. Standard strip wood flooring must be finished with a durable polyurethane to hold up to typical kitchen conditions. Be certain that the manufacturer recommends the finish for kitchen use. Finishes include oil and wax (penetrating finishes) and polyurethane (a surface finish). Polyurethane is better for kitchens because it forms a hard, clear coating that protects against scratches, spills, and wear. This flooring is best installed on sleepers when installed on a concrete slab (to prepare the room for installing wood flooring refer to Table 8.3 in Chapter 8). There are materials manufactured for a direct glue-down of wood flooring, but wood floors installed in this manner are very prone to moisture problems. Wood flooring must be sanded and finished before installing cabinets.

A good wood floor can last the lifetime of the house, and actually improve with age, but it needs to be refinished periodically. When the floor is worn, it can be easily refinished to look like new. However, there are a few disadvantages to a hardwood floor in the kitchen. This type of flooring may shrink in heat or swell in dampness. Inadequate floor substructure and moisture damage can be major problems. With proper preparation, these problems can be avoided (to prepare the room for installing wood flooring, refer to Table 8.3 in Chapter 8). As far as maintenance, some surfaces can simply be damp-mopped.

Masonry floors

Masonry materials, such as slate, flagstone, terrazzo, stone, marble, and brick, have been used as flooring for centuries. They offer beautiful textures,

Figure 11.6 Parquet wood flooring adds warmth and coziness to the kitchen. (*Bruce Hardwood Floors*)

are generally easy to maintain, are virtually indestructible, and provide quite a luxurious look. Today its use is even more practical, due to the development of sealers and finishes.

Unfortunately, there are still a few disadvantages to masonry flooring. Hard floors can be uncomfortable to stand on for long periods of time. They are heavier than other types of flooring and therefore require a very strong, well-supported subfloor. Natural stone can be very expensive, especially if it has to be shipped any distance. Many types (i.e., flagstone, slate, brick) can still be stained. Masonry floorings also have the disadvantage of being quite noisy, as well as cold and slippery underfoot.

Carpeting

Should carpet be installed in the kitchen? At one time carpeting in the kitchen was very popular, but could not hold up to all of the traffic and moisture. Today, kitchens can benefit from the development of hardwearing carpeting materials. However, kitchens still tend to be better suited to a more water- and stain-resistant type of flooring. Carpet can be installed directly over a well-prepared subfloor or over any type of old flooring that is clean, smooth, and free from moisture. Always remove carpeting before installing new floor coverings.

Walls

As the installer, typically, you will turn over the finish painting and wallcovering work to a subcontractor. This option may be decided by the size of the project. A major partition relocation may occur, and hiring a carpenter, framer, or drywall professional may become more time efficient. Offering these services expands your ability to fulfill the clients' desires. Specialists are available and better equipped to tackle the work more efficiently, allowing you more freedom to address your own specialty. Therefore, being knowledgeable about these options will prove to be important to you.

Note: Painting and wallcovering can not be started until the cabinet installation is totally complete. Each finished wall style demands various preliminary subsurface applications that must be dealt with before cabinet work may start.

One of the most important factors the installer should be aware of is that the finished appearance of the paint and wallcovering can only be as good as the wall surface under it. The actual wall preparation is almost always the responsibility of the installer, and it is essential for attractive, long-lasting results. The walls must be clean, dry, and smooth before they can be covered. While some wallcoverings are more susceptible to showing wall imperfections (dents, cracks, popped nails, etc.), all wallcoverings do to some extent. Be sure to review all wall surfaces with the subcontractor before the work begins, and repair or touch up any areas that he or she might find unacceptable.

Walls in older buildings are likely to be covered with plaster. If only minor repairs are needed, it is worthwhile to retain the plaster walls. However, if major reconstruction (or demolition) is going to take place, you should plan on replacing the plaster with drywall. The walls in kitchens can be drywalled and painted, or they can be drywalled and covered with a wallcovering such as tile, wallpaper, or paneling.

Drywall provides smooth surfaces that can be decorated with paint, wallpaper, textures, fabric, or vinyl wallcoverings. For satisfactory finishing results, care must be taken to prepare the surface properly to eliminate possible decorating problems. Plan to span the entire length of the ceiling or wall with single boards, if possible, to reduce the number of butt joints, which are more difficult to finish. Stagger butt joints and locate them as far from the center of the wall and ceiling as possible so they will be inconspicuous (Fig. 11.7a). When viewed in natural lighting, the joints and fasteners in painted walls and ceilings might be visible (Fig. 11.7b).

Regular drywall has a paper covering on each side and on the edges. The backs of the boards are surfaced with a gray liner paper. The facing is generally a light gray manila paper that extends over the long edges. The surface is smooth and takes a wide variety of finishes. They can be applied in one or more layers directly to wood framing members, to steel studs or channels, or to interior masonry surfaces. The edges of drywall are tapered so that when joint compound is applied, it firmly bonds the tape to the board and the panel "V" edges to each other making a strong, rigid joint.

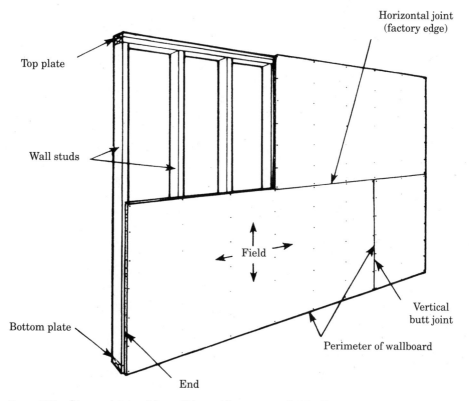

Figure 11.7a Stagger joints of drywall to avoid an uneven finish. (*Leon E. Korejwo*)

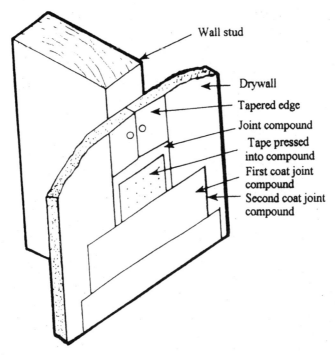

Wall stud

Drywall

Tapered edge

Joint compound

Tape pressed
into compound

First coat joint
compound

Second coat joint
compound

Figure 11.7b Preparing drywall joints for a flawless finish. (*Leon E. Korejwo*)

Drywall may be installed in a variety of ways to solve many existing remodeling problems. Simple projects may only require single-ply construction, which is adequate in the majority of situations (Fig. 11.8). Single-layer or single-ply construction is the most commonly used system in residential construction. The single-ply uses just one layer of drywall and is usually adequate to meet many fire-resistance and sound-control requirements.

Multi-ply or multilayer construction consists of a face layer of drywall applied over a base layer of drywall that is directly attached to the framing members. This construction can offer greater strength and higher resistance to fire and sound transmission compared to single-ply applications (Fig. 11.9). Double-ply construction is especially resistant to cracking and provides the finest, strongest wall available.

To prevent many decorating problems, it is usually recommended that a high-quality latex primer/sealer be applied prior to decoration. Color or surface variations are thereby minimized, and a more uniform texture for any surface covering is provided. The sealer allows the wallcoverings to be removed more easily without marring the surface. Glue size, shellac, and varnish are not suitable as sealers or primers.

As stated earlier, while the installer does not generally apply the final drywall decorative finish, you should know something about preparing the surface of drywall panels to receive the different types of final finishes.

Paint

By far the easiest, least expensive, and most popular of kitchen wall finishes is paint. Today it offers an almost unlimited variety of colors and hues. It can be mixed to the precise shade and color selected. Increasingly popular today are fanciful decorative paint finishes that lend new vitality and depth to painted walls. Techniques range from the very simple, such as ragging and sponging, to the more technically difficult methods that produce the appearance of other patterns or textures, such as marbling, wood graining, and modern art.

Alkyd or latex enamel paints are among the most popular house paints sold today. Paint finishes range from flat to high gloss. They offer moisture-, grease-, and stain-resistance and give a smooth, easy-to-clean finish to walls. Alkyd paint (made of synthetic resins) is considered the best type of paint for rooms that receive a lot of use, such as the kitchen. Any painted or wall-papered surface, or bare wood, can be covered with an alkyd paint. This type of paint adheres to bare masonry or plaster, but it should not be used on bare drywall because it will raise a nap on the drywall's paper covering. Alkyd tends to be more durable than latex paint. Although it is harder to apply and may take longer to dry, it is easily sanded.

Latex paints are easy to work with, dry quickly (in little more than an hour), and combine the long-lasting finish with a high-gloss finish. They are not as

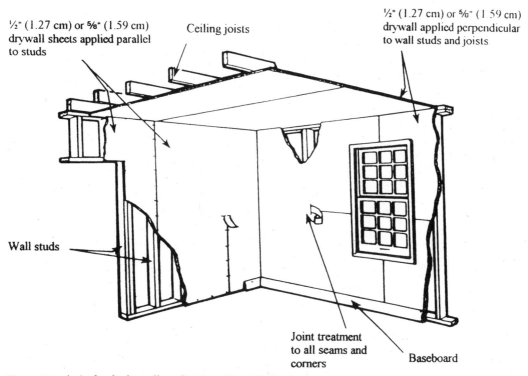

½" (1.27 cm) or ⅝" (1.59 cm) drywall sheets applied parallel to studs

Ceiling joists

½" (1.27 cm) or ⅝" (1.59 cm) drywall applied perpendicular to wall studs and joists

Wall studs

Joint treatment to all seams and corners

Baseboard

Figure 11.8 A single-ply drywall application. (*Leon E. Korejwo*)

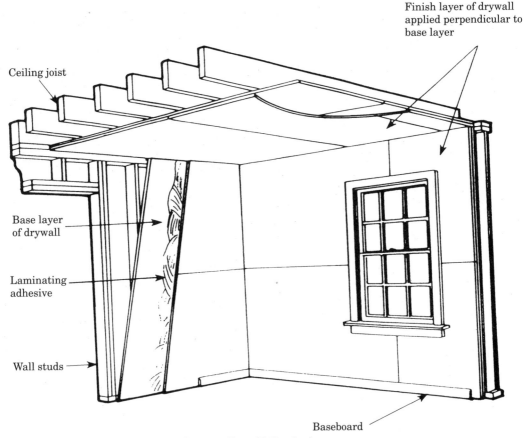

Finish layer of drywall
applied perpendicular to
base layer

Ceiling joist

Base layer
of drywall

Laminating
adhesive

Wall studs

Baseboard

Figure 11.9 A double-ply drywall application. (*Leon E. Korejwo*)

durable as alkyds. Latex tends to tear or melt when sanded. It is not recommended to use latex over unprimed wood, metal, or wallpaper. Latex quality is identified by the type of resin used. The highest-quality and most durable latex paint is one with 100-percent acrylic resin. In general, high resin content is the mark of durable, abrasion-resistant, flexible paint—the kind needed in the kitchen. The higher the resin content, the higher the gloss.

The glossier the finish, the more durable and washable it is. Kitchens are typically painted with high-gloss paints that can be easily washed or wiped down. High-gloss paint has the highest luster; it is a highly reflective finish for areas where washability is crucial, provides maximum durability, and is grease-resistant. Medium or semigloss paint also has a highly washable surface, but with a slightly less-reflective finish. It tends to show surface flaws, but not as obviously as high-gloss paint does.

Use an enamel undercoat on the surface and, as stated earlier, properly sand to help even the surface and reduce any visible flaws that will be more noticeable with the paint application. When remodeling, before removing or sanding any old paint, check to determine whether the paint contains lead. Lead may be harmful. Do not remove or sand paint without contacting local health officials for information on lead paint testing and safety precautions.

Remember—paint only covers the surface, it does not fill in defects. Unfortunately, wall imperfections that were not noticeable before the painting may now show through. Refer to Table 11.1 for proper preparation of the existing surface for paint. Preparation is the key to good-looking, long-lasting results. A properly prepared surface is clean, solid, and dry, and without cracks and imperfections.

As stated earlier, a good-quality, white latex primer/sealer formulated with higher binder solids, applied undiluted, is typically specified for new drywall surfaces prior to the application of texture materials and gloss, semigloss, and flat latex wall paints. An alkali- and moisture-resistant primer and a tinted enamel undercoat might be required under enamel paints. These products are claimed to make drywall strippable, to bind poor latex paint, to allow hanging of wallcoverings over glossy surfaces and existing vinyls, to hide wall colors, and to be water-washable.

After the painting job is complete, examine the job to verify that the specified number of coats of paint were applied to the walls. Indications of insufficient or uneven paint coverage include blotches in the finished surface or areas where the paint is not as glossy. Because the seams have been taped and spackled, the first coat of paint looks different on the seams than it does on bare drywall surfaces. The difference is usually not noticeable after the second coat is applied. To easily verify if the required number of coats of paint have been applied, ask the painter to tint the first coat of paint. When the second coat is applied, any missed spots will be noticeable.

Another important factor regarding paint in the kitchen is that latex and alkyd paints do not work on certain surfaces, such as acoustical panels, tile, and glass. Special acoustical ceiling paint is needed. Ordinary paints change the sound-deadening qualities of the panels. Epoxy paint is the best paint to use on hard, impermeable surfaces such as ceramic tile, plastic, porcelain, and glass.

Wallcoverings

Wallpaper provides an extraordinary variety in materials, styles, colors, patterns, and textures while opening up a world of decorating possibilities (Fig. 11.10a). Wallpapers are available as prepasted, washable, and strippable—features that make the hanging, cleaning, and removal easier. Choosing the right type of wallcovering is as important as the style. Here are some factors to consider when choosing.

Fabric-backed vinyl has a vinyl top layer and an undersurface of fiberglass or cheesecloth. The sturdiest kind of wallpaper, fabric-backed vinyl is washable, often scrubbable, and usually strippable. Compared with other papers, it is more moisture-resistant and less likely to tear if a wall cracks. It usually comes unpasted, because it is often too heavy to roll if prepasted.

Table 11.1   Guidelines for preparing an existing surface for paint

Surface	Preparation for painting
New plaster	Must be clean and completely cured. Seal surface with a vinyl acrylic wall primer; dry thoroughly. Follow with two coats of desired paint.
New drywall	Panels must be securely nailed or glued in place. All panel joints must be taped and filled before painting. When joint cement and/or patching materials are thoroughly dry, sand smooth, wipe away dust, then prime and paint. Seal surface with PVA (polyvinyl acetate) sealer; dry thoroughly. Apply two coats of desired paint. Do not use an alkyd primer—it will raise nap in paper.
Existing drywall or plaster	Treat small stains with a white-pigmented shellac, larger ones with a quick-drying alkyd primer. Spot prime patches with PVA sealer or finish paint (diluted 10%). If surface is more than five years old, or there is a big color change, using a vinyl acrylic wall primer over a latex finish; prime entire surface. Use an alkyd primer over an oil-base finish. Apply two coats of desired paint.
Bare wood	With patching paste, fill nail holes, joints, and cracks. Sand smooth and remove sanding dust with a tack cloth. Prime all bare wood and patched areas with an alkyd wood primer; allow to dry overnight. Follow with two coats of desired paint. (On fir, use a latex enamel undercoater since it does not bleed.) An enamel finish is usually recommended for wood.
Previously painted surfaces	Wash off dirt, grease, and oil buildup. Rinse thoroughly. Chip away loose, flaking paint. Patch holes and cracks with patching paste. Allow to dry and sand smooth. Prior to applying topcoat, spot-prime all bare spots with a white-pigmented shellac. Allow to dry for at least a half hour. An enamel finish is usually recommended for wood.
Bare wood to be stained	Fill holes with natural latex wood patch. Use a stain-controlling sealer for uniform stain absorption on soft woods. Stain in desired color; allow to dry overnight. If surface feels rough, apply a quick-drying sanding sealer and lightly sand. Apply varnish as applicable: thinned 10% with paint thinner; dry thoroughly; sand; apply second undiluted coat; and allow to dry for 24 hours.
Wallpaper	Completely remove existing wallpaper before painting. If necessary, use a chemical wallpaper remover. Once removed, wash off old adhesive. Rinse with water; allow wall to dry thoroughly before priming.
Metal	Remove dirt and rinse thoroughly. Sand off rust; prime. Use rust-inhibitive primer on new metal that will rust, latex metal primer on galvanized metal, conventional metal primer on aluminum. Apply two coats of desired paint. Do not sand between coats.
Masonry	Use acrylic or latex block filler. For a waterproof surface, follow with a hydrostatic coating. If going over a previous coating, consult dealer—all coatings are not compatible. Apply two coats of desired paint.

Paper-backed vinyl has a vinyl top layer with a paper rather than fabric backing. The paper makes the wallcovering lighter, so paper-backed vinyl can come prepasted. It is often peelable and washable.

Vinyl-coated paper can stain and tear more easily than other papers with vinyl content.

Vinyl's durability, strength, and stain-resistance make these papers relatively easy to install, easy to maintain, and terrific for high-traffic areas like the kitchen. Common untreated and vinyl-coated wallcoverings are susceptible to grease stains and abrasions, and pattern inks may run if washed. These products should be avoided.

Be sure that all special moldings have been installed before the wallcovering work begins, including any crown or chair rail molding over which the wall covering is to be applied.

Wallcovering work should be checked carefully when complete to ensure that the wallpaper is fully adhered without air bubbles. Check around doorways and openings to make sure that all edges are well bonded. Edge guards may be advisable if the opening does not have trim to protect the edges. Figure 11.10b shows the installation of a metal corner bead on drywall. Properly install screws and nails by dimpling the nail into the drywall, filling the dimple with joint compound and tape (Fig. 11.10c). This procedure ensures that the fastener won't ruin the finish by showing a circle or dent. As with glossy paint surfaces, shiny wallpaper finishes show imperfections (especially foil papers). Seams should be tightly butted without overlapping. Pattern matching is also very important, although some patterns can be more forgiving than others.

Tile

Tile for the kitchen walls may be ceramic or metal. Ceramic tile makes an attractive wall surface and is especially practical in a kitchen because it is

Figure 11.10a Wallpaper provides unlimited possibilities for kitchen walls. (*Wellborn Cabinet, Inc.*)

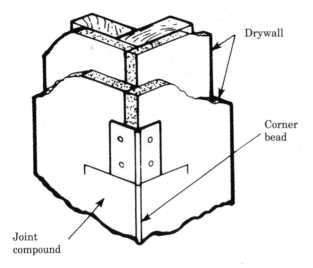

Drywall

Corner bead

Joint compound

Figure 11.10b Corner beads provide strong, durable protection for wall finishings. (*Leon E. Korejwo*)

waterproof, fireproof, durable, and easy to clean. It is highly resistant to food acids, impervious to grease, almost completely scratch-proof, and never needs waxing or painting. Cleaning can be done with a damp cloth, and most cleaning agents can be used. An extensive selection of colors and shapes of tiles offer infinite possibilities in pattern design on the wall. Wall tiles range in size from small 1-inch-square (3-cm-square) mosaics to impressive 12-inch (30-cm) squares and are available in high- or low-relief designs with colorful glazes or multicolored patterns.

The most important distinction between types of tile is whether they are glazed or unglazed. Glazed tile, available in matte or shiny finish, is impervious to stains but can be scratched; it is the standard tile used around sinks and above counters. Unglazed tile, made only in matte finish, picks up stains from grease and oil but resists scratching; it is the choice for floors. Basically, either type can be used on walls.

Installing tile on the walls between the top of the kitchen counter and the bottom of the kitchen cabinets, behind a range, or simply as a splash guard are a few suggestions for tile on the kitchen walls (Fig. 11.11). Metal tiles—copper, stainless steel, and aluminum—do have beauty and durability, but their high cost usually limits their use to the grease-catching walls (i.e., the backsplash behind the cooking area) of the kitchen.

Installation of tile Ceramic tiles can be applied to any drywall, plaster, or plywood surface that is smooth, sound, and firm. With existing walls, strip off flexible coverings (such as wallpaper) and scrape away loose paint. Moisture-resistant (MR) or water-resistant (WR) drywall is a specially processed drywall for use as a base for ceramic tile and other nonabsorbent finish materials in wet areas. The core, face paper, and back paper of MR board are treated to withstand the effects of moisture and high humidity. MR board is available in the standard 4-foot (122-cm) width and 8-foot (244 cm) and 12-foot (366 cm) lengths. It is applied to the studs in the same way as regular drywall (Fig. 11.12).

There are, however, a few limitations that should be kept in mind when considering the use and handling of MR or WR board:

☐ No vapor retarder should be placed behind the MR board where tile is to be applied to the face. A vapor retarder can be created on the face of the MR board by applying a skim coat of tile adhesive or by using a silicone grout for tile.

☐ The MR board should not be used on interior ceilings unless extra framing is provided. Consult the local building code for compliance.

☐ Do not apply joint compound to joints or fasteners to be tiled. This requirement might be waived if the applied tile in combination with the bonding adhesive employed fully protects the drywall and water-sensitive materials, if present (such as when a joint compound is used), from penetration of water.

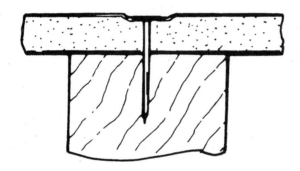

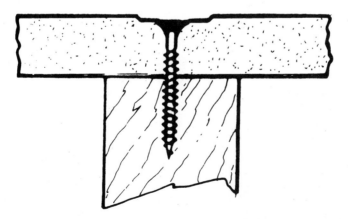

Figure 11.10c Proper depth of a nail dimple (Top) or driven screw dimple in drywall (Bottom). (*Leon E. Korejwo*)

Figure 11.11 Ceramic tile provides a distinctive, natural look for kitchen walls and backsplashes. (*Florida Tile*)

Responsibility for performance of completed installations should rest with the surfacing material manufacturer or the surfacing material applicator.

□ Adherence to the recommendations concerning sealing exposed edges, painting, tile adhesives, framing, and installation is necessary for satisfactory performance. For high-humidity or "wet" areas, cement board should be considered as an alternative.

Ceramic tiles are relatively brittle, but once they have been cemented to a solid backing and the joints grouted with special mortar, you have an exceptionally sturdy wall. Note that wall tiles are thinner and slicker than floor tiles. To piece together a smooth installation, two different types of tiles are needed. Field tiles ($4\frac{1}{4}$- and 6-inch, or 11- and 15-cm, squares are typical sizes) cover most of the surface. Trim tiles with rounded-off edges get around the corners. Smaller mosaic tiles come bonded to pieces of 1-by-1-foot or 1-by-2-foot paper or fabric mesh. They go up a little faster but require more grouting. Pregrouted tile sheets include $4\frac{1}{4}$-inch (11-cm) tiles and flexible synthetic grouting. You cement the sheets to the wall, and then seal edges with a caulking gun. These are the easiest to put up. Many professionals still prefer to mud-set ceramic tiles in cement-based mortar—a tricky masonry process. Fortunately, if you have decided to not hire a subcontractor to tile the walls, you can now choose from a number of mastic-like adhesives especially developed with less-demanding handling qualities. Grout between tile comes in colors and can be applied after the tile is set. Due to its porosity, grout must be coated with a sealer to prevent mildew or grease from penetrating and staining.

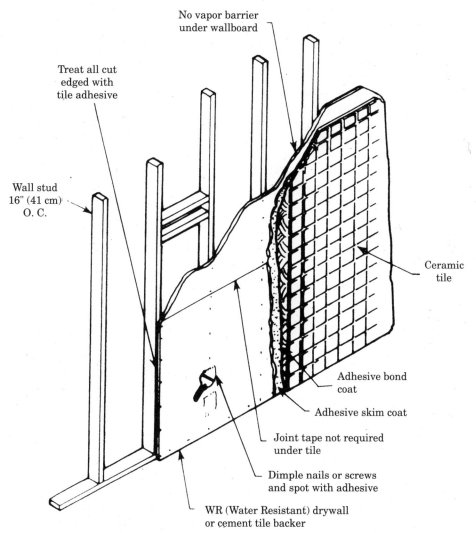

No vapor barrier
under wallboard

Treat all cut
edged with
tile adhesive

Wall stud
16" (41 cm)
O. C.

Ceramic
tile

Adhesive bond
coat

Adhesive skim coat

Joint tape not required
under tile

Dimple nails or screws
and spot with adhesive

WR (Water Resistant) drywall
or cement tile backer

Figure 11.12 Details of installing ceramic tile over MR board. (*Leon E. Korejwo*)

Of all the things that can be done with a wall, veneering it with ceramic tile makes the most lasting, water-resistant, and easily cleanable improvement. Unfortunately, it can also be one of the most expensive for your client.

Paneling

The two main types of paneling, sheet and solid boards, come in a variety of textures and finishes. What is the best? It depends on what the homemaker wants. Most paneling never needs painting, is easy to clean, and has a good resistance to dirt. The smoother the surface of the paneling, the higher the resistance. Sheet paneling is easier to apply over large, unbroken surfaces because of its dimensions. Solid boards (or strips), however, are easier to fit around openings and obstructions.

Sheet paneling   Sheet paneling comes in large, machine-made panels, most commonly 4 by 8 feet (122 by 244 cm). The two main types of sheet paneling are hardboard and plywood.

Hardboard paneling is manufactured for use as prefinished paneling and is specifically treated for resistance to stains, scrubbing, and moisture. It is also highly resistant to dents, mars, and scuffs. Hardboard is produced by reducing waste wood chips to fibers and then bonding the fibers back together under pressure with adhesives. In most cases, the material is prefinished in wood grains such as walnut, cherry, birch, oak, teak, and pecan, in a variety of shades. It may be smooth-surfaced or random-grooved. In addition, there are the decorative and work-saving plastic-surfaced hardboards that resist water, stains, and household chemicals exceptionally well. A typical surface consists of baked-on plastic. Most hardboard is sufficiently dense and moisture-resistant for use in kitchens and laundry rooms. The variety of laminated plastic finishes and sizes is extensive. A tough, pliable paneling, hardboard is sold in 4-by-8-foot (122-by-244 cm) sheets in thicknesses ranging from $\frac{3}{16}$ to $\frac{3}{8}$ inch (0.48 to 1.33 cm); $\frac{1}{4}$ inch (0.6 cm) is usual. It is usually less expensive than plywood, but it is also less durable and more subject to warping or moisture damage. The most common surface finishes are imitation wood; generally grooved to look like solid-board paneling. Wood imitations are available in highly polished, resawn, or coarser-brushed textures in a range of colors. Panels embossed with a pattern, such as basket weave, wicker, or louver are also available.

Plywood paneling is more popular in the kitchen than solid woods. The panels are manufactured from thin wood layers (veneer) peeled from the log and then glued together. The grain of each veneer runs perpendicular to adjacent veneers, making plywood strong in all directions. Less expensive than most solid-board strips, plywood panels are also less subject to warping or shrinkage. Any standard, unfinished plywood sheets may be used for wall paneling. In addition, there are sheets expressly intended for paneling. Just about any type of hardwood, and most of the major softwoods, laminated onto the surfaces of plywood panels can be purchased. Prefinished and vinyl-faced decorative styles are available, as well as resin-coated panels designed for painting.

Plywood face textures range from highly polished to resawn. Many types feature decorative grooves (often imitating solid-board patterns) or shiplap edges. There are two standard thicknesses—$\frac{1}{4}$ inch (.63 cm) and $\frac{5}{16}$ inch (.79 cm). Thinner panels are not very durable and difficult to work with. Many come prefinished and resist staining (to a degree). Plywoods faced with genuine hard and softwoods provide the most natural look. Often, no two panels are exactly the same.

Solid-board strip paneling Natural fragrance, texture, subtle variations in color and grain, and imperfections make solid-board strip paneling extremely warm and inviting. It is especially suitable around doors, windows, and other large openings; areas where extensive handling and cutting are required. Quite simply, solid-board strip paneling is any paneling made up of solid pieces of lumber positioned side-by-side. In some cases, standard, square-edged lumber is used—1-by-4s (2.5-by-10-cm), 1-by-6s (2.5-by-15-cm), and so forth. But generally, the boards have edges specially milled to overlap or interlock. The three basic millings are square edge, tongue and groove, and shiplap (Fig. 11.13a).

Various types and patterns of woods are available for application on kitchen walls to obtain the desired decorative effects. For informal treatment, pine, redwood, whitepocket Douglas fir, sound wormy chestnut, and pecky cypress,

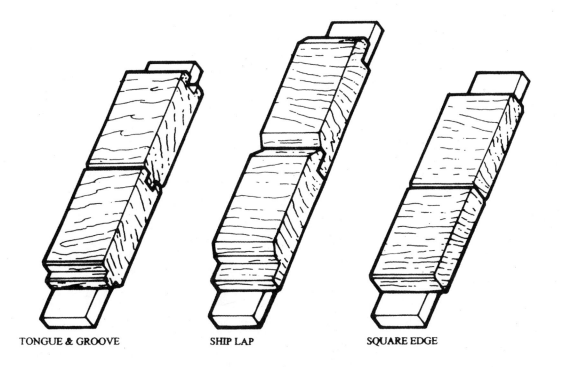

TONGUE & GROOVE   SHIP LAP   SQUARE EDGE

Figure 11.13a Solid-board strip paneling three basic millings: tongue-and-groove; shiplap; and square edge. (*Leon E. Korejwo*)

finished natural or stained and varnished may be used to cover one or more sides of a room. In addition, there are such desirable hardwoods as red oak, pecan, elm, walnut, white oak, and cherry also available for wall paneling. Most types of paneling come in thicknesses from $\frac{3}{8}$ to $\frac{3}{4}$ inch (1.33 to 1.9 cm); widths vary from 3 to 12 inches (7 to 30 cm), lengths from 3 to 10 feet (91 to 305 cm).

When planning a wood-paneled room, remember that when the wall is to be accented, use boards of random widths; subdue it by the use of equal-width boards. Small rooms can be given the illusion of increased size by applying the paneling horizontally. Of course, paneling can be applied vertically, horizontally, diagonally, or in combined directions. Solid-wood paneling must be finished with a good wood sealer and finished to resist staining.

Like sheet paneling, solid boards can be attached to new stud walls, to existing walls that are in good shape, or to a gridwork of furring strips applied over old, bumpy walls. Although solid boards are usually installed vertically or horizontally, consider a diagonal pattern. Before paneling vertically with solid boards, horizontal furring strips must be attached to the wall every 24 inches (61 cm) on center or install nailing blocks at those spacings between studs (Fig. 11.13b).

Generally you don't need to apply furring if attaching boards horizontally, unless the wall is badly damaged or out of plumb. Nail the boards to the studs directly or through existing wall coverings (Fig. 11.14a). To avoid ending with a very narrow board at the ceiling, calculate its size and split the difference so the lowest and highest boards are the same width. Start at the bottom of the wall and work toward the ceiling (Fig. 11.14b). Temporarily nail the first board at one end, $\frac{1}{4}$ inch (0.6 cm) above the floor. Then level the board and complete the nailing. If needed, scribe and trim the board as for sheet paneling. Minor inconsistencies can be covered with molding (Fig.11.14c).

Because you are working with lightweight, semirigid materials, each of which covers a lot of wall territory, paneling with hardboard or plywood is applied more quickly and easily than wallpaper on the same space. However, like wallpaper, paneling often tends to play up irregularities in its backing. If the sheets are nailed directly to studs, you could end up with an undulating wall as the lumber shrinks, swells, and warps. To prevent this, first put up a layer of drywall, then laminate the paneling to the drywall with adhesive (Fig. 11.15). The use of drywall as a substrate when applying wood paneling provides increased fire resistance and sound control. In new or existing construction, a $\frac{3}{8}$-inch or $\frac{1}{2}$-inch (0.95 or 1.27 cm) drywall substrate is recommended before applying combustible paneling.

Prefinished sheet paneling is probably the easiest type of wall finish. Sheets $\frac{1}{2}$ inch or thicker may be nailed or glued directly to studs. Thinner panels require a $\frac{1}{4}$- or $\frac{3}{8}$-inch (0.6 or 0.95 cm) drywall backing. Paneling hides flaws while providing a tough finish for heavy-traffic areas. Plywood sheets may be glued to furring when the wall is in poor condition (Fig. 11.16).

Like painting, wallpaper, and most other interior finishing approaches, a successful paneling job requires careful preparation of the surface to be cov-

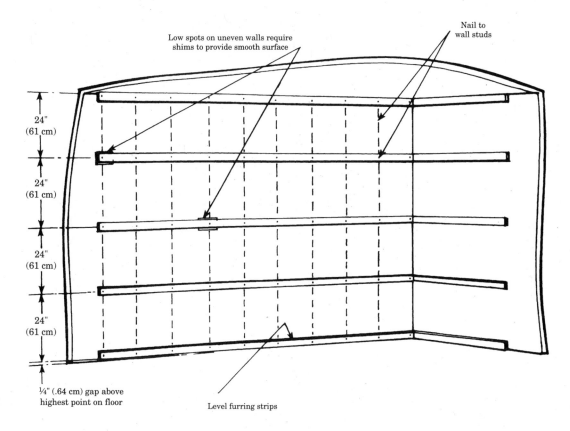

Low spots on uneven walls require
shims to provide smooth surface

Nail to
wall studs

24"
(61 cm)

24"
(61 cm)

24"
(61 cm)

24"
(61 cm)

¼" (.64 cm) gap above
highest point on floor

Level furring strips

Figure 11.13b   Vertical plank-board wall preparation. (*Leon E. Korejwo*)

ered. Be sure to complete all necessary preparations before installing any paneling.

Moldings

Molding details can set the theme of the kitchen's decor. Moldings can give the kitchen a whole new look, as well as help to conceal the room's architectural and carpentry faults. Even in the most basic room, moldings have their place along the base of walls and around door and window frames (Fig. 11.17).

Here is a sample of the more popular types of interior trim:

1. **Baseboard** protects the bottom of the wall from wear and tear and conceals irregularities at the wall-floor joints. Quarter-round or base shoe moldings may be used to complete the trim.

2. **Casing** is used to trim doors, windows, and other openings. It may also be used for chair rails, cabinet trim, and decorative purposes.

3. **Cornices**, whether of the crown or cove type, give a rich appearance wherever two planes, such as a wall and ceiling, meet. They are also used for

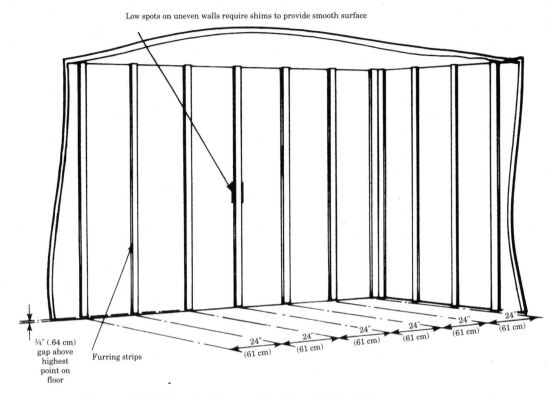

Low spots on uneven walls require shims to provide smooth surface

¼" (.64 cm) gap above highest point on floor

Furring strips

24" (61 cm)

Figure 11.14a   Horizontal plank-board wall preparation. (*Leon E. Korejwo*)

trimming exposed beams and, singly or in combination with other moldings, in decorative mantels and frames.

4. **Wainscot** caps are applied to the top of wainscoting. Some patterns have a wraparound lip to conceal craftsmanship defects. Others may be used to cap decorative baseboards.

5. **Chair rail** protects the wall in areas subject to chair-back damage. It is installed at a height appropriate to the furniture style.

The variety of moldings available seems endless. Traditional wood moldings come in many standard patterns and sizes. Integral molding systems include corner pieces that eliminate the need for tricky corner cuts and joints. They can be purchased prefinished (painted or stained), wrapped with printed vinyl, or natural. Lengths range from 3 to 20 feet (91 to 610 cm). While wood is still the most popular, vinyl-coated, plastic, or metal moldings, manufactured to look like wood, are available.

Brick and stone

Brick and stone can provide a wonderful feeling of warmth and create an outstanding highlight wall in a kitchen. Real brick and stone have permanence matched only by ceramic tile. But real bricks and stones do have certain disadvantages. They are porous, and grease sinks right in, as does dirt of any kind. In addition, bricks and stones add enormous weight to a wall and require the support of a foundation wall. These disadvantages have led manufacturers to develop veneers and imitations. Imitation bricks and stones are made of various materials; styrene, urethane, and rigid vinyl are the most common. All false bricks and stones are highly durable and come in a wide variety of colors and styles. Some are sold in sheet form, while others are installed individually, similar to ceramic tile. Real brick may be purchased cut thin to match an existing fireplace or interior wall. Some are fire-resistant and may be used near ranges and ovens.

We have covered the most important groups of wallcovering materials. Others might be considered. Special accents can enhance the appearance of the walls. Glass blocks, a decoupage wall, and mirrored walls can give "size" and "richness" to a small kitchen. In some cases, archways may be cut through an existing wall at countertop level and above to create a spacious atmosphere. The arch may be trimmed in brick or stone to enhance an informal country kitchen. Milled wood molding trimming an arch or opening can carry the environment of a formal family room or dining room into the kitchen. While not fully exposing the food preparation surfaces, it allows a smaller room a visual increase in dimension.

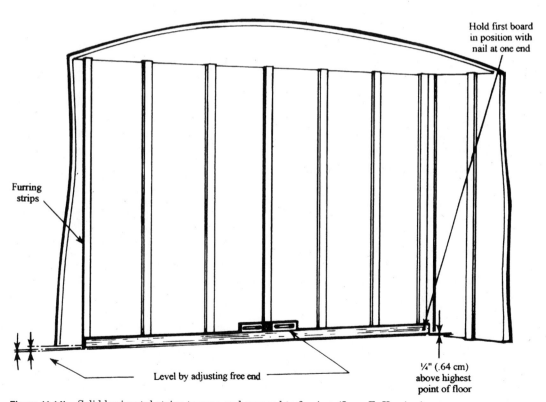

Figure 11.14b Solid horizontal strips tongue-and-grooved to furring. (*Leon E. Korejwo*)

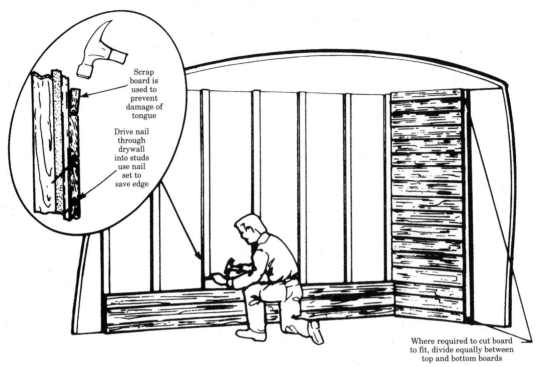

Scrap
board is
used to
prevent
damage of
tongue

Drive nail
through
drywall
into studs
use nail
set to
save edge

Where required to cut board
to fit, divide equally between
top and bottom boards

Figure 11.14c Applying horizontal solid-board strip paneling. (*Leon E. Korejwo*)

Ceilings

Almost everything that has been mentioned for use on walls can be applied to the kitchen's ceiling—paint, wallpaper, tile, or paneling—to give the broadest possible spectrum of colors, textures, and patterns. Kitchen ceilings can be made very interesting, as well as made to fit most any design or style of decoration, by using different angles (some drop beams and some raised ceilings), exposed beams, pitched ceilings, faux-finished ceilings, and even hammered-tin ceilings. Skylights help brighten the kitchen ceiling, too (Fig. 11.18). It is important to remember that most experts agree that ceilings should be relatively light in color, especially in a kitchen, to maximize the reflection of light to provide more even lighting.

Combine the framing of a floor with the covering materials used for walls, and you get the anatomy of a typical home ceiling. It begins with the same joists that support the subfloor above. Next the builder may level off the joists' bottom edges with furring strips, or—if the lumber is even to begin with—fasten drywall or plaster lath directly to the joists. There are exceptions to this construction cutaway, though. Sloping top-floor ceilings usually are attached to the roof framing and, in properly built homes, have insulation above them. You also may encounter lightweight tiles suspended below the joists of an old ceiling. An open-beam ceiling consists of nothing more than the underside of the roof decking above.

Suspended ceilings

A suspended ceiling can change the entire appearance of the room. Compared to installing drywall, putting up ceiling tiles or panels is a breeze. Instead of handling awkward 4-by-8-foot (122-by-244-cm) sheets, you work with lightweight materials. The most common panel size is 2 by 4 feet (61 by 122 cm), though panels are available in a variety of sizes. Many tiles on today's market won't break, stain, or chip, are easily cleaned with soap and water, and are available in a variety of colors, patterns, and natural wood grain finishes. In addition, many manufacturers have produced ceiling tiles that are highly insulated, water- and insect-proof, have flame-retardant metal backings, and allow complete accessibility to all above-ceiling utilities. Tiles are available that provide an acoustical, soundproofing

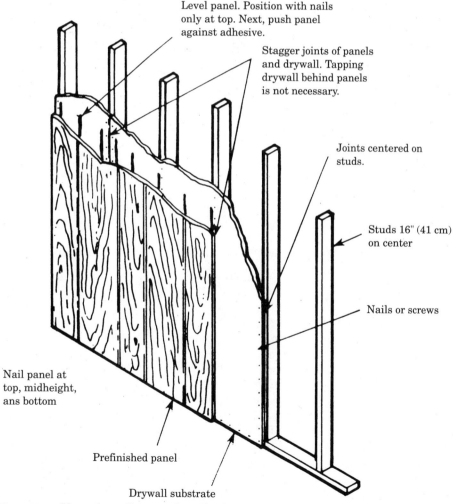

Level panel. Position with nails only at top. Next, push panel against adhesive.

Stagger joints of panels and drywall. Tapping drywall behind panels is not necessary.

Joints centered on studs.

Studs 16" (41 cm) on center

Nails or screws

Nail panel at top, midheight, ans bottom

Prefinished panel

Drywall substrate

Figure 11.15 Plywood paneling over drywall substrate. (*Leon E. Korejwo*)

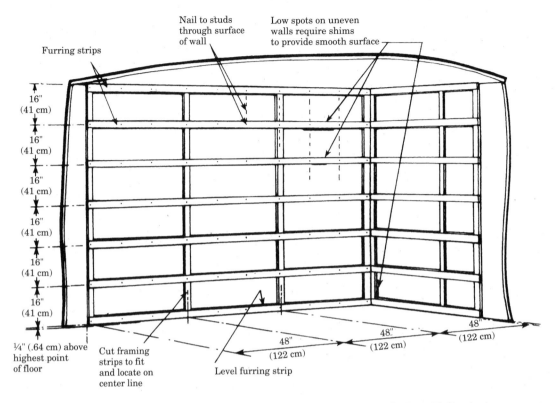

Figure 11.16a   Prefinished 4-by-8-foot plywood panels over furring on an old wall. (*Leon E. Korejwo*)

feature, as well as decorative patterns with or without the acoustical properties.

There are three popular ways to tile or panel a ceiling:

1. Fasten the tiles directly to the ceiling or to furring strips nailed directly across it.

2. Install a grid system directly to the exposed beams

3. Suspend the new ceiling from a grid that drops below the existing one.

The first method is the one most often employed for installing a new ceiling in an existing room. If the old ceiling is in fairly good condition, the new ceiling can be fastened directly to it with adhesive. But if the ceiling is rough, not uniformly level, or is unfinished, furring must be used.

Lay out a plan of the room, show location of the joists and choose the tile placement (Fig. 11.19). Furring a ceiling is much like furring a wall, although the placement of the furring strips is different (Fig. 11.20a). It is recommended to use 1-by-4s (2.5-by-10-cm) for furring; they are usually straighter and easier to space for the tiles. Tile may be nailed to furring, stapled in place (Fig. 11.20b), or held by nailer clips (Fig. 11.20c).

The use of grids, either fastened directly to the beams or suspended below them, offers many interesting ceiling treatments. Transparent and translucent panels and egg-crate grilles are made to fit the gridwork to admit light from above. Recessed lighting panels that exactly replace one panel are also available.

A dropped ceiling consisting of removable tiles allows installation work (wiring, pipes, and heating and air-conditioning components) to remain easily accessible (Fig. 11.21). Ceiling tiles are available in mineral-fiber (noncombustible) and fiberglass (may be fire resistant). Mineral-fiber and fiberglass are drop-in panels that work with a grid system. If the homeowner dislikes the look of drop-in tiles, or if protruding ducts or beams would require a lower ceiling than desired, box around them using a wood

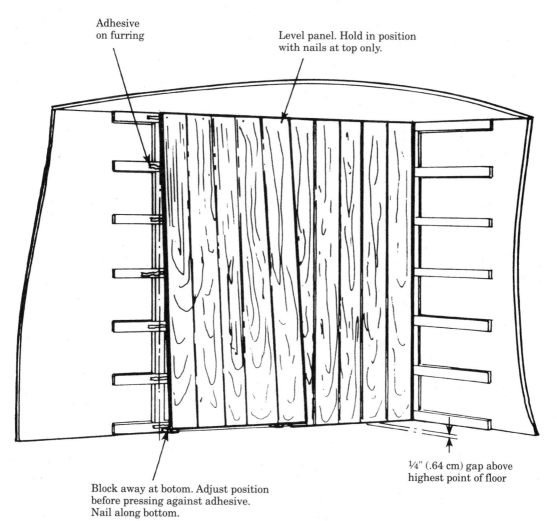

Adhesive on furring

Level panel. Hold in position with nails at top only.

¼" (.64 cm) gap above highest point of floor

Block away at botom. Adjust position before pressing against adhesive. Nail along bottom.

Figure 11-16b Plywood panels applied to furring strips. (*Leon E. Korejwo*)

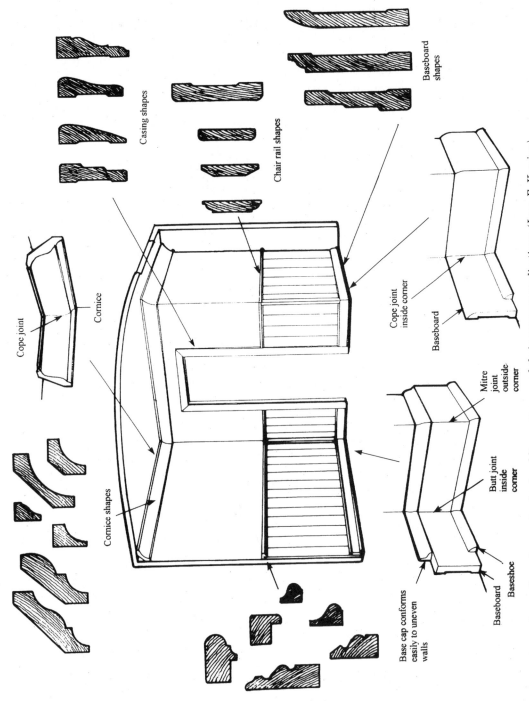

Casing shapes

Chair rail shapes

Baseboard shapes

Cope joint

Cornice

Cope joint inside corner

Baseboard

Mitre joint outside corner

Butt joint inside corner

Cornice shapes

Baseboard

Baseshoe

Base cap conforms easily to uneven walls

Figure 11-17 A sampling of some standard molding patterns and their common applications. (*Leon E. Korejwo*)

Figure 11-18 Ceiling skylights, casement, and roof windows let in sunlight—an important feature in kitchens. (*Anderson Windows, Inc.*)

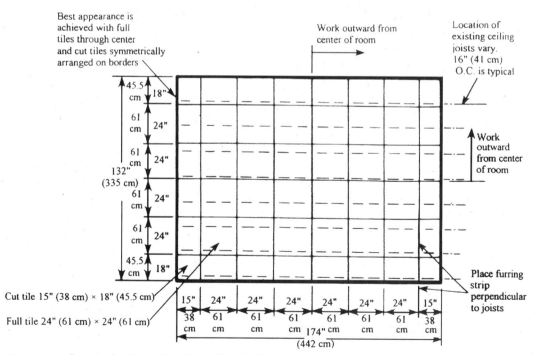

Figure 11-19 Layout plan for a tile ceiling. (*Leon E. Korejwo*)

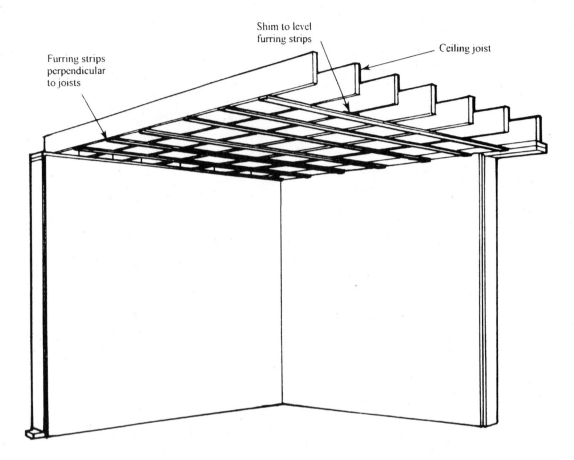

Figure 11-20a Furring strips for a tile ceiling. *(Leon E. Korejwo)*

frame. Cover it with acoustic tile and install the remaining ceiling at a more desirable height (Fig. 11.22).

Drywall

Painted drywall is the most common type of ceiling covering in use today. It is relatively easy to install, inexpensive, and provides a good-looking finished product. The same drywall panels used on walls are used for ceilings. However, it is harder to install on ceilings than on walls because working overhead is awkward, and the weight of the full panels can also be awkward.

Sheets larger than 4 by 8 feet (122 by 244 cm) are rarely used for ceiling construction. Drywall ceiling installation is usually a two-person job. If a one-person ceiling installation is necessary, use temporary support, as seen in Fig. 11.23a, and a T-brace ceiling installation, as in Fig. 11.23b. Mechanical cradle lifters are available (Fig. 11.23c). The installation can be accomplished in the same basic manner as wall installation except that you are working in a difficult position. If covering both surfaces, do the ceiling first.

Water-resistant drywall is designed for areas where heavy moisture might collect. MR or WR board should not be used on interior ceilings unless extra framing is provided. Spacing of framing members should not exceed 12 inches (30 cm) on center, including use as a substrate when ceramic tile or similar materials are to be applied to the ceiling. Drywall may be installed perpendicular to joists, Fig. 11.24a, or parallel to joists, Fig. 11.24b. Fig. 11.25 shows a method of floating corners at ceilings to minimize cracking due to expansion.

Drywall ceiling panels have a noncombustible drywall core with high light reflection and attractive predecorated surfaces. The panels install easily in standard exposed grid systems. Because ceiling panels have a rigid drywall core, they resist sagging and warping, and they do not require clips to offset bowing. Most drywall ceiling panels are accepted by the U.S. Department of Agriculture (USDA) for use in food-processing areas. Most drywall ceiling panels are $\frac{1}{2}$ inch (1.27 cm) thick with a 2-mil, white, stipple-textured vinyl laminate for high light reflection and easy cleaning.

There is a good chance that some of the ceiling might have to be opened up for plumbers, electricians, heating crews, or other trades. Drywall is easy to cut and patch. New drywall can be blended into existing drywall very well. A coat of sealer, followed by a coat of paint, is all that is needed to camouflage any disruption of the ceiling. However, as previously stated, not all kitchens have existing drywall in the ceiling.

Figure 11-20b   Tile stapled in place. (*Leon E. Korejwo*)

Figure 11-20c Tile held by clips. (*Leon E. Korejwo*)

Beams

While beams are not a part of the ceiling, they often add to the beauty of it. The ceiling can be made very interesting with a beam arrangement—real or artificial. Certain styles incorporate patterned beams and enclosed ceiling bays (usually painted drywall). What would an early American kitchen be without a beamed ceiling? Today, ready-made, ready-to-install beams are available at most lumber dealers for this purpose. They are made of solid lumber, plywood, polyurethane plastic foam, or metal. The number of beams and patterns employed depends on the size of the room and personal preference. Some room ceilings look best with the beams running in one direction only, while others look fine with crossing beams. The beams themselves can be finished or painted to match any color scheme. If you have removed a bearing-wall and added a structural beam, disguise its presence by flanking it with fakes. The inside provides efficient space to hide electrical and plumbing lines or heating ducts.

Open it up. If a one-story house has an attic or crawlspace above, you may be able to remove ceiling joists or add more widely spaced beams and forget the ceiling material. The underside of the roof deck needs to be finished off with tongue-and-groove wood planks or drywall and paint. This is also the

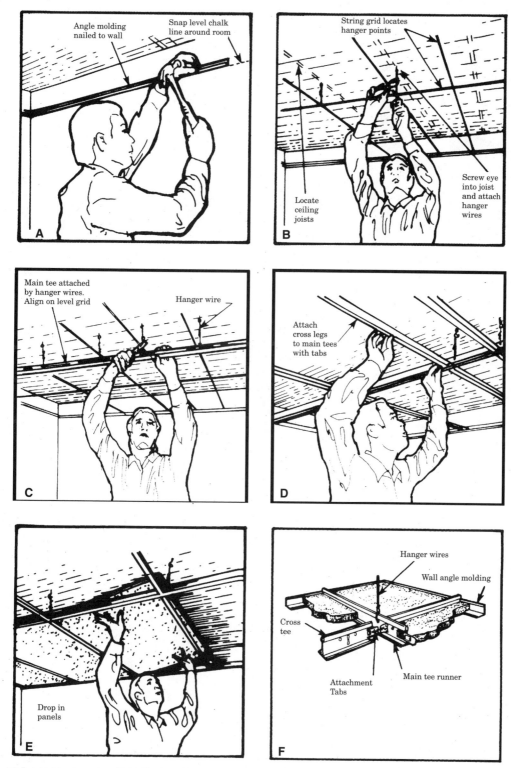

Figure 11-21a through f Installation of a suspended ceiling. (*Leon E. Korejwo*)
Figure 11-21f Suspended ceiling components. (*Leon E. Korejwo*)

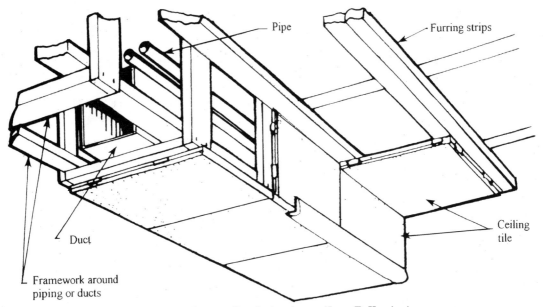

Figure 11-22 Drop ceiling boxed around protruding duct or beam. (*Leon E. Korejwo*)

time to add a skylight or special windows to fit the gables revealed through the cathedral ceiling.

Wood paneling

Sheet paneling, plank paneling, and tongue-and-groove lumber—natural, stained, bleached, or painted, provides a charming accent. Wood paneling is not difficult to install on a ceiling and can be very attractive in the proper setting, especially on a vaulted ceiling. Installation of the sheets or planks on the ceiling is the same as application to the wall surface. Flooring strips can be nailed to the ceilings, but it is faster and easier to use hardboard panels with the desired wood facing.

Plaster

Plaster was a common covering for ceilings at one time. It is more difficult to repair cuts, cracks, and holes in plaster than with drywall. Plaster has rises and depressions that create an uneven surface. It is not easy to match drywall to a surface like this, however, it can be done with some skill. The wood strips and wire used in conjunction with plaster makes any job harder for plumbers, electricians, and others to do their installation work. If an existing kitchen has plaster, be sure to factor the extra work into the overall production schedule and budget.

Tin-stamped ceiling panels and cornice patterns drawn from the past enliven contemporary and traditional styling with finely crafted detail. Lightweight, noncombustible panels are manufactured from tin-free steel 0.010 inch thick and properly coated for superior paint retention (for little or no maintenance and cleaning with soap and water). Conventional suspension methods are recommended for the ceiling installation of lay-in panels. Heavy-duty safety gloves should be worn when handling panels and cornices—edges can be sharp.

Kitchen lighting

Because the kitchen is both a work area and a gathering place, the selection of lighting is important. No kitchen is complete without adequate lighting,

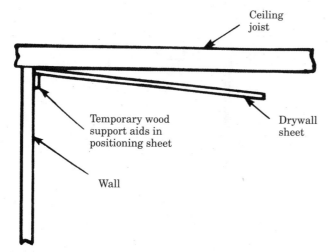

Figure 11-23a The use of temporary support when installing a ceiling. (*Leon E. Korejwo*)

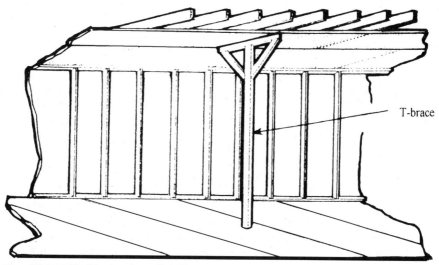

Figure 11-23b Positioning drywall ceiling panels with a temporary wall (T-brace) installation. (*Leon E. Korejwo*)

Figure 11-23c Cradle lifter for installing ceiling drywall.
(*Leon E. Korejwo*)

and, of all the components in the kitchen, lighting is the one the homeowner can least afford to skimp on. Not only can poor lighting make the cheeriest kitchen seem dreary, it can also promote fatigue and even cause accidents. Proper lighting has long been established as being essential to safety, good health, and a general feeling of well-being. The need for effective lighting is at its greatest in the kitchen where meal preparation, entertaining, and a variety of other tasks all require proper levels of illumination.

The amount of light needed in a kitchen may be obtained from one source or from a combination of several (Fig. 11.26). Factors to take into consideration

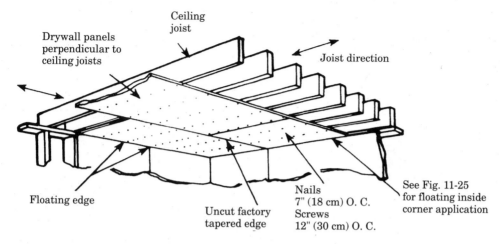

Ceiling
joist

Drywall panels
perpendicular to
ceiling joists

Joist direction

Floating edge

Uncut factory
tapered edge

Nails
7" (18 cm) O. C.
Screws
12" (30 cm) O. C.

See Fig. 11-25
for floating inside
corner application

Figure 11-24a Method of mounting drywall perpendicular to ceiling joists. (*Leon E. Korejwo*)

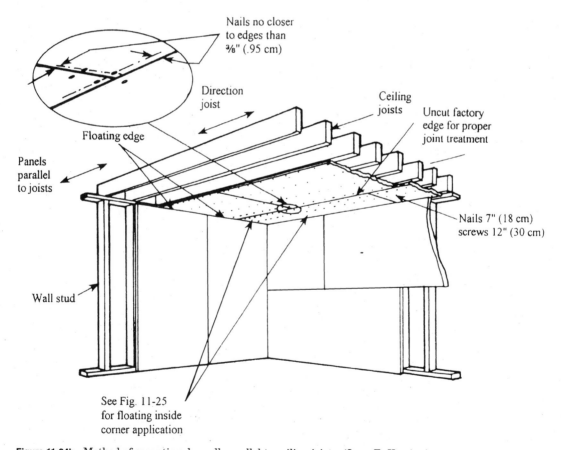

Nails no closer
to edges than
³⁄₈" (.95 cm)

Direction
joist

Ceiling
joists

Uncut factory
edge for proper
joint treatment

Floating edge

Panels
parallel
to joists

Nails 7" (18 cm)
screws 12" (30 cm)

Wall stud

See Fig. 11-25
for floating inside
corner application

Figure 11-24b Method of mounting drywall parallel to ceiling joists. (*Leon E. Korejwo*)

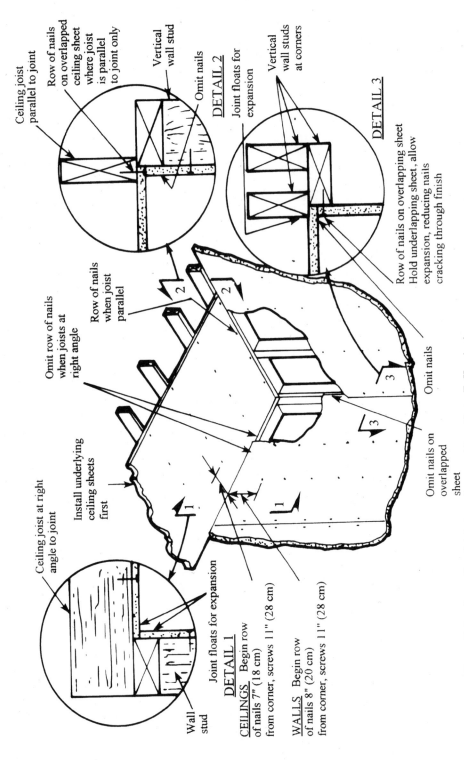

Ceiling joist parallel to joint

Row of nails on overlapped ceiling sheet where joist is parallel to joint only

Vertical wall stud

Omit nails

DETAIL 2

Joint floats for expansion

Vertical wall studs at corners

DETAIL 3

Row of nails on overlapping sheet Hold underlapping sheet, allow expansion, reducing nails cracking through finish

Omit row of nails when joists at right angle

Row of nails when joist parallel

Install underlying ceiling sheets first

Omit nails on overlapped sheet

Ceiling joist at right angle to joint

Joint floats for expansion

DETAIL 1

CEILINGS Begin row of nails 7" (18 cm) from corner, screws 11" (28 cm)

WALLS Begin row of nails 8" (20 cm) from corner, screws 11" (28 cm)

Wall stud

Figure 11-25 Drywall-floating ceiling joists to reduce cracking. *(Leon E. Korejwo)*

430

Figure 11-26   Natural light, recessed lighting fixtures, and hanging pendants provide for overall effective lighting here. (*Cooper Lighting*)

are the size of the room and its use. The larger the room, the greater its requirements. The lighting fixtures chosen highlight your client's new kitchen and control the atmosphere of the room, as well as help them to perform their tasks more efficiently.

Note: You want to establish a good, workable lighting plan early in the kitchen project to allow for the proper placement of wiring, fixture boxes, and switches as the kitchen is being worked on.

Some of the choices you must decide on with your client are

1. Structural (built-in as part of the home's finished structure) lighting or store-bought fixtures

2. Incandescent bulbs, fluorescent tubes, or halogen lights

3. Level of intensity (brightness) for general and for task lighting

4. Fixture types: luminous ceiling, luminous panels, recessed, tracks, spots, wall brackets, etc.

Most kitchens have four basic kinds of lighting: natural daylight resulting from windows and skylights; general lighting that gives overall illumination; local or task lighting that provides proper illumination at specific work centers; and decorative lighting that emphasizes the color and decorative theme of the kitchen.

Natural light

Everybody wants more natural light in the kitchen (Fig. 11.27). Natural light can enter a kitchen through windows, doors, and skylights—all a part of connecting the kitchen to the outdoors. Windows furnish the natural light to brighten work areas and supplement the artificial light (Fig. 11.28). Minimum property standards of the HUB/FHA specify a window area totaling, in square feet, 10 percent of the floor area. (Good kitchens should have at least 15 percent; an area equal to 20 to 30 percent of the room's area is still better.) Whenever possible, two window areas should be planned. Glass panels in outside doors and skylights count as part of the total window area. Incidentally, skylights give a great deal of natural light without cutting down on wall cabinet space. Each square foot of skylight supplies natural lighting to 20 square feet of floor space. A skylight can also help avoid that claustrophobic feeling in kitchens by opening up spaces to the outdoors where there are no windows. In addition, the venting models, placed near the roof ridge, can also greatly improve natural ventilation.

Glass doors versus wood doors can double the natural lighting potential. For safety and security, be sure that the new door has tempered glass or shatterproof plastic. In addition, energy conservation is a prime consideration; double- or triple-glazed panes can help conserve energy.

Discuss with your client the exposure of the kitchen to sunlight. Have them consider the view from the window, the need for privacy, and the possible glare from uncovered windows at night. Also, keep in mind that natural light may illuminate the kitchen unevenly. A single window in the middle of a wall often creates a strong contrast with the surrounding area, causing a glare. For more even light, use two windows on adjacent walls, add a skylight, or compensate for the glare by illuminating the surrounding area with artificial light. If your client wants a window to provide natural task light for a kitchen sink or work surface, its sill should be 3 to 6 inches (8 to 15 cm) above the countertop.

It is also very important to consider which way the windows should face. A good northern window usually provides an almost constant brightness throughout the day. Because it is from an open sky, without direct sun, it does not create glaring hot spots or deep shadows in work areas, which also keeps the sun from shining directly into the homeowners eyes while they work. Eastern windows are not too bad because they brighten the morning yet rarely have the intensity of the hot afternoon sun. Skylights on north and east-facing roofs lessen heat gain in the summer. Southern light warms a

Figure 11-27 Turn the kitchen into a sunroom with abundant natural lighting. (*Four Seasons*)

Figure 11-28 Natural light supplements artificial lighting. (*Anderson Windows, Inc.*)

kitchen in winter but may require an overhang or awning to cut summer heat gains. A skylight on a southern or western exposure will capture solar heat during the winter. Western light subjects a kitchen to the hot, direct rays of late-afternoon sun, which can make a room uncomfortable until far into the night. Southern and western windows may provide more sunshine than desired (as long as seven hours).

General (ambient) lighting

General lighting fills in the undefined areas of a room with a soft level of light and illuminates the room as a whole. General lighting can come from hanging fixtures, fixtures mounted on the ceiling, or recessed downlights mounted in the ceiling. It also can come from a "luminous ceiling" where light floods through an entire translucent suspended ceiling, from valance lighting, or illuminated soffits. General lighting should be indirect, aimed to bounce off pale walls or ceiling and then into the room. Bouncing diffuses the light for even illumination over the entire area.

How much general light is needed in the kitchen? Table 11.2 shows some general lighting guidelines for the kitchen. It is difficult to tell in terms of light bulb wattages, because wattages are not accurate measures of light intensity. However, the following system serves as a rough guide.

Measure the kitchen to find how many square feet of floor space is there. Then simply apply the following minimums. If the homeowner is comfortable with more light, do not hesitate to move up from the minimums.

☐ For ceiling-mounted or suspended fixtures in the kitchen, 2 watts per square foot for incandescent and 1 watt per square foot for fluorescent fixtures are needed.

☐ For valance, cornice, or wall-bracket types of indirect lighting, 5 to 6 watts of fluorescent lighting per square foot of floor space are needed.

☐ For nondirectional recessed lighting (fixtures mounted in the ceiling), the minimums are 3 watts per square foot for incandescent and $1\frac{1}{2}$ watts per square foot for fluorescent fixtures.

Task lighting

Task lighting is the local illumination needed for the special task areas of the kitchen. This type of lighting should be directly from the fixture to the work surface, providing brighter lighting for close work. It should be provided over the

Table 11.2   General lighting guidelines for the kitchen

Kitchen size	Incandescent	Fluorescent
Small kitchen (under 75 sq. ft.)	150 watts min.	60–80 watts min.
Average kitchen (75–120 sq.ft.)	150–200 watts total	60–80 watts, min.
Large kitchen (over 120 sq.ft.)	2 watts/sq. ft.	1 watt/sq. ft.

Figure 11-29 Task lighting is needed for the various work areas. (*Cooper Lighting*)

work surfaces (counters) and over the cook and sink/cleanup work centers (Fig. 11.29). If an informal dining area is planned for the kitchen, lighting for this area must also be considered. Table 11.3 summarizes task lighting for the kitchen.

The lighting fixtures for task areas may be mounted under cabinets, on the ceiling, on the walls, or in soffits, and are often pendant, track, or recessed lights. Regardless of where the source is located, it must fall on the area in front of the homeowners to prevent them from working in their own shadows. With well-lighted work areas, time is saved and the possibility of accidents is greatly reduced. Remember to have proper shielding for each light, so no one ever has to look directly at a bare bulb or fluorescent tube.

Decorative (accent) lighting

While general lighting and task lighting provide required illumination for the work areas of the kitchen, decorative lighting focuses attention on an area or

Table 11.3 Summary of task lighting for the kitchen

Kitchen area	Fixture placement	Incandescent lamping	Fluorescent lamping
Counter lighting	*Under cabinets:* Mount fluorescent fixtures as close to the front of the cabinet as possible		Tubes long enough to extend two-thirds length of counter; e.g., 36-inch 30W or 48-inch 40W
	Counters with no overhead cabinets: Hang pendant 24–27 inches above the counter	60–75W for every 20 inches of counter	Same as above
	Recessed or surface-mounted units: 16–24 inches apart, centered over the counter	75W reflector lamp	Two-tube fixture extending two-thirds length of counter
At the range	Built-in hood light	60W bulb	
	With no hood, place recessed or surface-mounted units 15–18 inches apart over the center of the range	Minimum of two 75W reflector floods	Two 36-inch 30W or three 24-inch 20W
At the sink	Same as for range	Same as for range	Same as for range
Eating area	Pendant centered 30 inches above table or counter; multiple pendants over counters 4' or longer	One 100W or two 60W or three 40W or 50/100/150W	

Courtesy of the American Home Lighting Institute, 1992

object, such as artwork, with a concentrated beam of light. Today, fixtures are available from manufacturers for every possible decorating theme or budget and are used to highlight architectural features, to set a mood, or to provide drama. Proper accent lighting should be three to five times ambient lighting levels. Control a light with a dimmer switch, and it can also serve as general lighting. Accent lighting is usually provided by recessed, track, or wall-mounted fixtures. In the kitchen, use it to wash a wall, play up interesting textures, spotlight a unique kitchen fireplace, or dramatize a dining table.

Light bulbs and tubes

It is important to know the general characteristics of light sources to choose types appropriate for each kitchen application. By knowing how to work successfully with each type individually, or with a combination, it is possible to achieve lighting results that are visually comfortable, functional, and beautiful. Light sources can be grouped in categories according to the way they produce light. The key to good lighting starts with the right bulbs.

Incandescent lights Incandescent bulbs have filaments of tungsten that give off light when heated by electric current. The light from incandescent bulbs (clear, frosted, or tinted)—the oldest of our artificial light sources—is warm in color quality and imparts a friendly, homelike feeling to interiors. Under this light source, the warm colors (oranges, reds, brown, etc.) are enhanced, while the cool colors (blues and greens) are subdued. Generally, when the overall atmosphere of a room is on the warm side, the full values of the warm colors are acceptable when lighted with filament bulbs. Inside-frosted standard-service incandescent bulbs, ranging in wattages from 15 to 300, are the most commonly used types. Some of these wattages are available in other finishes. Due to a fine white inner coating in these bulbs, light is distributed over the entire surface, eliminating a bright spot near the center. Tinted bulbs are recommended for locations where the specific effect of a warm or cool mood or atmosphere is desired. In general, tinted light subtly accents like colors and subdues complementary colors.

Incandescent reflector (R-type) and projector bulbs (PAR-type) can be used where a very definite application of either a moderately concentrated spotlight or a more widespread floodlight beam is desired (Fig. 11.30b). Remember that for the best results and the most economical operation, all incandescent bulbs should have the same voltage rating as that of the power supply line. Low-voltage incandescent lighting is especially useful for accent lighting. Operating on 12 or 24 volts, these lights require transformers to step down the voltage from standard 120-volt household circuits.

Fluorescent lights Fluorescent light sources offer a higher light efficiency, cooler operating temperatures, and longer life than incandescent bulbs of comparable wattage (Fig. 11.30a). The average fluorescent tube delivers three times as much light as an incandescent for the same wattage and lasts 20 times longer. For these reasons, they are most adaptable to and very commonly used in custom-designed, built-in installations, as well as in surface-mounted and recessed fixtures.

Fluorescent light sources commonly used in the home range in size from the 4-watt, T-05, 6-inch (15 cm) length to the 40-watt, T-12, 48-inch (122 cm) length. ("T" means tubular and the number equals the diameter in eighths of an inch.) The length of the tube given in all cases is the overall nominal length, including the required sockets.

All fluorescent light sources require a current-limiting and control device called a ballast. The ballast limits the amount of current used by the lamp and provides the proper starting voltage. For the operation of some fluorescent tubes, an automatic switch known as a starter is required, in addition to the normal wall switch. The starter is a small metal can inserted into the fixture body or channel and is a replaceable part.

The choice of fluorescent tubes is simply a matter of deciding first whether a warm or a cool atmosphere is desired for the kitchen. When the kitchen atmosphere and color rendition are of prime importance, the corresponding warm or cool tube would be chosen. Otherwise, warm white or cool white tubes serve if light output is the more important requirement.

Manufacturers have developed fluorescent lamps in many color balances, from cool white to natural, or soft white. Deluxe cool white strengthens all colors nearly equally and gives the best overall color rendition. Daylight tubes strengthen green and blue. Deluxe warm white tubes simulate incandescent

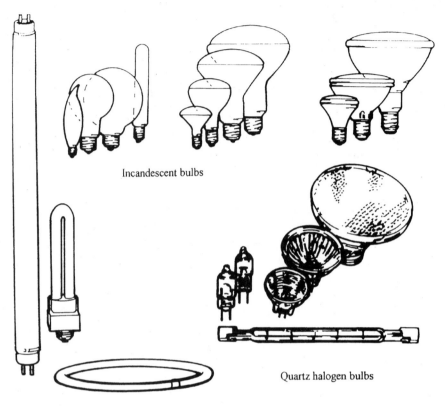

Incandescent bulbs

Quartz halogen bulbs

Fluorescent bulbs

Fig. 11-30a Fluorescent bulbs
Fig. 11-30b Incandescent bulbs
Fig. 11-30c Quartz halogen bulbs. (*Leon E. Korejwo*)

light and is recommended when these two kinds of lights are used in the same room. New types of fluorescent tubes are constantly being introduced. Consult a lighting specialist to keep up with the new lighting sources.

Quartz halogen bulbs Bright white quartz halogen bulbs are excellent for task lighting, pinpoint accenting, and other dramatic accents (Fig. 11.30c). Halogen is usually low-voltage, but it may be standard line current. Halogen bulbs are tiny bulbs that give out a lot of lumens on 12 to 20 volts. It is actually a form of incandescence, but the lamp is filled with halogen, and the tungsten filament is a different design. It produces as much light as an incandescent bulb 10 times its size, consumes half as much power, and lasts up to seven times longer.

The popular MR-16, or mini reflector, bulb creates the tightest beam and creates drama by varying the levels of lighting intensity on objects and areas. For a longer reach and wider coverage, choose a PAR bulb. An abundance of smaller bulb shapes and sizes fit pendants and under-cabinet strip lights. They are used in strings hidden in toekicks for general illumination, under stair steps, in glass-door cabinets to illuminate glassware, on top of wall cabinets, and under wall cabinets for task lighting.

Low-voltage lighting takes a transformer that is usually quite small. The system requires a transformer to convert the normal household current to the lower voltage requirement. Low-voltage systems also require special rheostat dimming controls. Overall, the miniature size, halogen gas, and precise beam result in superior optical efficiency and long life. However, be aware that halogen does have very high heat and a high initial cost.

Tungsten-halogen   Tungsten-halogen is a special type of incandescent lamp that has its filament in a quartz enclosure filled with a halogen gas. Due to a regenerative cycle, the evaporated tungsten is removed from the bulb wall and redeposited on the filament. As a result, lamp blackening is eliminated and lamp life is roughly doubled compared to standard incandescent lamps.

Xenon lamps   Xenon lamps are widely used for their increased brightness and lumen output over similar wedge base incandescent lamps. An excellent alternative to halogen, xenon lamps are cooler with comparable color rendition.

Increased energy conservation is constantly creating new lighting with lower power consumption. Clients may want to know which kind uses less power.

Types of fixtures

Kitchen lighting requirements depend on and vary according to the size of the space, ceiling height, ceiling color, overall kitchen color scheme, finishes used, and the general layout of the kitchen. A room designed with dark cabinets, deep-toned countertops, and a cathedral ceiling requires more light than a white-on-white kitchen. A light color on the ceiling reflects 60 to 80 percent of the light. Light colors reflect nearly twice as much light as deep, dark colors.

Ceiling and recessed fixtures   The easiest way to obtain good general lighting is by using one or more ceiling-mounted fixtures. Recessed housings are installed during "rough-in" construction and are one of the most permanent elements of the kitchen design. Therefore, a lighting plan must be considered early.

Today a wide range of well-designed ceiling fixtures of the surface-mounted or pendant type are on the market. Recessed (downlights) fixtures are the most popular choice in today's kitchens (Fig. 11.31). These fixtures can handle ambient, task, and accent needs, as long as they are fitted with the right baffle or shield. Recessed incandescent downlights should be placed 6 to 8 feet (183 to 244 cm) apart. Use two 48-inch (122-cm) lamps for kitchens up to 120 square feet. Remind your client that the type and size of the shielding material, the position of the fixture relative to viewing angles of occupant, and light distribution are all important. Ceiling-mounted fixtures in the kitchen are more readily seen within one's field of view and consequently should have only half the surface brightness of similar fixtures in utility areas. Similarly, pendant fixtures (directly in the line of sight) should have half the surface brightness of ceiling-mounted fixtures. Most luminous small-diameter pendant fixtures should use 25 watts or less, as they serve only for decorative accent. However, pendant fixtures of very dense or opaque materials may use much higher wattage without causing discomfort.

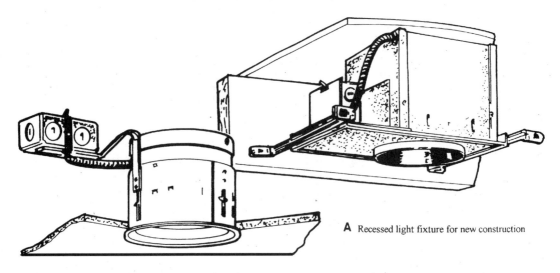

A Recessed light fixture for new construction

B Recessed light fixture for remodeling

Figure 11-31a   Recessed light fixture for new construction. (*Leon E. Korejwo*)
Figure 11-31b   Recessed light fixture for remodeled kitchen. (*Leon E. Korejwo*)

If the homeowner chooses a pendant light, the fixture should be positioned 30 to 36 inches (76 to 91 cm) above the table, so as to not obstruct the view of the diners (Table 11.4). Use a 100-watt bulb, two 60-watt, or three 40-watt. For an eating counter, place the fixture at least 6 feet (183 cm) above the floor. A well-shielded fixture distributing light up, down, and to the sides (by itself) creates a much more pleasing atmosphere in a room than one that produces only downward light. Fixtures suspended from the ceiling and reflecting all or nearly all of their light to the ceiling for redistribution produce very comfortable (though bland) general lighting.

Recessed fixtures provide no light on the ceiling. Because most of the light is directed downward, the wattage and the number of fixtures needed should be at least double the number of surface-mounted or pendant fixtures used to light the same area. Keep in mind that while any down lighting done in recessed fixtures holding R- or PAR-type bulbs may be most decorative in the light patterns and shadows they cast, they are not suitable for general lighting because they are not diffused enough for comfortable visibility throughout the kitchen. Typically, however, downlights follow countertops or shine on the sink or island. Recessed lighting is integral to the architecture, economical, and applicable to all architectural styles and interior designs.

Luminous panels and ceilings   A ceiling can be fully illuminated, or it can have just one panel of light (Fig. 11.32). In the latter case, a panel of louvered or diffusing material—usually about 4 by 6 feet (122 by 183 cm) in size—is mounted on the ceiling or suspended from it, with the light source concealed behind it. Most of the major fixture manufacturers offer kits consisting of plastic diffusers in pans, sheets, or rolls; a suspension system with a gridwork

Table 11.4   Kitchen lighting fixture types and uses

Fixture type	Fixture use
Surface-mount	Mounted directly on the ceiling's surface; distributes very even, shadowless general lighting. To minimize glare, shield with translucent material. To distribute more even lighting, have sockets for several smaller bulbs rather than one or two large bulbs.
Dropped (suspended)	To light a room or table: chandeliers, pendants, globes, and other suspended fixtures. Hang fixtures 12–20 inches (30–51 cm) below an 8-foot ceiling or 30–36 inches (76–91 cm) above table height. Use low-wattage bulbs or a dimmer switch to prevent glare. Shares many of the characteristics of flush-mounted fixtures.
Recessed	Mounted flush with the ceiling or soffit; include fixed and aimable downlight, incandescent or fluorescent bulbs (shielded by plastic diffusers), and totally luminous ceilings. Require more wattage— up to twice as much as suspended and surface-mounts.
Wall bracket	Serve as task or accent lighting. Conserve space in tight quarters. Mounted either high or low.
Track	Used for general, task, accent lighting, or a combination. Track lighting is very versatile—add, subtract, or rearrange fixtures, aiming in any direction desired. To minimize shadows, locate tracks 12 inches (30 cm) out from edges of wall cabinets.
Cornice	Mounted at the intersection of the wall and ceiling; provides the wall with soft, downward light. Mount tube so its center is 6 inches (15 cm) from the wall. Good in rooms with low ceilings. May also be used over windows where space above the window doesn't accommodate valance lighting. Because the wall is emphasized, it gives an impression of greater ceiling height.
Valance	Similar to cornices, but installed lower on the wall, often over a window. A lighted valance above a window provides both up-light, which reflects off the ceiling for overall lighting in a room, and down-light, which accents the window treatment. Should not be mounted closer to the ceiling than 10 inches (25 cm) to avoid annoying ceiling brightness.
Cove	Reflects upward toward the ceiling. Creates smooth, even general lighting or dramatic architectural effects. Ceiling should always be painted flat white or a near-white, natural color. Mount on top of wall cabinets, about a foot below ceiling level.
Under-cabinet	For efficient, simple task lighting, fluorescent or incandescent fixtures are mounted to the undersides of wall cabinets, as near as possible to the front, not back by the wall. Fluorescent tubes and halogen lamps are best because their low profile makes it easier to shield the eyes from direct light. Add shielding strips or valances to cabinet fronts to hide fixtures and prevent glare. Use the longest tubes that fit, and fill at least two-thirds of the counter's total length: ☐ 15W (18 inches/46 cm) for counters less than 24 inches (61 cm) ☐ 20W (24 inches/61 cm) for counters less than 36 inches (91 cm) ☐ 30W (36 inches/91 cm) for counters less than 48 inches (122 cm) ☐ 40W (48 inches/122 cm) for counters less than 72 inches (183 cm)

Figure 11-32 Luminous ceiling panels can fully illuminate the kitchen. (*Leon E. Korejwo*)

of wood, metal, or plastic; and fluorescent fixtures to mount above the suspended luminous panel.

While more than one luminous panel may be used in a kitchen, a single unit is best where it is to be employed as an architectural feature, for example, suspended over an island work center. When used in this manner, the luminous panel should be the same size as the island and about 6 inches (15 cm) deep. If it is hung at least a foot from the ceiling, it will provide good general illumination as well as shadowless light for the work center island.

For shadow-free general lighting throughout the kitchen, the luminous ceiling is the answer. But such a lighting plan is usually only possible where there is a cavity of at least 10 to 12 inches (25 to 30 cm) above the desired ceiling line. Frequently, the cavities between the ceiling joints are used as a part of or the whole space needed for the luminous ceiling fixtures. For uniform lighting, use a minimum of one 40-watt fluorescent tube for every 12 square feet of room area, or one 60-watt incandescent bulb for every 4 square feet of panel. For proper light diffusion, all surfaces of the cavity must be highly reflective and must have a matte texture. White lampholders for incandescent tubes are also needed. In addition, any other materials within the cavity, such as joists, conduits, pipes, etc., should be painted white. The ceiling panels themselves are usually made of translucent plastic, with decorative metal gridwork providing the necessary support. Manufacturers of luminous ceilings can provide design and installation instructions for their particular products.

Track lighting   Of all the kitchen lighting options, track lighting offers complete decorating freedom and provides endless opportunities to achieve function and accent lighting. Track lighting is a concept of accent light that can be as flexible as the imagination. It can be recessed, surface-mounted, or suspended by pendant stems. Most of these movable, adjustable track systems can be swiveled, angled, and pointed in any direction or grouped for every functional design effect. Fixtures come in myriad styles and give the look of built-in lighting without the installation hassle. Tracks mount on ceilings or walls for task lighting at work centers or general kitchen illumination. For task lighting, fit track fixtures with spotlight bulbs; for general illumination, install more diffuse floodlight bulbs. For best illumination, tracks or downlights should be installed at least 2 feet (61 cm) away from a wall. To avoid shadows on the work surface, point lamps directly over the counters or place them so light comes from the side.

Cove lighting   Coves are particularly suited to rooms with two ceiling levels. In these applications they should be placed right at the line where a flat, low-ceiling area breaks away to a higher-ceiling space. The upward light emphasizes this change of level and is very effective in rooms with slant or cathedral-type ceilings. However, the lighting efficiency of coves is low in comparison with that of valances and wall brackets. Coves that are usually mounted high on the wall direct all of their light upward to the ceiling where it is, in turn, reflected back in to the room. The illumination effect produced by cove lighting is soft, uniform, and comfortable. Cove lighting should be supplemented by other lamps and lighting fixtures to give the room interest and provide lighting for tasks (Fig. 11.33).

Valance and soffit lighting   A valance board installed between the cabinet tops and ceiling is an easy and effective way to obtain general kitchen light (Fig. 11.34a). The valance faceboards, which shield a continuous row of fluorescent tubes running the full length of the cabinets, can be simple and unobtrusive, or they can be as decorative and stylish as the imagination

Figure 11-33 Typical cove lighting. (*Cooper Lighting*)

allows. A wide variety of faceboard materials are available that can be trimmed, scalloped, notched, perforated, papered, or painted. Faceboards should have a minimum width of 5 inches (13 cm) and seldom should be wider than 10 inches (25 cm). They should not interfere with the opening and closing of cabinet doors. To obtain proper upward diffusion of the light, the top of the fluorescent fixtures and the top of the faceboard should be at least 10 inches (25 cm) from the ceiling. The faceboard should also be mounted a minimum of 6 inches (15 cm) from the wall or soffit face to permit air to circulate freely around the fluorescent tubes. It is a good idea to tilt the faceboard in 15 to 20 degrees to shield the light source from anyone sitting in the kitchen. The inside of the valance board should always be painted flat white.

Where the space above the wall cabinets is not enclosed and there is at least 12 inches (30 cm) between them and the ceiling, the bare fluorescent channels can be mounted on the top of the cabinets at the rear (Fig. 11.34b). The channel should be tilted on 45-degree blocks for upward diffusion of light. If the fluorescent tubes can be seen from across the room, a 3- to 5-inch (8- to 13-cm) board can be fastened on top of the cabinets at the front. This faceboard can be finished to match the wall cabinets.

The underside of an architectural member is known as a soffit. If the soffit area above the cabinet is at least 12 inches (30 cm) high, it is possible to illuminate the perimeter of the entire kitchen. The continuous two-tube fluorescent fixtures are fastened to the wall at the back of the cavity, and the soffit is faced with either translucent plastic or glass (Fig. 11.34c). The inside of the cavity should be painted flat white or lined with a reflecting material. To be most effective, the ceiling itself should be white or a very light color.

Another way to have an illuminated soffit is to extend it out from the cabinets 16 to 24 inches (41 to 61 cm) and, on the ceiling of this extension, install continuous, two-tube fluorescent fixtures (Fig. 11.34d). The bottom of the extension is then covered with either translucent plastic or glass. The result is a soft, intimate light all around the cabinet walls. This extended soffit method

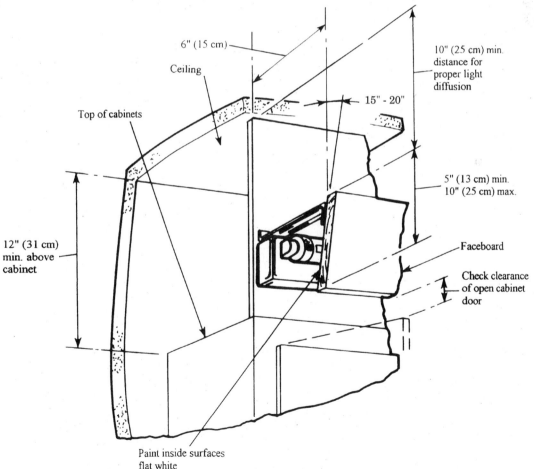

Figure 11-34a Valance lighting installed in front of a cabinet. (*Leon E. Korejwo*)

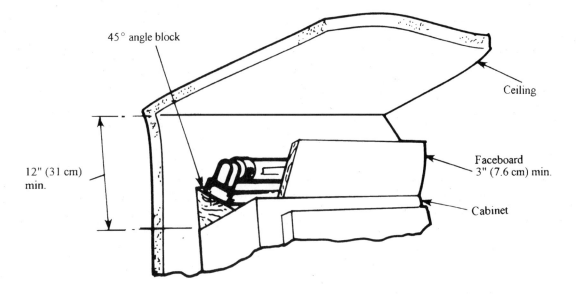

Figure 11-34b Valance light above the cabinet. (*Leon E. Korejwo*)

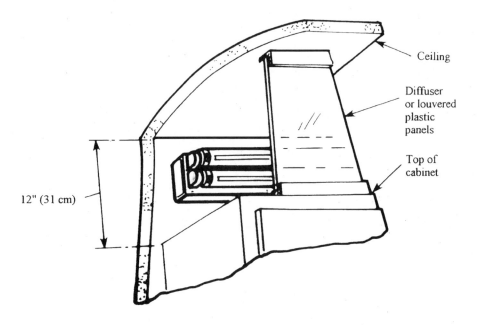

Figure 11-34c Soffit lighting above cabinets. (*Leon E. Korejwo*)

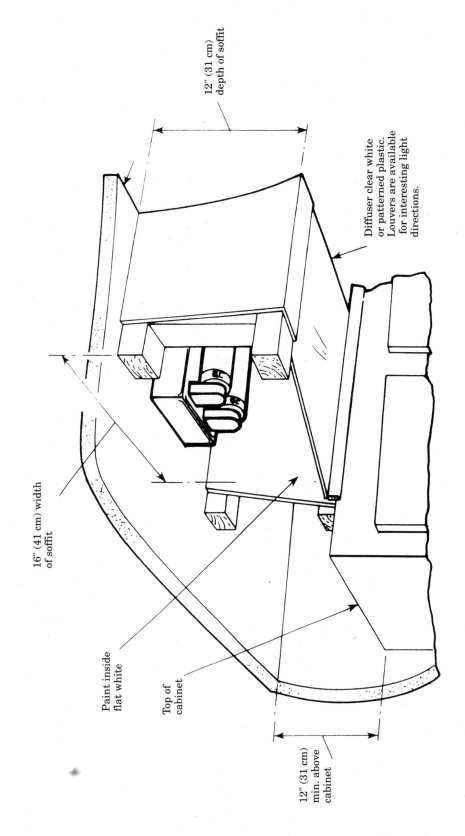

12" (31 cm) depth of soffit

Diffuser clear white or patterned plastic. Louvers are available for interesting light directions.

16" (41 cm) width of soffit

Paint inside flat white

Top of cabinet

12" (31 cm) min. above cabinet

Figure 11-34d Soffit lighting extended in front of the cabinet. (*Leon E. Korejwo*)

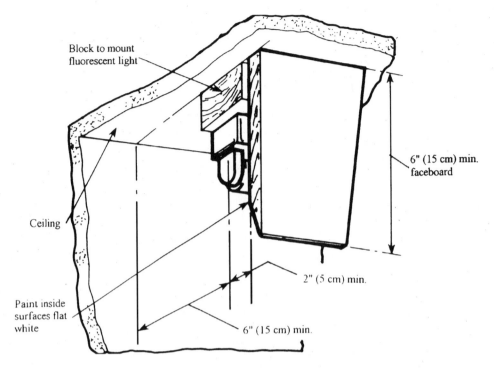

Block to mount
fluorescent light

6" (15 cm) min.
faceboard

Ceiling

2" (5 cm) min.

Paint inside
surfaces flat
white

6" (15 cm) min.

Figure 11-34e Typical cornice lighting. (*Leon E. Korejwo*)

is very similar to cornice lighting that is often used in other rooms of the house to dramatize wall textures and wall coverings (Fig. 11.34e). Also, strips of small lights can be installed in the toekick or soffit.

Under-cabinet lighting   When overhead cabinets are present, the best method of lighting the counter surface below is to fasten fluorescent channels either to the bottom of the cabinets at the front (preferable) or directly under the cabinet on the back wall. The channels may be painted to match the wall, if desired. An opaque shield is recommended, and it may be of metal, wood, or laminated plastic to match the cabinets or countertops. The wattage used depends on the total length of counter areas. Use the longest tube that fits, and fill at least two-thirds of the counter length. These wattages are good for up to 22 inches (56 cm) above the countertop, a height for wall cabinets that is rather uncomfortable for most householders. A two-socket incandescent bracket with 60 watts in each socket is the equivalent for each 3 feet of countertop, but incandescent light is not as popular for under-cabinet applications as fluorescents.

For a right-angle counter with cabinets above, use a fluorescent channel mounted lengthwise under the center of the upper cabinets, with an opaque shielding on both sides of the tube. The tube length and wattage needed are the same as given for under-cabinet installations.

For a right-angle counter with no cabinets above, use either louvered, sur-face-mounted, or semirecessed cylinders or recessed, louvered "high-hats" with 75-watt R-30 floodlights, spaced 20 inches (51 cm) from center to center.

In framed cabinets, the face frame helps shield the eyes from under-cabinet lights. Frameless cabinets do not have a protecting shield, so a 2-inch (5-cm) bar has to be added along the front of the cabinet. Some frameless cabinets have recesses built into the cabinet bottoms where lights can be mounted. Some halogen under-cabinet lights plug into a track attached to the cabinet bottom that shields the eyes.

Cook and sink/cleanup centers   The cooking and sink/cleanup work centers need special lighting. The sink/cleanup center is best lighted by a hanging fixture, soffit lighting over the sink, or valance or cornice lighting that uses fluorescent strips. If there are cabinets above the sink, the area can be illuminated in the same manner as already described for under-cabinet lighting. If the sink has wall cabinets on either side of it, the light source can be mounted in or on the soffit as mentioned earlier. If the sink is not located under a soffit, wall brackets mounted from 14 to 22 inches (36 to 56 cm) above the sink can provide some general upward light, as well as the necessary task lighting. The minimum for such a bracket unit is one 3-watt fluorescent tube or a multiple-socket incandescent in a box at least 18 inches (46 cm) long with 60 or 75 watts in each socket. A surface ceiling-mounted fixture or recessed downlights can also be used. When using a fluorescent surface-mounted fixture, it is usually a good idea to shield it with a faceboard at least 8 inches (20 cm) deep, installed between the cabinets. Recessed downlights should have lenses or louvers to provide the necessary shielding.

Modern ranges usually have their own lighting and ventilating hoods that provide task lighting for cooking surfaces. If the vent hood does not have its own lighting fixtures, one or two sockets can usually be mounted inside it to hold one 100-watt or two 60-watt incandescent bulbs. (To obtain optimum reflectance of light, the inside of the hood should be painted flat white.) Some larger vent hoods contain fluorescent tubes shielded by plastic lenses. The number of tubes and their wattage, generally anywhere from 15 to 40 watts, depends on the size of the hood.

Should additional illumination be needed, the cook/work area can be lighted in many of the same ways as the sink/cleanup center—soffit lighting, a ceiling-mounted fixture, under-cabinet lighting, recessed lighting, or hanging fixtures. When using hanging fixtures, it is an interesting touch to drop a heat lamp from the ceiling over the cutting board, providing both illumination and warmth to keep foods hot while they are being readied for serving.

If the task lighting over the work surface is in the ceiling or soffit, sufficient illumination can be provided by 75-watt incandescent bulbs in fixtures located 2 feet on center or equivalent fluorescent tubes. An equivalent amount of lighting could also be provided by recessed 100-watt incandescent fixtures with inner reflectors or pendant or bullet-shaped fixtures equipped with 75-watt reflector flood lamps. If wall-mounted fixtures are preferred over the sink or range, they should be mounted 14 to 22 inches (36 to 56 cm) above the surface with at least one 30-watt fluorescent tube.

Light for informal dining   While the kitchen is seldom used for formal dining, many people like to eat snacks, brunches, and lunches there. Usually, it is best to provide some kind of fixture, pendant or swag, over the center of the table (or snack bar) to provide downlight that is shadowless yet relatively low in intensity. The fixture can be hung from the ceiling or from a swinging bracket mounted on the wall. Several pendant fixtures may be mounted 16 to 20 inches (41 to 51 cm) apart at the same or varying heights; the bottom of the lowest fixture should be a minimum of 72 inches (183 cm) from the floor. If a dropped fixture is used over the eating area, be certain that the size complements the table and harmonizes in brightness with the rest of the kitchen.

A pulley lamp can be used to allow for adjusting the light level over the table. Be careful, though, in locating the fixture on the ceiling. The furniture arrangement of such things as hutches or buffets can necessitate the placing of the table off center; make sure the light is centered 30 inches (76 cm) over the table. The softness and warmth of incandescent lighting are usually thought to be more comfortable for dining than the brighter illumination of fluorescent lighting.

The lighting levels of the kitchen can be controlled by both switching and dimming. As a bare minimum, the general illumination, the task lighting, and the dining area should have their own separate switches. It is always best, of course, to have each lighting arrangement on its own switch. Be sure to plan and show which switch operates which light source and indicate whether it is a single or multiple switch or a dimmer. Plan to add a three- or four-way switch if the kitchen has more than one entrance. Switches are usually placed 44 inches (112 cm) above the floor on the open (or latch) side of doorways.

For special lighting effects in the kitchen, dimmer controls can be used on both incandescent and fluorescent light sources. This type of control is especially desirable in combination kitchen-family rooms where variable lighting levels would be useful for creating different moods. Several sizes and varieties of dimmer controls are available for incandescent bulbs. The smaller-sized controls can be used in place of a standard switch, which makes installation extremely simple. For larger-sized dimmers, a double box may be needed. For uniform dimming, all tubes controlled by one dimmer should be of the same size.

With states and communities now providing energy-conserving guidelines in their building codes, the Illuminating Engineering Society (IES) developed a residential lighting power budget or limit formula, which is based on well-organized task lighting and the use of energy-efficient lighting equipment. Table 11.5 is a wattage table for residential kitchens prepared in accordance with procedures outlined by the National Conference of States on Building Codes and Standards (NCSBCS) in the lighting power budget (limit) determination procedure. Calculations were based on levels of illumination for specific tasks and for normal vision requirements listed in the IES Lighting Handbook, which was the basis of many of the kitchen lighting suggestions given in this book. Table 11.5 provides an upper limit of power for the budget proposed and should be used as such for developing the actual lighting design. In using the power-budget tables, if dimensions of a kitchen fall in between two categories, use the next higher category. The length of a kitchen is defined as the longer dimension of the room. New motion sensing switches activate the lighting and shut off when an occupant leaves the area, thus lowering power usage.

Table 11.5   **Power-budget table**

Kitchen dimensions (feet)		Power budget
Width	Length	(watts)
4	4–12	280
5	5–11	280
	12 and over	320
	6–9	280
6	10–12	320
	13 and over	360
	7–8	280
7	9–10	320
	11–14	360
	15 and over	400
	8–9	320
8	10–11	360
	12 and over	400
9	9–11	360
	12 and over	400
10 and over	10–11	360
	12 and over	400

Completion

With the majority of the kitchen project complete, the installer must complete all touch-ups, repairs, alignments, and adjustments that are inevitably needed at the end of a project. Typically, this work represents a small portion of the total project. This last portion of the installation project is the most difficult to complete and is also the most crucial in ensuring client satisfaction.

Perhaps the most effective way to complete a job quickly and neatly with a satisfied client is by using a quality control precompletion checklist. Table 11.6 is an example form for a final inspection with the customer. When you feel the project is complete, and you have completed any lists you may have prepared for yourself, as well as for any subcontractors, you should schedule a meeting between the homeowner, dealer or designer, and architect (if applicable). The focus of this meeting is to produce a list of items to be completed before final payment. Any items not on this list that need work or repair may be done under the warranty. The final bill must be paid before any warranty work can be done to prevent the never-ending list where final payment is postponed over and over again as more items are added to the checklist.

Due to excellent quality control during the installation, the checklist does not need to contain more than 10 items. Be aware that the client will point out most of these items. To assure your client that you are committed to the highest level of client satisfaction, you should also add a few items to the list.

As the end of the installation project approaches, you need to step back and review the installation to see if everything is up to the level of quality you have

Table 11.6   Form for final inspection before punchlist with customer

Job name: _____

Job address: _____

Cabinetry	☐ Kitchen cabinets installed properly; enough screws. Must be plumb and level.
	☐ Doors and drawer fronts aligned properly.
	☐ Door and drawer hardware functioning properly.
	☐ Cabinet surfaces free of scratches and nicks.
	☐ Glass or plastic door inserts installed securely and unbroken.
	☐ Accessory items installed as specified.
	☐ Cabinet shelves installed with proper number of chips.
Countertops	☐ Countertops securely attached to base cabinets or means of support.
	☐ Backsplash installed if required.
	☐ Backsplash scribed to wall if irregular wall surface exists.
	☐ Countertop caulked as required.
	☐ Countertop free of surface scratches.
	☐ Solid-surface and stone countertops correctly finished.
	☐ Countertop joints fit tightly and sealed.
Appliances	☐ Dishwasher screwed in; air gap installed.
	☐ Food waste disposer clear and operating properly.
	☐ Water connection made to refrigerator; ice maker functioning properly.
	☐ Range/cooktop installed; all burners functioning properly.
	☐ Oven securely installed in opening; oven and broiler functioning properly.
	☐ Microwave oven securely mounted, if required, and operating properly.
	☐ Hood securely mounted, connected to ductwork, and operating properly.
Electrical	☐ All outlets and switches work; wall plates straight and level.
	☐ All outlet wall plates installed tight to drywall or finish surface.
	☐ Electric panel labeled for new circuits.
	☐ All light fixtures have lamps installed and are working properly.
	☐ Batteries in smoke detector, if required.
	☐ Final inspection completed.
Plumbing	☐ All fixtures working properly.
	☐ Fixtures not chipped or showing signs of finish damage.
	☐ Fixtures secure to wall, floor, or countertop surface.
	☐ Aerators and escutcheon plates installed.
	☐ All gas connections working properly.
	☐ Final inspection completed.
HVAC	☐ Diffusers installed and working; return air grille installed.
	☐ System operable; thermostat properly installed, if required.
	☐ Filters replaced after construction completed.
Drywall/paint/ wallpaper	☐ Drywall finished properly and sanded.
	☐ No loose nails or nail pops apparent.
	☐ No damage to drywall from electrical or plumbing work.
	☐ Painting complete; sufficient coats applied.
	☐ Wallpaper installed, if required, and properly adhered.
Finish flooring	☐ Grout cleaned from tile.
	☐ Mastic cleaned from sheet goods and vinyl tile.
	☐ Base materials (shoe molding, vinyl base, etc.) installed.
	☐ Seams in sheet flooring sealed correctly.
	☐ Thresholds and reducer strips installed.
General	☐ Job broom-clean or maid-clean as appropriate.
	☐ Windows and newly installed glass cleaned.

☐ Trash removed.
☐ Countertops and cabinet surfaces, interior and exterior, cleaned.
☐ Yard and staging or storage areas returned to good order.
☐ Owner received all warranty and operating literature for materials and equipment installed.

Comments:

_____

_____

_____

_____

_____

_____

_____

_____

_____

Signature _____ Date _____

Home Tech Information Systems, Inc.

established. You, as the installer, are responsible for deciding whether work is acceptable or is defective. This step is extremely necessary.

You need to completely review the cabinet installation, all of the flooring, drywall, and carpentry work, as well as the electrical, plumbing, and HVAC work done by the subcontractors. Check every outlet and light switch, every door and every drawer to be certain they operate correctly and are free from visual defects. Although you can come back to take care of any problems after a project is finished, it is much easier to get mistakes corrected while the crew is still onsite. You and your client both want the job finished quickly and professionally. To make sure all goes well, Table 11.7, prepared by Home Tech Information Systems, Inc. is a list of things to check before making the final payment. Preparation for this final inspection is helpful. This list helps you review every segment of the installation job. In addition, keep a written list of items that need attention as you see or think of them. When the items on this list have been completed, you are ready for the quality control pre-completion punchlist with the client

Final repairs

Immediately take care of any items that need to be corrected. Schedule any subcontractors who need to return to the job. Since they may not be able to handle the job right away, give the subcontractors a deadline for completion of the work. Although it seems like common sense, assemble all items for the subcontractors into one list so they can take care of all of the items in one visit.

Be aware that nicks and scratches on countertops and cabinets are unavoidable. Plan on making a few surface repairs to cabinets on every kitchen

Table 11.7   Quality Control Pre-Completion Punchlist

Sheet _____ of _____

Owner _____

Address _____

City, State, Zip _____

Telephone _____

Job location
_____
_____
_____

List of items to be completed prior to final payment of $ _____

_____

_____

_____

_____

_____

_____

Amount to be retained in escrow pending completion of above items: $ _____

It is agreed that when the above list of items is completed, approval for final payment will be authorized. Any omitted or defective items noted after final payment will be covered by the warranty.

_____   _____
Submitted by (Installer)   Date   Owner   Date

_____   _____
Dealer/Designer   Date   Owner   Date

Home Tech Information Systems, Inc., 1997

installation project. Countertop and cabinet dealers sell stains and touch-up kits to match the finish of the cabinetry and nicks and scratches can be repaired so that they are unnoticeable. Never try to blame a subcontractor for having scratched a cabinet. It is more efficient to make the repair with the touch-up kits rather than trying to place blame. Be sure to order the touch-up kit from the cabinet manufacturer when the cabinets are ordered. Do not wait until it is time to make repairs to try to obtain the touch-up kit.

Final adjustments to cabinet doors, drawers, and hardware may also need to be made at this stage of the project. Anything more than minor adjustments should have been done at an earlier point in the installation. Frameless cabinets with fully adjustable hinges make door alignment relatively easy, but framed cabinets are difficult to adjust once they are installed and the countertop attached, because door and drawer misalignment in framed cabinets is generally due to the twisting or racking of a cabinet unit that was not installed plumb and level.

Check to make sure that all cabinet hardware is properly aligned and securely attached to the cabinets. Although this detail may seem minor, imagine how unhappy your clients will be when they open a drawer and the handle comes off. In addition, be certain that drawers slide out smoothly, and the safety features that keep the drawer from being pulled all the way out are working correctly.

Final cleanup

Cleanup procedures need to be reviewed at this stage of the installation. Upon completion, you are responsible for leaving the kitchen spotless and ready for use. The installation estimate should include a line item for a final professional cleaning of the job. Several items need to be taken care of before the cleaning service can begin:

1. All debris must be removed from the job site, including areas that were used for staging and assembly, interior and exterior. If installers have their own cleanup and trash truck, the driver must keep time not only for cleaning up, but also for the time spent traveling to dispose of the refuse.

2. Remove all canvas runners and polyurethane covers. If possible, these can be saved for the next job.

3. If you relocated any of the client's personal possessions during the installation project, they should now be returned to their original locations.

The professional cleaning service should clean all appliances, floors, windows, light fixtures, plumbing fixtures, countertops, cabinets (inside and out), as well as any areas adjacent to the kitchen that may have been affected by the installation, i.e., dusting, wiping down woodwork, and shampooing carpets. When this work is completed, the kitchen must be ready to use.

Client education

Once the major appliances have been installed according to the manufacturer's instructions, the installer is usually provided with an excellent chance to answer questions about the operation and use of the appliances. In addition to replying to direct questions, the kitchen remodeler can offer information that will increase the value of the appliance to the customer and perhaps avoid a future service call.

Today's kitchen equipment and appliances are more advanced than ever. Most of them work the same as they did in the past, but all of the buttons, indicator lights, and computer readouts can be overwhelming to the homeowner. To help your clients understand the features of their equipment and appliances and learn how to use them safely, you should schedule time for instruction. Therefore, you should be familiar with and feel comfortable with the appliance and equipment operating techniques. It is important to review the manual and warranty papers with the owner.

Another important factor to be covered is the routine maintenance and cleaning of appliances and equipment. If sections of the oven or range can be disassembled for cleaning, let the client know. Try to answer some of the inevitable questions that every client with a new kitchen wants to know. Your experience will tell you what some of these commonly asked questions are. Always present this information in a positive way.

At this point, cabinet maintenance is another important topic to be addressed. Make sure that the clients are aware of the type of finish on the cabinets and that they know how best to care for it. Enamel and polyurethane finishes can be wiped clean, but they also withstand cleaning with a mild, nonabrasive cleaner. Streaks on high-gloss surfaces can be prevented when cleaning by adding a little vinegar to the cleaning water.

Woods and veneers with natural-looking, penetrating finishes can be waxed to provide added protection against stains. These finishes should be polished periodically to renew their luster, and wax should be applied once or twice a year. Nonpenetrating wood finishes, such as polyurethane, can be wiped clean with a damp cloth. Wax should not be used.

Decorative laminates and melamine cabinet finishes can be wiped with a wet cloth or a sponge. These surfaces can handle most nonabrasive cleaning products without surface damage. However, clients should understand that water should not be left on the edges of laminate cabinets or near seams where it could penetrate into the substrate and cause the laminate to lift as the core material expands. Wax or oil-based polishes are never to be used on a laminate finish.

As a final word on the subject of customer education, it must be pointed out again that it is the responsibility of the kitchen installer to establish and maintain cordial relations with customers. If you are able to establish pleasant relationships, the associations with your customers will not only be more agreeable, but immeasurably more effective.

Post-completion meeting

Some installers find it beneficial to set up a post-completion meeting with the dealer or designer to review the installation. This meeting can be held a few days after the kitchen installation project is completed. It provides an excellent opportunity to discuss areas of the work that you feel were successful or those which may have been problematic (be constructive). The communication should help to strengthen your relationship with the dealer or designer.

The following terms are used in the planning and remodeling procedures necessary for the installation of a good kitchen.

ANSI American National Standards Institute—an association that sets construction and performance standards.

ASME American Society of Mechanical Engineers.

ASTM American Society for Testing and Materials, a U.S. agency universally recognized for testing methods and performance criteria of materials, including plumbing pipe, fittings, and solvent cements.

accountability Responsibility for one's actions or work.

actual dimension The true size of a piece of lumber, after milling and drying; see also *nominal dimension.*

air gap In a water supply system, the distance between the faucet outlet and the flood rim of the basin it discharges into. Air gaps are used to prevent contamination of the water supply by back-siphonage. *Air gap* is also used to describe a unique fitting installed to prevent back-siphoning in residential dishwasher drain lines.

ampere (amps) The measurement of the rate of flow of electricity.

appliance The range, refrigerator, dishwasher, garbage disposer, trash compactor, and kitchen sink are considered major or large appliances. Small or portable appliances include toaster, blender, mixer, coffee maker, can opener, etc.

apron The trim board placed immediately below the window stool.

asbestos A material comprising several minerals used as a noncombustible, nonconducting, or chemically resistant construction material.

backflow The term used for negative water pressure.

back panel The piece of material making up the back of a cabinet.

back rail The top and bottom horizontal members on the back of a cabinet. They are used for mounting the cabinet to the wall with screws.

backsplash The portion of a countertop extends up the wall at the rear of a base cabinet.

bar graph A graph or chart that lists categories of work down the left side and dates for the duration of the project across the top of the schedule. The projected start and completion of each category of work is plotted on the chart.

baseboard The finishing board covering a finished wall where it meets the floor.

base cabinet Those cabinets that rest on the floor and support the countertop to provide a working surface; commonly has a single drawer over a single door.

base corner filler A special accessory item used to turn a corner in the most economical way.

base drawer unit A base cabinet composed entirely of drawers.

base drawers and bread board A base cabinet composed of several drawers and a bread board.

base end panel An accessory item used to provide a finished end next to items such as dishwashers and lazy Susans.

base filler An accessory used to fill spaces resulting from odd wall dimensions that cannot be filled using standard base cabinets.

base pantry A deluxe base cabinet that includes a bread board, bread box, and cutlery divider.

base slide-out shelves A base cabinet with shelves that can be pulled out of the inside of the cabinet.

base tray unit A base cabinet designed to accommodate large items, such as cookie sheets, trays, etc. This cabinet contains no shelves or dividers.

beam A horizontal loadbearing part of the building.

bearing capacity How much weight a floor can support or bear.

bearing wall A wall carrying more than its own weight, usually supporting floor or ceiling joists.

blind cabinet A cabinet that has a shelf, pull-out, or swing-out apparatus to provide access to a corner.

blind corner base A base cabinet used to turn a corner and use all available storage space.

blind corner wall A wall cabinet used to turn a corner and use all available storage space.

blueprints A general term for a set of plans for a project.

bottom panel The material used in the bottom of a base or wall cabinet.

branch Any part of a piping system other than a riser, main, or stack.

bread board A solid maple board included in certain base cabinets used as a cutting surface.

Btuh British thermal units per hour; a measure of heating or cooling.

building drain The lowest house piping that receives discharge from waste and soil stacks and other sanitary drainage pipe and carries it to the building sewer outside the house.

building official The person who is legally responsible for enforcing building codes.

butt joint A joint framed by butting the ends of two pieces together without overlapping.

CFM Cubic feet per minute.

CKD Certified kitchen designer. To qualify to use the initials CKD after his or her name, a designer must pass a rigid examination and submit affidavits of competence and integrity to an accrediting body known as the Council of Certified Kitchen Designers of the AIKD.

CPM (critical path method) schedule Type of schedule that uses a diagram or network to show the sequence and interrelation of the categories of work for a project.

cantilevered Two beams or trusses that project from piers toward another that when joined, directly or by a suspended connecting member, form a span.

cast iron A manufacturing process used to mold metal when it is so hot it is in a liquid state.

cast polymer A fixture and surfacing material created by pouring a mixture of ground marble and polyester resin into a treated mold, where curing takes place at room temperature or in a curing oven.

center stile The vertical member that separates doors on two-door base and wall cabinets. Also used on blind corner wall and base cabinets.

certificate of occupancy A written certificate issued by the local building official at the completion of the project that permits owners to occupy the home or a portion thereof.

certified cabinet A cabinet that, through testing, has been verified as meeting the quality and performance requirements of the standards program of the NKCA.

change order An amendment to the original contract to perform changes or additional work.

check-valve A one-way valve for water flow.

circuit breaker A safety device that keeps the electrical branch circuits and anything connected to them from overheating and catching fire.

cleanout Accessible opening in the drainage system used to provide access for removing obstructions.

clearance space A term used to describe the space necessary for the safe and convenient use of appliances, working surfaces, and storage, and for passage in the kitchen.

close grain Small, closely spaced pores or fine texture.

concrete slab Used as a base for building when there is no basement or crawl space.

conduit A hollow metal tube that contains electrical wiring.

convection heat Air heated by an element and then circulated.

coped cut A profile cut made in the face of a piece of molding that allows for butting it against another piece at an inside corner.

corner block The piece of wood used to square and strengthen the corners of base cabinets.

corner cabinet A cabinet designed to fit specifically in a corner.

counter frontage The measurement of the counter along its front edge.

countertop fasteners The mechanical items used to secure the countertop pieces together.

crawl space A 2- to 3-foot-high space under the floor.

credentials Evidence of training and education that establish an installer as a qualified professional.

cubic footage A measure of volume, the product of length times width times height.

customer analysis An examination of a customer to determine if the customer is likely to be difficult to work for and cause the installer to lose money on the project.

cutlery divider An accessory item used in certain cabinet drawers to provide separate storage areas for silver and cutlery items.

dado joint A joint formed when the end of one member fits into a groove cut partway through the face of another.

dead load Weight on the floor that is not moving.

diagonal sink front A cabinet front designed to enclose a sink in a corner area.

diagonal wall cabinet A wall cabinet designed to turn a corner on an angle.

diffuser A device used to deflect air from an outlet in various directions. Also a device for distributing the light of a lamp evenly.

dimensional stock Solid lumber pieces of specified length, width, and thickness.

dimmer control (rheostat switch) A device used to vary the light output of an electric lamp.

dishwasher front An accessory item used to cover a dishwasher door. It is often finished to match the cabinets.

door cabinet component used to cover an opening in the cabinet base frame. Also an entryway to the kitchen.

door frame The outside portion of a frame-and-panel door.

door panel The inside portion of a frame-and-panel door.

dovetail A projecting part that fits into a corresponding indentation to form a joint.

drain Any pipe that carries wastewater or waterborne wastes in a building drainage system.

drainage system All the piping that carries sewage, rainwater, or other liquid wastes to the point of disposal or sewer.

drain-waste-vent (DWV) system A system that carries water and waste out of the house.

drain stack The portion of the plumbing in the kitchen used to remove wastewater.

drawer cabinet A cabinet that features two, three, four, or five drawers.

drawer front   The exposed front portion of the drawer box. Same as drawer face.

drawer pull   The hardware on a drawer used to pull it open.

drawer suspension   The mechanism used to guide the drawer box when it is being opened and closed.

elbow   Fitting that changes the angle of two lengths of pipe.

electrical system   Includes the electric meter, the wiring, the receptacles, switches, and lights, to provide electricity.

elevations   Drawings of the walls of a kitchen, made as though the observer were looking straight at the wall.

encapsulation   In reference to asbestos, wrapping asbestos material in a material or paint that prevents loose fibers from floating in the air.

end panel   The material that makes up the sides of a cabinet.

endsplash   The portion of the countertop used to protect walls and tall cabinet sides where the countertop joins them.

equipment   The appliances, plumbing fixtures, storage units, and counters used in a kitchen.

escutcheon   A protective plate used to enclose a pipe or fitting at a wall or floor covering.

exhaust system   A system in which air is captured by a ventilation unit and is exhausted outdoors.

face frame   The exposed front portion of the cabinet, which provides shape and strength to the cabinet.

fascia   The front of a soffit area.

faucet   Fixture valve that opens and closes the flow of water to the fixture.

fiber-core   Sheet core material made from compressed wood fibers that have been impregnated with a waterproof phenolic resin. Density of material develops great dimensional stability and resistance to warping and twisting. Core is veneered on both sides for use as doors, drawer panels, and overlays.

filler   An accessory used to fill spaces resulting from odd wall dimensions that cannot be filled using standard cabinets.

filter   Device for removing sediment from water and improving taste and odor. Sometimes used in a potable water system.

finish coat   The final covering applied to cabinets to provide a tough durable surface.

finished paneling   Prefinished sheets that can be used to provide a matching surface for walls, cabinet ends, and peninsula backs.

firestop   A 2 by 4 nailed horizontally between studs to inhibit the spread of a fire between uninsulated stud cavities.

fitting   The term used for any device that controls water entering or leaving a fixture. Faucets, spouts, drain controls, water supply lines, and diverter valves all are fittings.

fixture supply   Water supply pipe that connects a fixture to a branch water-supply pipe or directly to a main water-supply pipe.

flat or slab doors   Doors made of flat pieces of lumber, particleboard, or plywood with a veneer surface.

floor plan   The horizontal section of a kitchen showing the size, doors, windows, etc., in the walls.

flowchart   A written, week-by-week schedule of the crucial tasks to be accomplished on a job.

flue gas   A toxic gas produced during the combustion of fossil fuels.

flush-mounted   A method of installation whereby a sink is installed completely even with the surrounding deck.

frame construction   A method of construction where thin component parts form the sides, back, top, and bottom of a cabinet.

framed doors   Doors made of laminate or wood that can be of a slab configuration, with a wood, thick PVC edging, or a metal frame around the doors and drawers.

frame-and-panel door   A type of door constructed of several separate pieces fastened together.

frameless construction   A method of construction where the core material sides of a cabinet are connected with either a mechanical fastening or dowel system.

frontage   A measure equaling the horizontal dimension across the face of an appliance, base cabinet, counter, or wall cabinet. To be credited as having *accessible frontage,* a component must be directly accessible from the area located in front of the component. In accordance with this definition, cabinet space (base or wall) and counter surface located in corners are not credited in counting accessible frontage.

full-shelf option   Base cabinets are manufactured with half-shelves standard. A full shelf can be added at an additional cost.

furring   The building out of a wall or ceiling with wood strips.

furring strips   Narrow strips of nonstructural wood.

fuse   Safety devices that keep electrical branch circuits and anything connected to them from overheating and catching fire.

grain   Pattern arrangement of wood fibers.

granite   An igneous rock with visible coarse grains, used as a surfacing material.

ground-fault circuit interrupter (GFCI)   A device that monitors the electrical circuit at all times and, upon detecting a ground fault (a leakage on the line to ground), removes power to the circuit.

ground wire   A bare copper wire or a coated green wire that does not carry current.

gusset   Brace installed in corners of base cabinets to provide strength and a means of attaching the countertop. Also used in peninsula wall cabinets to ensure positive installation.

hanging strip   A portion of the cabinet back through which screws are attached to the wall.

header   The framing component spanning a door or window opening in a wall. A header supports the weight above it and serves as a nailing surface for the door or window frame.

heat gain   Heat that is gained from the sun's rays.

hinges (self-closing)   A type of hardware used to attach doors to the face frame. They are spring-loaded and, as a result, close automatically if the door is partially closed.

hinging   A term applied to single-door cabinets to identify whether the hinges attaching the door to the face frame are on the right or left side.

hollow core   A panel construction with plywood, hardboard, etc., bonded to both sides of a frame assembly.

independent contractor   A self-employed person offering services to other companies. To be classified as an independent contractor, an individual must be truly independent from the companies for which the person does work.

installation delivery system   An established system for providing installation services that describes the roles of manufacturers, distributors, retailer/dealer/designers, and installers.

integral   A method of installation whereby a fixture is fabricated from the same piece of material as the countertop material.

island   A cabinet or group of cabinets that are freestanding (not attached to a wall).

island corner base cabinet   A base cabinet used to form an island.

jamb   The inside vertical finished face of a door or window frame.

joists   The framing members that are the direct support of a floor.

joist span   The distance a joist is allowed to stretch between walls.

junction boxes   Rectangular, octagonal, or circular boxes made of plastic or metal, which house connections between wires.

kiln-dry   To remove excessive moisture from green lumber by means of a kiln or chamber, usually to a moisture content of 6 to 12 percent.

L shape   One of the basic kitchen shapes where two sides join to form an L shape.

laminate   A surfacing material used on wall areas, countertops, cabinet interiors, and cabinet doors.

lamp   An artificial light source, such as an incandescent or halogen bulb or fluorescent tube.

lap joint   Two pieces of dimensional stock overlapping each other and bonded together.

layout   A drawing showing how cabinets and appliances will be arranged within a kitchen.

lazy Susan corner base cabinet   A base cabinet, installed in a corner, that contains revolving shelves.

lead installer   An installer who has been trained to handle most installation projects as a one-person crew and given responsibility for the job from start to finish.

ledger   A horizontal strip (quite often wood) used to provide support for the ends or edges of other members.

level   Exactly flat in the horizontal plane, or the instrument used to determine horizontal flatness.

liberal kitchen   A kitchen that meets liberal-ample cabinet space requirements.

liberal-minimum kitchen   A kitchen that meets limited-ample or liberal-minimum cabinet space requirements.

lineal footage   A measurement of distance of lengths, such as the distance around the perimeter of a room.

line-a-wall   Another name for a one-wall kitchen, which positions all work centers and appliances along one wall.

lintel   The horizontal structural member supporting the wall over an opening.

live load   The load in a room from people walking on the floor.

loadbearing walls   Walls on the exterior or interior of the home that carry the load of the roof down to the foundation.

M₂   Square meters.

magic triangle   A technique used in a kitchen design to ensure efficiency.

main   Principal pipe to which branches are connected.

marble   Recrystallized limestone. A brittle stone used for countertops and flooring.

mechanical systems   Include heating and cooling equipment, any fans that provide ventilation, and the water heater.

meter   Measurement device to measure utility usage such as electric, gas, and water service.

millwork   Finished woodwork, machined, and partly assembled at the mill.

minimum kitchen   A kitchen that meets the limited-minimum requirement of cabinet space.

model code   A standard building code published by one of several model code organizations.

modular   Pertaining to a module or a standard or unit of measurement. Kitchen cabinets are generally manufactured in 3-inch modules.

molding   Prefinished pieces of various shapes used to improve visual appearance of areas where cabinets meet walls, ceilings, etc.

monorail   Metal center guide or rail for supporting and guiding a drawer roller.

mortise   A notch or hole cut into a piece of wood to receive a projecting part (tenon) shaped to fit.

mounting rail   Top and bottom horizontal members on back of cabinet, used for mounting cabinet to the wall with screws.

mullion   A vertical piece that separates and supports windows, doors, or panels.

muntins   The small members that divide the glass in a window.

NKCA   National Kitchen Cabinet Association—an association of cabinet manufacturers.

NKBA   National Kitchen and Bathroom Association.

nailer   A structural support.

nominal dimension   The stated size of a piece of lumber, such as a 2 by 4 or a 1 by 12. The *actual dimension* is somewhat smaller.

obstructions   A term applying to exposed obstacles, such as pipes, in a kitchen area.

on center (OC)   A phrase used to designate the distance from the center of one regularly spaced framing member to the center of the next.

one-wall   One of the basic kitchen shapes. It is composed of a single row of cabinets.

open-end shelves   An accessory attached to the end of base and wall cabinets to provide exposed shelf space. Primarily used for decorative purposes.

open grain   Large pores or coarse texture.

out-to-out measurements   Measured from one outside edge of a window, opening, or space to the other outside edge.

oven cabinet   A tall cabinet designed and engineered to house a built-in oven.

overlay   Decorative panel fixed to a door or drawer panel surface rather than inlaid.

panel doors   Doors that have a frame composed of two horizontal rails and two vertical stiles, with a thinner panel floating in between.

partition walls   Walls that divide spaces within the home but are not loadbearing.

peninsula   A group of cabinets that extend at right angles to a wall to create a divider. A peninsula is usually used to separate the food preparation from the dining area.

peninsula base cabinet   A base cabinet used to create the base extension of a peninsula.

peninsula cabinet   A wall cabinet that is accessible from two sides and is installed above a cabinet that juts out into the room.

peninsula blind corner wall cabinet   A base cabinet used to turn a corner on an angle and begin the extension of a peninsula. This cabinet usually has access from two sides.

peninsula diagonal wall cabinet   A wall cabinet used to turn a corner on an angle and begin the extension of a peninsula. This cabinet usually has access from two sides.

peninsula wall cabinet   A wall cabinet used to create the upper portion of a peninsula extension.

pie-cut cabinet   A type of corner cabinet that abuts two walls, with both sides usually, but not necessarily, equal.

pipe, fixture supply   The pipe bringing water from behind a wall to a fixture.

pipe, vent   The pipe installed to provide air flow to or from a drainage system or to provide air circulation within the system.

pipe, waste   The line that carries away the discharge from any fixture except toilets, conveying it to the building drain.

plumb   A term applying to a perfectly vertical measurement— straight up and down. The vertical equivalent of *level*.

plumbing appliance   Special class of plumbing fixtures, such as water heaters, water softeners, filters, etc.

plumbing code   A set of rules that govern the installation of plumbing.

plumbing fixture   A receptacle or device that demands a supply of water, either temporary or permanently connected to the water supply system. It discharges used water or wastes directly or indirectly into the drainage system.

plumbing system   Includes potable water supply and distribution pipe; plumbing fixtures and traps; drain, waste, and vent pipe; and building drains, including their respective joints and connections, devices, receptacles, and appurtenances within the property lines of the premises. It also includes water-treating equipment, gas piping, and water heaters and their vents.

plywood   A panel made of layers (plies) of veneer bonded by an adhesive. The grain of adjoining piles is usually laid at right angles, and an odd number of piles are bonded to obtain balance.

preconstruction conference   A meeting attended by the retailer/dealer/designer, installer, and customer, among others, prior to the start of construction, for the purpose of informing the customer what is to occur during construction.

preliminary preparation   Steps to be taken in advance of cabinet installation.

pressure-treated wood   Lumber and sheet goods impregnated with one of several solutions to make the wood virtually impervious to moisture and weather.

productivity   The level of efficiency in project production; increased productivity can create increased profit.

pull   Knob or handle on doors and drawers.

pullman   Another name for a one-wall kitchen, which positions all work centers and their appliances along one wall.

quality control   The process of working to ensure that projects are completed in a quality manner in keeping with established industry standards and customer requirements.

RPM   Rotations per minute.

R-value   A rating system given to insulating material that rates heat-retention capability.

racking   Twisting or warping out of square, due to uneven floors, walls, or pressure being applied as a result of forcing into position.

radiant heat   Heat transferred through invisible electromagnetic waves from an infrared energy source.

radon   An invisible pollutant that comes from trace amounts of uranium in the soil.

rail   The horizontal members of a cabinet face frame.

range   A cooking appliance. It can be free-standing, drop-in, or built-in.

range cutout   The opening cut into the countertop so that the cooktop unit fits below and rests on the countertop.

range panel   A base cabinet front engineered to accept drop-in ranges.

reveal   The offset or exposure of the jamb edge that is not covered by the trim.

right of recision   A customer's right under federal law to cancel a transaction within three business days of contract signing.

rimmed   A mounting method whereby a sink sits slightly above the countertop with the joint between the sink and the countertop concealed by a metal trim.

roll-out shelf cabinet A base cabinet unit with interior shelves installed on drawer glides.

rough-in The installation of parts of the plumbing system that must be done before installing of the fixtures, including drainage, water supply, vent piping, and installation of the necessary in-the-wall fixture supports. Usually done before the walls are closed in. The term is also applied to the dimensions used for roughing in.

roughing-in The framing stage of a carpentry project. This framework later is concealed in the finishing stages.

rough opening The rough-framed opening into which window and door frames are installed.

sash The framework that holds glass in the window.

sash size The overall measurement of a window sash.

scribe allowance Small extension of sides beyond the back and frame stiles beyond sides for trimming to ensure proper fit.

scribing A technique used to adjust for wall irregularities. Scribing is used in fitting countertops, molding, and fillers.

sealer A protective coating applied to the cabinet during the finishing process.

self-closing A term applied to hinges that assist in closing a door.

self-rimming A mounting method whereby a sink is designed to sit on top of the countertop.

separate oven An oven unit that is not part of a range.

service box The control center for the home's electrical service.

shelf clip An item used to support the adjustable shelves in wall cabinets.

shim A small angular piece of wood used to compensate for unevenness in wall and floor surfaces.

shimming A technique using small wood pieces to compensate for unevenness in wall and floor surfaces.

sink/cooktop cabinet Base cabinet with shallow top drawers or a tilt-down front that houses a plastic or stainless steel container.

sink cutout The opening cut into the countertop so that a sink unit fits below and rests on the countertop.

sink front A cabinet front, with no sides, back, or bottom, used as a low-cost substitute for a sink base cabinet.

sink-front bottom A piece of material used to cover the floor area behind a sink front.

sink/range base cabinet A base cabinet designed to house and support either a sink or cooktop unit.

slab-on-grade A foundation or floor created by pouring concrete directly on prepared ground.

sleepers A series of 2 by 4s laid flat on a slab, used under a finished floor made of wood strips.

soffits (bulkhead and fascia)   The area between the cabinet top and ceiling. When this area is closed off, it appears to be part of the ceiling.

softwood lumber   Produced from coniferous or evergreen trees that have needles or scalelike leaves and remain green throughout the year.

specifications   Descriptions, in words, of the materials to be used on a project, as well as how to install them.

square   At a right angle or perpendicular to an adjacent surface.

square footage   A measure of area equal to the product of length times width.

stack   General term for the main vertical pipe of soil, waste, or vent pipe systems.

stack, vent   Vertical pipe providing circulation of air to and from any part of the drainage system.

staging area   A space that is reserved for the receiving, unpacking, and initial assembly of materials for a project.

stile   The vertical members of a cabinet face frame.

storage   A general term used to denote any space that is used for storing food, utensils, small appliances, etc., in the kitchen. Generally the refrigerator space is excluded from this term, although in reality it provides cold storage. Most storage space is provided in cabinets, but other special storage devices such as closets, pantries, etc., may be employed.

strip kitchen   Another name for a one-wall kitchen, which positions all work centers and their appliances along one wall.

structural system   The shell of the house: the foundation, walls, and roof.

studs   Wall framing members, generally 2 by 4s to which cabinets are to be attached.

style   A term referring to the design and decorator effect of the cabinet door.

subcontractor   A person or business that agrees to render services or provide materials necessary for the performance of a project under contract to another contractor, retailer, dealer, designer, or installer.

subrail   Horizontal interior framing member, not a part of the front frame assembly.

substrate   A material that lies below a finish material and provides support for the finish material.

subtraction method   A technique used to check and ensure layout accuracy.

supply branches   The two sets of pipes that run horizontally through the floor joists or concrete slab and deliver water from the supply main.

supply risers   Water supply branches that run vertically in the wall to an upper story.

surface cooking unit   Cooking units that are not part of a range, but are usually installed in a counter work surface. Same as a cooktop.

tailpiece   The short pipe leading from a faucet or sink drain.

tall cabinets   Large cabinets that are 84 inches high. These cabinets are designed to supply a large quantity of storage space or house built-in oven units.

tee   A T-shaped pipe. Joins a branch pipe into a run of pipe.

tenon   A projecting part cut on the end of a piece of wood for insertion into a mortise to make a joint.

top panel   A piece of material making up the top of a wall cabinet.

toe space   The open area at the lower portion of the base cabinet. This space allows a person to stand close to the cabinet without kicking it.

trap   Bent pipe section or device that holds a water deposit and forms a seal against the passage of sewer gases from a waste pipe into the room.

trim   The finishing frame around an opening such as doors and windows.

trusses   An assembly of members, such as beams, which frame a rigid framework.

turnkey installation services   Installation services provided as an integrated part of the sale of a kitchen project.

twin-track hardware   The item that guides and directs the drawer when it is being opened and closed.

U shape   One of the basic kitchen shapes where three rows of cabinets are joined together to form a U.

underlayment   Nonstructural subflooring that lies under the sheet flooring. Also, a form of particleboard that has a low density and low resin content.

under-mounted   Installed from below the counter surfaces.

union   A fitting that joins pipes with a method for future disassembly without cutting them apart.

unit dimension   The overall outside dimensions of a manufactured window.

utility cabinet   A tall cabinet providing large amounts of storage space.

utility filler   An accessory used next to tall cabinets to fill spaces resulting from odd wall dimensions that cannot be filled by using standard cabinets.

vacuum breaker   A device to prevent wastewater backflow into a water supply line.

valance   An accessory usually used to fill the area between wall cabinets next to a sink.

valve   Fitting for opening and closing the passage for water.

vapor retarder   A plastic sheet, 4 mil thick, that is unrolled to cover soil, primarily for crawlspace systems where there is no basement. A vapor retarder is also used to provide a barrier over unfaced wall insulation.

veneer   A thinly cut slice of decoratively marked wood. Produced by sawing, slicing, or rotary process.

vitreous china   A form of ceramic or porcelain that is glasslike.

volt   The measurement of electrical pressure.

wall cabinets   Those cabinets that attach to the walls and provide storage space above the countertop work area.

wall filler   An accessory used to fill the space resulting from odd wall dimensions that cannot be filled with standard wall cabinets.

wall units   Cabinets fixed to walls with screws.

warp   Any of several lumber defects caused by uneven shrinkage of wood cells.

warranty   A written guarantee offered to the purchaser of a product or service that the product or service will be free from defects for a specified period.

water heater   Plumbing appliance used to heat water.

water meter   A device used to measure the amount of water that flows through it.

water softener   A plumbing appliance used to reduce the hardness of water, usually through the ion-exchange process.

water-supply system   A water service entry pipe; water distribution pipes; and the necessary connecting pipes and fittings, control valves, and appurtenances in or adjacent to the building or premises.

watts (wattage)   The measurement of the amount of energy consumed by a light source.

well   A driven, bored, or dug hole in the ground from which water is pumped.

wide bottom rails   A wide bottom rail is used to replace the toe space on base cabinets.

window rough opening   The opening left in a stud wall for the window.

window sill   The exterior slanting ledge of a window frame.

window stool   The interior horizontal ledge that seals a window at the bottom.

wood floor system   Used when a home has a basement or crawlspace.

workers compensation insurance   A program that provides funds to employees that are injured during the course of performing work for the company by which they are employed. The funds for this program are derived from premiums that are paid by employers.

zoned heat   A division of the house into two or more parts with a separate heat source for each part.

B

Metric conversion charts

1. Actual metric conversion to centimeters is 1 inch = 2.54 cm.

2. Actual metric conversion to millimeters is 1 inch = 25.4 mm. To facilitate conversions between imperial and metric dimensioning for calculations under 1 inch, 24 mm is used.

3. To facilitate conversions between imperial and metric for calculations over 1 inch, 25 mm is typically used.

Inches	Centimeters	Inches	Centimeters
$\frac{1}{8}$"	0.32	69"	175.26
$\frac{1}{4}$"	0.64	72"	182.88
$\frac{1}{2}$"	1.27	75"	190.5
$\frac{3}{4}$"	1.91	78"	198.12
1"	2.54	81"	205.74
3"	7.62	84"	213.36
6"	15.24	87"	220.98
9"	22.86	90"	228.6
12"	30.48	93"	236.22
15"	38.1	96"	243.84
18"	45.72	99"	251.46
21"	53.34	102"	259.08
24"	60.96	105"	266.7
27"	68.58	108"	274.32
30"	76.2	111"	281.94
33"	83.82	114"	289.56
36"	91.44	117"	297.18
39"	99.06	120"	304.8
42"	106.68	123"	312.42
45"	114.3	126"	320.04
48"	121.92	129"	327.66
51"	129.54	132"	335.28
54"	137.16	135"	342.9
57"	144.78	138"	350.52
60"	152.4	141"	358.14
63"	160.02	144"	365.76
66"	167.64		

Area

100 sq. millimeters = 1 sq. centimeter
100 sq. centimeter = 1 sq. decimeter
100 sq. decimeters = 1 sq. meter
100 sq. meters = 1 are
10,000 sq. meters = 1 hectare
100 hectares = 1 sq. kilometer

Square measure

1 sq. inch = 6.4516 sq. centimeters
1 sq. foot = 9.29034 sq. decimeters
1 sq. yard = 0.836131 sq. meter
1 acre = 0.40469 hectare
1 sq. mile = 2.59 sq. kilometers

Long measure

1 inch = 25.4 millimeters
1 foot = 0.3 meter
1 yard = 0.914401 meter
1 mile = 1.609347 kilometers

Length

10 millimeters = 1 centimeter (cm)
10 centimeters = 1 decimeter
10 decimeters = 1 meter (m)
10 meters = 1 decameter
100 meters = 1 hectometer
1000 meters = 1 kilometer

Liquid measure

1 pint = 0.473167 liter
1 quart = 0.946332 liter
1 gallon = 3.785329 liters

Linear drawing measurements

1 millimeter (mm) = 0.03937"   1" = 25.4 mm   12" = 304.8 mm
1 centimeter (cm) = 0.3937"   1" = 2.54 cm   12" = 30.48 cm
1 meter (m) = 39.37"   1" = 0.0254 M   12" = 0.3048 M

Quantity	Metric (SI) unit	Metric (SI) symbol	U.S. equivalent (nom.)[1]
Length	millimeter	mm	0.039 in.
	meter	m	3.281 ft.
			1.094 yd.
Area	meter	m^2	10.763 ft.^{2}
			1.195 yd.^{2}

Quantity	Metric (SI) unit	Metric (SI) symbol	U.S. equivalent (nom.)[1]
Volume	meter	m^3	35.314 ft.^{3}
			1.307 yd.^{3}
Volume (fluid)	liter	L	33.815 oz.
			0.264 gal.
Mass (weight)	gram	g	0.035 oz.
	kilogram	kg	2.205 lb.
	ton	t	2,204.600 lb.
			1.102 tons
Force	newton	N	0.225 lbf.
Temperature (interval)	kelvin	K	1.8°F
	degree Celsius	°C	1.8°F
Temperature	Celsius	°C	(°F–32) × $\frac{5}{9}$
Thermal resistance		$K \cdot m^2/W$	5.679 ft$^2 \cdot$hr$\cdot$°F/Btu
Heat transfer	watt	W	3.412 Btu/hr.
Pressure	kilopascal	kPa	0.145 lb./in.^{2} (psi)
	pascal	Pa	20.890 lb./ft.^{2} (psf)

Decimal equivalents of inch fractions

Fraction	Decimal	Fraction	Decimal	Fraction	Decimal	Fraction	Decimal
$\frac{1}{64}$	0.015625	$\frac{17}{64}$	0.265625	$\frac{33}{64}$	0.515625	$\frac{49}{64}$	0.765625
$\frac{1}{32}$	0.03125	$\frac{9}{32}$	0.28125	$\frac{17}{32}$	0.53125	$\frac{25}{32}$	0.78125
$\frac{3}{64}$	0.046875	$\frac{19}{64}$	0.296875	$\frac{35}{64}$	0.546875	$\frac{51}{64}$	0.796875
$\frac{1}{16}$	0.0625	$\frac{5}{16}$	0.3125	$\frac{9}{16}$	0.5625	$\frac{13}{16}$	0.8125
$\frac{5}{64}$	0.078125	$\frac{21}{64}$	0.328125	$\frac{37}{64}$	0.578125	$\frac{53}{64}$	0.828125
$\frac{3}{32}$	0.09375	$\frac{11}{32}$	0.34375	$\frac{19}{32}$	0.59375	$\frac{27}{32}$	0.84375
$\frac{7}{64}$	0.109375	$\frac{23}{64}$	0.359375	$\frac{39}{64}$	0.609375	$\frac{55}{64}$	0.859375
$\frac{1}{8}$	0.1250	$\frac{3}{8}$	0.3750	$\frac{5}{8}$	0.6250	$\frac{7}{8}$	0.8750
$\frac{9}{64}$	0.140625	$\frac{25}{64}$	0.390625	$\frac{41}{64}$	0.640625	$\frac{57}{64}$	0.890625
$\frac{5}{32}$	0.15625	$\frac{13}{32}$	0.40625	$\frac{21}{32}$	0.65625	$\frac{29}{32}$	0.90625
$\frac{11}{64}$	0.171875	$\frac{27}{64}$	0.421875	$\frac{43}{64}$	0.671875	$\frac{59}{64}$	0.921875
$\frac{3}{16}$	0.1875	$\frac{7}{16}$	0.4375	$\frac{11}{16}$	0.6875	$\frac{15}{16}$	0.9375
$\frac{13}{64}$	0.203125	$\frac{29}{64}$	0.453125	$\frac{45}{64}$	0.703125	$\frac{61}{64}$	0.953125
$\frac{7}{32}$	0.21875	$\frac{15}{32}$	0.46875	$\frac{23}{32}$	0.71875	$\frac{31}{32}$	0.96875
$\frac{15}{64}$	0.234375	$\frac{31}{64}$	0.484375	$\frac{47}{64}$	0.734375	$\frac{63}{64}$	0.984375
$\frac{1}{4}$	0.2500	$\frac{1}{2}$	0.5000	$\frac{3}{4}$	0.7500	1	1.0000

Bibliography

Better Homes & Gardens. 1980. *Complete Guide to Home Repair, Maintenance and Improvement*. Des Moines, IA: Meredith Corp.

Better Homes and Gardens Special Interest Publications—Kitchen and Bath Ideas. Summer 1995. Des Moines, IA: Magazine Group of Meredith Corp.

Better Homes and Gardens Special Interest Publications—Remodeling Ideas for Your Home. Summer 1995. Des Moines, IA: Magazine Group of Meredith Corp.

Better Homes and Gardens Special Interest Publications—Kitchen & Bath Products Guide. Spring/Summer 1995. Des Moines, IA: Magazine Group of Meredith Corp.

Bianchina, Paul. 1988. *Kitchen Remodeling, A Do-It-Yourselfer's Guide*. Blue Ridge Summit, Pennsylvania: TAB Books, Inc.

Black & Decker Home Improvement Library—Kitchen Remodeling. 1989. Minnetonka, Minnesota: Cy DeCosse, Inc.

Black & Decker Home Plumbing Projects & Repairs. 1990. Minnetonka, Minnesota: Cy DeCosse, Inc.

Cheever, Ellen, CKD, CBD. 1992. *Beyond the Basics—Advanced Kitchen Design*. Hackettstown, N.J.: National Kitchen and Bath Association.

Creative Homeowner Press. 1991. *Kitchens—Design, Remodel, Build*. Upper Saddle River, N.J., CHP.

Day, Richard. *The Practical Handbook of Plumbing & Heating*. New York, N.Y.: ARCO Publishing Company, Inc.

Galvin, Patrick and Ellen Cheever. 1995. *National Kitchen & Bath Association, Kitchen Basics—A Training Primer for Kitchen Specialists*. E. Windsor, N.J.: Galvin Publications.

Hemp, Peter. 1994. *Installing & Repairing Plumbing Fixtures*. Newton, CT: The Taunton Press, Inc.

Hilts, Len. 1981. *The Dremel Guide to Compact Power Tools*. Dremel, Division of Emerson Electric Co.

House Beautiful—Kitchens/Baths. Vol. 18, No. 1, Spring 1997. New York, NY: The Hearst Corporation.

Hughes, Herb. 1980. *Adding Space Without Adding On*. Upper Saddle River, N.J.: Creative Homeowner Press.

Kitchen & Bath Design News. May 1995, October 1996, November 1996, December 1996, January 1997 issues. Melville, NY: PTN Publishing Co.

Kitchen & Source Book, 1995. New York, N.Y.: Sweet's Group/McGraw-Hill, Inc.

Lewis, Darrel L. and Walter W. Stoeppelwerth. 1995. *Kitchen & Bathroom Installation Manual, Volume 2*. Hackettstown, N.J.: National Kitchen & Bath Association.

Love, T. W. 1976. *Construction Manual: Rough Carpentry*. Carlsbad, CA: Craftsman Book Co.

Russell, James E. 1981. *Advanced Kitchens*. Upper Saddle River, N.J.: Creative Homeowner Press.

Scharff, Robert, and the Editors of *Walls & Ceilings* Magazine. 1995. *Drywall Construction Handbook*. New York, N.Y.: McGraw-Hill, Inc.

Scharff, Robert. 1975. *The Complete Book of Home Remodeling*. New York, N.Y.: McGraw-Hill, Inc.

———. 1978. *The House & Home Kitchen Planning Guide*. New York, N.Y.: McGraw-Hill, Inc.

Scharff, Robert and Kenneth M. Swezey. 1979. *Popular Science, Formulas, Methods, Tips & Data for Home and Workshop.* New York: Harper & Row.

Stoeppelwerth, Walter W. 1994. *Kitchen & Bathroom Installation Manual, Volume 1.* Hackettstown, N.J.: National Kitchen & Bath Association.

Stoeppelwerth, Walter W. 1997. *How to Implement the Lead Installer Concept.* Bethesda, MD: Home Tech Information Systems, Inc.

Sunset Books, Editors. 1991. *Decorating with Paint & Wall Coverings.* Menlo Park, California: Sunset Publishing Corporation.

———. 1994. *Kitchens—Planning & Remodeling.* Menlo Park, California: Sunset Publishing Corporation.

———. 1995. *Ideas for Great Kitchens.* Menlo Park, California: Sunset Publishing Corporation.

———. 1995. *Basic Plumbing.* Menlo Park, California: Sunset Publishing Corporation.

———. 1996. *Creating Beautiful Floors.* Menlo Park, California: Sunset Publishing Corporation.

———. 1992. *Flooring—How to Plan, Install, & Repair.* Menlo Park, California: Sunset Publishing Corporation.

Time-Life Books, Editors. 1987. *Kitchen & Bathroom Plumbing.* Alexandria, Virginia: Time Life Books.

———. 1989. *Kitchens and Bathrooms,* Alexandria, Virginia: Time Life Books.

Woman's Day Innovation Series—Kitchens & Baths. 1996. Vol. VI, No. 4. New York, NY: Hachette Filipacchi Magazines, Inc.

Woodson, R. Dodge. 1993. *Home Plumbing Illustrated.* Blue Ridge Summit, Pennsylvania: TAB Books.

The following is a complete list of all of the manufacturers, corporations, etc., that have contributed photographs, illustrations, research, literature, etc., to this book:

Aristokraft, Inc.
P.O. Box 420
Jasper, IN 47547-0420

Armstrong World Industries
P.O. Box 3001
Lancaster, PA 17604

Blanco America
1001 Lower Landing Rd.
Suite 607
Blackwood, NJ 08012

Broan Mfg. Co., Inc.
P.O. Box 140
Hartford, WI 53207-0140

Bruce Hardwood Floors
16803 Dallas Pkwy.
Dallas, TX 75248

Elkay Manufacturing Co.
2222 Camden Ct.
Oak Brook, IL 60521

Five Star
P.O. Box 2490
Cleveland, TN 37320

Florida Tile Industries, Inc.
P.O. Box 447
Lakeland, Florida 33802

Formica Corp.
c/o Dan Pinger Public Relations
2245 Gilbert Ave.
Cincinnati, OH 45206

Four Seasons Sunrooms
5005 Veterans Memorial Hwy.
Holbrook, NY 11741

General Electric Co.
Building 3, Room 232
Appliance Park
Louisville, KY 40225

Halo Lighting
Cooper Industries
400 Busse Rd.
Elk Grove, IL 60007

Harris-Tarkett, Inc. Hardwood
 Flooring
P.O. Box 300
Johnson City, TN 37605-0300

Home Tech Information
Systems, Inc.
5161 River Road
Bethesda, MD 20816

Juno Lighting, Inc.
2001 S. Mount Prospect Rd.
Des Plaines, IL 60017

Kohler Co.
444 Highland Dr.
Kohler, WI 53044

Kraftmaid Cabinetry, Inc.
1422 Euclid Ave.
Suite 239
Cleveland, OH 44115

Mannington Resilient Floors
P.O. Box 30
Salem, NJ 08079-0030

Merillat Industries, Inc.
P.O. Box 1946
Adrian, MI 49221

National Kitchen & Bath Association
687 Willow Grove St.
Hackettstown, NJ 07840

Perstorp Flooring, Inc.
8750 N. Central Expressway
Suite 1200
Dallas, TX 75231

Sub-Zero Freezer Co., Inc.
P.O. Box 44130
Madison, WI 53744-4130

Thomson Consumer Electrics
P.O. Box 1976 - INH 430
Indianapolis, IN 46206-1976

Wellborn Cabinet, Inc.
P.O. Box 1210
Ashland, AL 36251

Index

Illustrations are indicated in **boldface**.

About the Author

Pamela Korejwo (Mohnton, PA) has written for and about the construction industry for many years. She contributed to a number of popular McGraw-Hill books, including *Drywall Construction Handbook, Roofing,* and *Residential Steel Framing Handbook*. The daughter of a former home contractor, she recently became a principal associate of her father's consulting company, L.E.K. Illustration, which does technical illustration and design.

About the Illustrator

Leon Korejwo (Mohnton, PA) has more than 30 years' experience as a mechanical designer, and has won numerous national and international awards. He has illustrated a number of books, among them *Diesel Engines, Heavy Duty Trucks, Motorcycle Maintenance Handbook,* and *Residential Steel Framing Handbook*. For many years he operated his own home construction business. He is president and owner of L.E.K. Illustration.